# ENVIRONMENTAL RADON: OCCURRENCE, CONTROL, AND HEALTH HAZARDS

The Pennsylvania Academy of Science Publications

*Books and Journal*

Editor: Shyamal K. Majumdar
Professor of Biology
Lafayette College
Easton, Pennsylvania 18042

1. *Energy, Environment, and the Economy,* 1981. ISBN: 0-9606670-0-8. Editor: Shyamal K. Majumdar.
2. *Pennsylvania Coal: Resources, Technology and Utilization,* 1983. ISBN: 0-9606670-1-6. Editors: Shyamal K. Majumdar and E. Willard Miller.
3. *Hazardous and Toxic Wastes: Technology, Management and Health Effects,* 1984. ISBN: 0-9606670-2-4. Editors: Shyamal K. Majumdar and E. Willard Miller.
4. *Solid and Liquid Wastes: Management, Methods and Socioeconomic Considerations,* 1984. ISBN: 0-9606670-3-2. Editors: Shyamal K. Majumdar and E. Willard Miller.
5. *Management of Radioactive Materials and Wastes: Issues and Progress,* 1985. ISBN: 0-9606670-4-0. Editors: Shyamal K. Majumdar and E. Willard Miller.
6. *Endangered and Threatened Species Programs in Pennsylvania and Other States: Causes, Issues and Management,* 1986. ISBN: 0-9606670-5-9. Editors: Shyamal K. Majumdar, Fred J. Brenner, and Ann F. Rhoads.
7. *Environmental Consequences of Energy Production: Problems and Prospects,* 1987. ISBN:0-9606670-6-7. Editors: Shyamal K. Majumdar, Fred J. Brenner and E. Willard Miller.
8. *Contaminant Problems and Management of Living Chesapeake Bay Resources,* 1987. ISBN: 0-9606670-7-5. Editors: Shyamal K. Majumdar, Lenwood W. Hall, Jr. and Herbert M. Austin.
9. *Ecology and Restoration of The Delaware River Basin,* 1988. ISBN: 0-9606670-8-3. Editors: Shyamal K. Majumdar, E. Willard Miller and Louis E. Sage.
10. *Management of Hazardous Materials and Wastes: Treatment, Minimization and Environmental Impacts,* 1989. ISBN 0-9606670-9-1. Editors: Shyamal K. Majumdar, E. Willard Miller and Robert F. Schmalz.
11. *Wetlands Ecology and Conservation: Emphasis in Pennsylvania,* 1989. ISBN 0-945809-01-8. Editors: Shyamal K. Majumdar, Robert P. Brooks, Fred J. Brenner and Ralph W. Tiner, Jr.
12. *Water Resources in Pennsylvania: Availability, Quality and Management,* 1990. ISBN 0-945809-02-6. Editors: Shyamal K. Majumdar, E. Willard Miller and Richard R. Parizek.
13. *Environmental Radon: Occurrence, Control and Health Hazards*, 1990. ISBN 0-945809-03-4. Editors: Shyamal K. Majumdar, Robert F. Schmalz and E. Willard Miller.

# ENVIRONMENTAL RADON: OCCURRENCE, CONTROL, AND HEALTH HAZARDS

**EDITED BY**
SHYAMAL K. MAJUMDAR, Ph.D.
Professor of Biology
Lafayette College
Easton, Pennsylvania 18042

ROBERT F. SCHMALZ, Ph.D.
Professor of Geology
The Pennsylvania State University
University Park, Pennsylvania 16802

E. WILLARD MILLER, Ph.D.
Professor of Geography and Associate Dean
for Resident Instruction (Emeritus)
College of Earth and Mineral Sciences
The Pennsylvania State University
University Park, Pennsylvania 16802

Founded on April 18, 1924

**A Publication of**
**The Pennsylvania Academy of Science**

Library of Congress Cataloging in Publication Data

Bibliography
Index
Majumdar, Shyamal K. 1938-, ed.

Library of Congress Catalog Card No.: 90-61965

ISBN-0-945809-03-4


Printed in the United States of America by

Typehouse of Easton
Phillipsburg, New Jersey 08865

# PREFACE

Radon—the silent killer—is now recognized as a world-wide threat to health. While radon has long been known as a health hazard to miners in the uranium industry, it was only in the 1980s that it was recognized that radon contamination of buildings was widely distributed over the earth. The seriousness of long-term exposure to radon cannot be overstated. The great difficulty of alerting the general public to the devastating effects of radon is that it requires years of exposure before a health problem develops. Then it is too late. The objective of this volume is to not only present a review of the occurrences and properties of radon but to provide insights into its effects on human health. The technology is available to control radon, but before there is effective control, society must recognize the danger it poses.

This volume is divided into six parts beginning with a historical perspective from the time of the discovery of radon in uranium mines to the modern recognition of radon contamination in the Watras house in eastern Pennsylvania in the mid-1980s and concluding with the recognition that radon is a universal health problem. The second part deals with the geologic aspect of radon. The third part treats the detection and measurement of radon. Part four is devoted to radon health problems with emphasis on lung dosimetry and cancer. The final section considers the regulation and policies of radon control, legal aspects, and economic impacts.

This book is not only for the experts who are evaluating the radon problem, but will be of value to a wide audience who are awakening to the perils of radon contamination. It will also provide information to specialists who want a broader perspective of the radon problem. Because the management of radon is a universal problem, this volume has special value to the general public.

We express our deep appreciation for the cooperation and dedication of contributors, who recognize the importance of solving this critical environmental and health problem. In addition to the authors, many other individuals in the Pennsylvania Academy of Sciences made contributions, and we are appreciative to these persons. Thanks are extended to Drs. Robert S. Chase, Head, Department of Biology, Lafayette College, George C. Shoffstall, Executive Secretary of the Academy and Kurt C. Schreiber, President of the Academy, for their encouragement and help in the preparation of this book. Gratitude is extended to Lafayette College and The Pennsylvania State University for providing facilities to the editors for this work. The editors extend heartfelt thanks to their wives for their encouragement and help in the preparation of this book.

Shyamal K. Majumdar
Robert F. Schmalz
E. Willard Miller
Editors, July 1990

## ACKNOWLEDGMENTS

The publication of this book was aided by contributions from The Pennsylvania Power and Light Company, Allentown, Pennsylvania; and Consolidation Coal Company, Pittsburgh, Pennsylvania.

# Environmental Radon: Occurrence, Control, and Health Hazards

# Table of Contents

**Part Three: Detection, Measurement and Mitigation**



**Part Four: Radon and Health**

# FOREWORD

Since the discovery of radioactivity around the turn of the century, we have had some awareness that this manifestation brings with it some problems of health and safety. In the years between the two World Wars little attention was paid to this. Thus, workers put radioactive substances on watchfaces without being aware of the health problems they were being exposed to.

After World War II, we focused our attention on the radioactivity generated by processes devised by man, first, on the testing of atomic weapons, and, later, on the operations of nuclear power plants and the disposal of their waste products.

Only in recent years has the general public become conscious and concerned about radioactivity generated by the environment, with special focus on the gas, radon, leaking from the interior of the earth to the surface, especially, where that surface is covered by our homes.

This book has been developed as a follow-up to a symposium on "Environmental Radon" held at the 65th annual meeting of the Pennsylvania Academy of Science at Mt. Airy Lodge on April 8, 1989. The book gives a comprehensive review of this topic from a variety of viewpoints and, thus, brings together information from a number of different disciplines. It is most important that these facts be brought together in order to promote rational and meaningful solutions to problems which are significant to our Commonwealth.

As President of the Academy, I extend my personal thanks to the participants and organizers of the symposium as well as to the authors who have contributed chapters to this book. In addition, I extend my sincere thanks to the Society for Analytical Chemists of Pittsburgh for their contribution to defray the costs of the Radon symposium. Special thanks are hereby conveyed to the editors who have worked hard and long to make this volume an outstanding addition to the books published by the Academy.

Dr. Kurt C. Schreiber
Immediate Past President
The Pennsylvania Academy of Science

*Environmental Radon: Occurrence, Control and Health Hazards.* Edited by S. K. Majumdar, R. F. Schmalz and E. Willard Miller © 1990, The Pennsylvania Academy of Science.

*Chapter One*

# RADON HAZARD: A HISTORICAL PERSPECTIVE

**E. WILLARD MILLER**

Professor of Geography and Associate Dean
for Resident Instruction (Emeritus)
College of Earth and Mineral Sciences
The Pennsylvania State University
University Park, PA 16802

The hazardous health effects of radon have long been recognized for miners in uranium mines. It has, however, only been realized recently that radon is widely distributed and presents a severe potential health hazard to millions of people. It is the purpose of this paper to trace a brief history of the hazards of radon to miners, the recognition of its widespread distribution, and the passage of national legislation in an attempt to mitigate the problem.

## RADON GENERATION

The natural decay of uranium -238, the principal form of uranium found in the earth's crust, leads to the formation of radon and its daughters. When uranium occurs as an ore body the intensity of the release of radon is increased. One of the principal forms of radon is radon-222, a noble gas, which escapes from exposed rock surfaces. Radon-222 has a half-life of 3.8 days and decays by the emission of alpha ($\alpha$), beta ($\beta$), and gamma ($\gamma$) rays. In the decay of radon-222 a number of short-lived nuclides (radon daughters) are produced which exhibit a half-life of about 30 minutes before reaching lead-210.

Radon daughters, either as free ions or absorbed in water droplets or dust particles, may be inhaled, deposited and retained in the respiratory system of individuals (Union of Concerned Scientists, 1975). Consequently, the respiratory epithelium, the thin layer of cells forming a tissue that covers the surface of the lungs, can be exposed to radiation doses that far exceed whole body doses from internally dissolved radon. Lung exposures from radon daughters are far more harmful than from radon itself. The bulk of tissue damage is from alpha particles, which have short ranges but high linear energy transfer.

Lung cancer believed to arise from exposure to radon daughters usually appears in the hilar region of the lung. This area is ciliated and covered with a moving sheet of mucus. Since this surface layer of cells develops from a layer of basal cells, the health of the epithelium depends on the continued integrity of the basal cells, thus, the nuclei of these basal cells are taken as the relevant biological target.

When the radon daughters enter the epithelium they will deliver a significant fraction of energy of the alpha particles to the basal cell region. With a 30 minute half-life, the radon daughters will decay mainly on the epithelium before they can be removed by normal clearance mechanisms of mucus transport. The actual radiation dosage to the lungs depends upon a number of factors: concentration of radon daughters in the atmosphere, respiratory volume and exchange rate, distribution of radon daughters between free ions and ions absorbed by dust particles, size of dust particles and respiratory clearance rate.

## HISTORICAL FINDINGS OF RADON HAZARD IN MINING

As early as the sixteenth century it was reported that miners in the pitchblende (radium) mines at Schneeberg in the Erz Mountains of Germany and at Joachimsthal in Czechoslovakia were plaqued with pulmonary diseases. Many of the workers died in middle age of lung infections which was known locally as "Bergkrankheit" or mountain sickness. The pulmonary infections normally occurred in miners who worked for more than 10 years in the mines.

In the nineteenth century studies of the diseases of the Schneeberg miners continued. In a study of 650 miners during the years 1869-1877, it was found that the average life expectancy after entering the mines was about 20 years. Further, the study revealed that lung diseases accounted for 75 percent of all deaths of underground miners. It was assumed, incorrectly, that the lung malignancies were caused by inhaled arsenic.

Between 1900 and 1940 a number of studies verified the earlier findings of the excessive rate of lung cancer among the miners of Schneeberg and Joachimsthal. The statistical data compiled by Peller and Sikl on lung cancer rates of Schneeberg and Joachimsthal miners compared to the lung cancer rate of the male population of Vienna indicated the health hazards of the two areas (Table 1). Estimates of the radioactivity in the air in various mines in the Schneeberg region were first made in 1924 and again in 1937 (Ludwig and Lorenser, 1924; Rajewsky, 1939). The values ranged from 3.6-180 Working Levels (WL). (A working level is defined as any combination of short-lived radon daughters in 1 liter of air that will result in the ultimate emission of $1.3 \times 10^5$

MeV of potential alpha energy. MeV—Million Electron Volts—A unit of energy used to describe radioactive emissions. It takes about 2.3 billion billion MeV 23 x $10^{18}$ to produce one kilowatt hour) (Holaday and others, 1957). When it is assumed that the average miner works 17 years with an average exposure of 30 working levels per month (WLM), the average exposure was 6,120 or equivalent to more than 12,000 rad. An analogous average exposure at Joachimsthal was 3,840 WLM or about 8,000 rad (Ludwig and Lorenser, 1924). Radiation and its radon daughters was clearly the cause of cancer in the radium miners of Schneeberg and Joachimsthal.

Other studies have confirmed the findings of cancer incidences in other mining areas of the world. In the Newfoundland flourspar mines, where radon entered the mine in ground water, de Villiers and Windish measured radon daughter concentrations. The mine air radon daughter concentrations ranged between 2.5 to 10 WL. The concentration radiation exposure over a 14-year-period of the flourspar miner was estimated to have ranged from 415 to 1,660 WLM. The health of 630 miners who had spent 12 months or more working in the mines over a 29-year-period was surveyed in the study. Of the deaths reported in 1961, 26 of the 29 were attributable to lung cancer. This is from 25 to 40 times the incidences found in the adult male population of Newfoundland.

Numerous studies have investigated the health hazards of the underground mines in many types of ore or energy deposits. One of the dangers to the underground miner is the high risk of nonmalignant respiratory diseases. However, an excess mortality from lung cancer has not usually been found in the absence of significant exposure to radioactive materials. For example, excess lung cancer risk has not been shown by studies of the health problems among British or American coal miners, United States potash workers or Russian manganese miners.

## URANIUM MINING IN THE UNITED STATES

### *Radon Hazards*

The mining of uranium in the southwestern states of Utah, Nevada, Colorado,

TABLE 1

*Cancer Mortality Per 1,000 Persons Per Year*

| | Joachimsthal Mines | Schneeberg Mines | | Vienna Males 15 to 39 years of age |
|---|---|---|---|---|
| | 1929-1938 | 1875-1894 | 1895-1912 | 1932-1936 |
| Cancer of Lungs | 9.8 ± 1.5 | 12.7 ± 1.0 | 16.5 ± 1.5 | 0.34 ± 0.01 |
| All Other Organs | 0.45 ± 0.1 | 2.4 ± 0.4 | 2.1 ± 0.6 | 2.1 |

Source: Peller, S. 1939. *Human Biology,* 11:130.
Sikl, H. 1950. *Acta Unio Internationalis Contra Cancrum* 6: 1366.

Arizona, and New Mexico prior to 1930 was to secure radium. From 1930 to the early 1940s vanadium output dominated. Since the 1940s uranium has been produced primarily for weapon production and for nuclear power plants.

The danger from radon in the mines was long ignored. The first measurements of radon were made in 1949, but the initial survey of radon daughters was not completed until 1951. Until the 1960s there was no established plan to measure radon concentrations in mines. Efforts appear to have concentrated on gathering information.

The first study to recognize the feasible health hazard of radon to uranium miners was prepared by the Colorado Department of Health in 1949. Other studies followed by the U.S. Public Health Service. In 1957 the USPHS, based on radon daughter surveys, predicted a significant increase in mortality from lung cancer among the miners (Doyle, 1957). In the period from 1950 to 1960, the health of more than 4,000 uranium miners was surveyed. The conclusions were based on radon and radon daughters concentrations compared with epidemiological and mortality data.

As an initial response to these studies radon daughter control programs began to be introduced in 1961 in uranium mines. Follow up studies of the potential hazard of radon and radon daughters indicated, however, a major problem still existed. A preliminary study of 907 white underground uranium miners, with more than three years underground experience, indicated that death from respiratory cancer was occurring at nearly 5 times the normal rate (Archer and others, 1962). A 1964 survey of 3,656 white uranium miners indicated that there were 218 deaths compared with 148.7 expected (Wagoner and others, 1964). Health categories in which deaths significantly exceeded that expected were: respiratory neoplasms (11 observed against 1.1 expected), pulmonary fibrosis and accidents. Saccomanno and others reported in 1971 that lung cancer among the miners occurs at an earlier age and with a strikingly higher ratio of undifferentiated, small cell type carcinomas.

The first quantitative relationships between the incidence of radiation cancer and cumulative dosage of radiation was presented by Wagoner and others in 1965. Using WLMs as cumulative radiation exposure estimates, it was revealed that the incidence of lung cancer increases from 3.1 per 10,000 miners per year to 116 per 10,000 miners per year over 6 exposure categories ranging from less than 120 WLM to more than 3,270 WLM. By 1965 sufficient data had been collected and analyzed which revealed that the relationship between airborne radiation and respiratory cancer in U.S. uranium miners was virtually certain. In 1971 a study by Lundin confirmed a statistically significant excess of respiratory cancer among white uranium miners down to and including the range of 120-359 WLM.

It has now been determined that the highest rates of lung cancer occur in miners who worked in mines in which the average concentration of radon daughters was usually greater than 10 WL, whose total concentration exposure exceeded 1,000 WLM, who were in the mines a minimum of 15 years, and who were modest to heavy cigarette smokers (Federal Radiation Council, 1967).

*Control of Radon and Radon Daughters*

Radon and its daughters can be removed from mines by proper ventilation. Nevertheless, the protection of the miner was extremely slow to develop. Although a health hazard was recognized as early as 1949 the discussion of authorities delayed action. In addition to the state agencies responsible for mining practice and health and safety standards, a number of federal agencies including the Atomic Energy Commission and the Department of Interior under the Federal Metal and Nonmetallic Mine Safety Act had health and safety responsibilities.

The initial methodical air monitoring programs began in 1954. However, radon control programs were not very effective or extensive until the 1960s. Over time the federal agencies withdrew from programs of radon control leaving the protection and ultimate health of the miners to the state. Radiation protection in the mines remained confused until 1967 when uniform Federal Regulatory Commission standards came into effect.

The Atomic Energy Commission failed to protect the uranium miner due to a peculiar interpretation of the Atomic Energy Act. The Act was interpreted by the AEC itself not to include uranium mining but to include all other aspects of the mining process. The view was held by the AEC that it could not regulate operations in the ground, but only after the ore was removed from the earth could AEC controls begin. The AEC considered the mined ore rather than the ore body as the 'place' of deposit in nature. This position was not tested by the courts and the ruling remained intact.

## UNIVERSAL RADON HAZARD

It was not until 1984 that it was recognized that health hazards from radon were extremely widely distributed over the earth. The modern recognition was triggered by the now famous incident near Boyerstown, Pennsylvania. On a December morning Stanley Watras set off the radiation alarm when he went to work at the Limerick nuclear power plant. It was soon proved that the radiation did not come from the plant but from his home. On testing for radiation in the Watras home, 3,200 picocuries (A trillionth of a curie. A curie is equal to 37 billion ($37 \times 10^9$) disintegrations of radioactivity per second.) were discovered in the living room. Up to this time no one realized that radon could be a danger to anyone except uranium miners and people who lived close to uranium mines. A totally new concept came into existence that radon could be constantly released from the soil and find its way into buildings and accumulate to life threatening concentrations.

As a response to the very high picocurie readings, the testing for radon contamination spread across the nation. It was found that some of the highest levels of radon were in Pennsylvania, New Jersey, Wisconsin, Minnesota, North Dakota, and Colorado. The Environmental Protection Agency and the Public Health Service estimate radon poses a health problem in 8,000,000 American homes. Further the EPA estimates that between 5,000 and 20,000 lung cancer deaths annually are due to radon exposure. Only cigarette smoking causes more deaths.

It is difficult to identify buildings where radon will be found. Originally it was thought that radon was only associated with igneous rocks. This has proved to be wrong. Radon is found in all types of rock. For example, in New Jersey at Whispering Hills one basement had a level of 3,000 picocuries that was located on sedimentary rock. In addition radon concentrations vary tremendously from house to house. Thus, every house in an area must be tested if a study of local distribution is to be accurate. It is now estimated that every square meter of land emits half a picocurie per second. Because of dilution in the atmosphere however, outdoor levels average only 0.2 picocuries per liter of air.

Although no one knowns how much radon in a house provides a hazardous condition, the Environmental Protection Agency has set the safe level at four picocuries. This was an arbitrary decision for it is relatively easy to reduce the level to four picocuries in all buildings, but extremely difficult to reduce the four to one. As more information is acquired, this standard will be adjusted either upward or downward.

Although radon is widely distributed, it is quite easy to control in buildings. When a building has a high radon presence two things need to be done to control the radon and radon daughters. The basement needs to be sealed so that no gas can penetrate the walls from the outside and vents need to be installed in the basement to direct the radon outside. Ventilation is the key to the control of indoor radon. If ventilation is not developed the concentrations of radon will gradually increase. In drafty, old houses the danger from radon is greatly reduced. In contrast, modern houses are highly insulated in order to prevent the escape of heat, and, of course, radon.

## NATIONAL RADON LEGISLATION

By the late 1980s radon as a hazardous gas was being recognized by the Congress of the United States. In 1987 hearings by the Committee on Energy and Commerce of the House of Representatives were held on the proposed Radon Pollution Control Act of 1987 and the Committee on Environment and Public Works of the United States Senate held hearings on the Indoor Air Quality Act of 1987. Although these bills were not enacted, Title III Indoor Radon Abatement was passed in 1988 as an amendment to the Toxic Substance Control Act.

This Act established the national long-term goal of the United States with respect to radon levels in buildings so "air within buildings in the United States would be as free from radon as the ambient air outside of buildings."

*Model Construction Standards and Techniques*

The recognition that radon is hazardous to the general public is so recent that little basic information has been developed to date. In order to secure this information the Act authorizes that:

> "The Administration of the Environmental Protection Agency shall develop model construction standards and techniques for controlling radon levels within new buildings. To the maximum extent possible, these standards and techniques should be developed with the assistance of organizations involved in establishing national building construction standards and techniques. The Adminstration shall make a draft of the document containing the model standards and techniques available for public view and comment. The model standards and techniques shall provide for geographic differences in construction types and materials, geology, weather and other variables that may affect radon levels in new buildings. The Administration shall make final model standards and techniques available to the public by June 1, 1990. The Administration shall work to ensure that organizations responsible for developing national model building codes, and authorities which regulate building construction within states and political subdivisions within states, adopt the Agency's model standards and techniques."

*Technical Assistance to States for Radon Programs*

The Act gave the Administration the authority to assist the states in implementing the radon program. These activities included:

1. Establishement of a clearinghouse of radon related information, including mitigation studies, public information materials, surveys of radon levels, and other relevant information.
2. Operation of a voluntary proficiency program for rating the effectiveness of radon measurement devices and methods, the effectiveness or radon mitigation devices and methods, and the effectiveness of private firms and individuals offering radon-related architecture, design, engineering, measurements and mitigation services.
3. Design and implementation of training seminars for state and local officials and private and professional firms dealing with radon and addressing topics such as monitoring, analysis, mitigation, health effects, public information and program design.
4. Publication of public information material concerning radon health risks and methods of radon mitigation.

5. Operation of cooperative projects between the Environmental Protection Agency's Radon Action Program and the state's radon program. Such projects shall include the Home Evaluation Program, in which the Environmental Protection Agency evaluates homes and states demonstrate mitigation methods in these homes. Whenever possible, homes of low-income persons should be selected for evaluation and demonstration.
6. Demonstration of radon mitigation methods in various types of structures and in various geographic settings and publication of findings. In the case of demonstration of such methods in houses, the Administration should select homes of low-income people.
7. Establishment of a national data base organized by states concerning the location and amount of radon.
8. Development and demonstration of methods of radon measurement and mitigation that take into account unique characteristics, if any, of nonresidential buildings housing child care facilities.

The Act also recommended discretionary assistance upon request of a state for technical assistance in the development or implementation of programs. These activities included:

1. Design and implementation of surveys of the location and occurrence of radon within a state.
2. Design and implementation of public information and education programs.
3. Design and implementation of state programs to control radon in existing or new structures.
4. Assessment of mitigation alternatives in unusual or unconventional structures.
5. Design and implementation of methods for radon measurement and mitigation for nonresidential buildings housing child care facilities.

The Act also has a provision to "provide appropriate information concerning technology and methods of radon assessment and mitigation to professional organizations representing private firms involved in building design, engineering, and construction."

The Act also authorizes the establishment of a proficiency rating program and training seminars. An initial sum of $1,500,000 was authorized. This money is to be recovered, however, by charges on persons applying for a proficiency rating and on private and professional firms participating in training seminars. Funds from these sources are to be placed in a special account in the U.S. Treasury.

*Grant Assistance to States for Radon Programs*

The states under the Act may apply for grants in order to assist in the development and implementation of programs in order to assess and mitigate the radon problem.

In order to secure a grant the state applications must include each of the following:

1. A description of the seriousness and extent of radon exposure in the state.
2. An identification of the state agency which has the primary responsibility for radon programs and which will receive the grant, a description of the rules and responsibilities of the lead state agency and other state agencies involved in radon programs and description of the rules and responsibilities of any municipal, district or areawide organization involved in radon programs.
3. A description of the activities and program related to radon which the state proposes in such year.
4. A budget specifying Federal and state funding of each element of activity of the grant application.
5. A 3-year plan which outlines long range program goals and objectives, tasks necessary to achieve them and resource requirements for the entire 3-year period, including anticipated state funding levels and desired Federal funding levels.

Activities eligible for grant assistance included:

1. Survey of radon levels, including special surveys of geographic areas or classes of buildings such as public buildings, school buildings, and high-risk residential construction.
2. Development of public information and educational materials concerning radon assessment, mitigation, and control programs.
3. Implementation of programs to control radon in existing and new structures.
4. Purchase by the state of radon measurement equipment and devices.
5. Purchase and maintenance of analytical equipment connected to radon measurement and analysis, including costs of calibration of such equipment.
6. Payment of costs of Environmental Protection Agency approved training programs related to radon for permanent state or local employees.
7. Payment of general overhead and program administration costs.
8. Development of a data storage and management system for information concerning radon occurrence, levels, and programs.
9. Payment of costs of demonstration of radon mitigation methods and

technologies as approved by the Administration, including state participation in the Environmental Protection Agency Home Evaluation Program.
10. A toll-free radon hotline to provide information and technical assistance.

Beginning in 1991 the EPA Administration will give a preference for grant assistance to those states that made reasonable efforts to build new structures according to the model construction standards. The Federal share of the cost of radon activities shall not exceed 75 percent of the costs incurred by the state in the first year, 60 percent in the second, and 50 percent in the third year. The state may use part of the grant funds to assist local governments.

The development of an information system is provided by this Act. The Administration may request such information, data and reports developed by the state in order to determine continuing eligibility for grants to the state. A state receiving funds shall make available to the public a list of firms and individuals within the state that have received a passing rating under the Environmental Protection Agency proficiency rating system.

*Radon in Schools*

A special feature of the Act is to conduct a study for the purpose of determining the extent of radon contamination in the nation's school buildings. In order to accomplish this the Administration is required to identify and compile a list of areas in the United States determined to have a high probability including schools which have elevated levels of radon.

The basic list is to be determined on the following:

1. Geological data.
2. Data on high levels in homes and other structures near any such schools.
3. Physical characteristics of the school buildings.

The survey of radon contamination in schools is to provide the basis for state action in controlling radon. The high-risk areas within each state are to be identified and the EPA Administration shall provide guidance and data detailing the risks associated with high radon levels. Technical guidance and information is to be provided to schools. After a problem school has been identified efforts are to be carried out to develop technology in order to solve the problem.

*Regional Radon Training Center*

The purpose of a regional radon training center is to develop information and provide training to Federal and state officials, professional and private firms and the public regarding the health risks posed by radon and demonstrated methods of radon measurement and mitigation. Any college, university, institution of higher learning or consortium of institutions may request funding for a center.

The criteria for selection of a center includes:

1. The extent to which the applicant's program will develop information and methods of solving radon problems.
2. The demonstrated expertise of the applicant regarding radon measurement and mitigation methods.
3. The demonstrated expertise of the applicant in radon training.
4. The seriousness of the radon problem in the region.
5. The geographical coverage of the proposed center.

*Study of Radon in Federal Buildings.*

The Act requires that radon measurements be taken in all Federal buildings. The high-risk buildings are to be identified, based on:

1. Geological data.
2. Data on high radon levels in homes and other structures near such Federal buildings.
3. Physical characteristics of the Federal building.

If a radon problem is identified the Act requires that studies be carried out based on design criteria specified by the EPA Administration, according to the most recent EPA guidance documents. The study design shall also include additional testing data to the extent resources are available, including any reliable data supplied by Federal agencies, states or other parties.

In order to provide information on risks and testing, the EPA Administration is required to provide the following:

1. Guidelines and data detailing the risks associated with high radon levels.
2. Technical guidance and related information concerning testing for radon within Federal buildings and water supplies.
3. Technical guidance and related information concerning methods for reducing radon levels.

## REFERENCES

Archer, V.E., H.J. Manguson, D.A. Holaday and P.A. Lawrence. 1962. *Journal of Occupational Medicine,* 4:55.

Doyle, H.N. 1957. *Radiation Hazards in Uranium Mining,* Washington, DC.

Federal Radiation Council. 1967. *Guidance for the Control of Radiation Hazards in Uranium Mining.* Report No. 8, Revised, Washington, DC: Government Printing Office.

Holaday, D.A., D.E. Rushing, R.D. Coleman, D.F. Woolrich, H.L. Kusnetz and

W.F. Bale. 1957. *Control of Radon and Daughters in Uranium Mines and Calculations on Biologic Effects.* PHS Publication 494, Washington, DC: Government Printing Office.

Ludwig, P. and E. Lorenser. 1924. *Zischr. Phys.,* 22:178.

Peller, S. 1939. *Human Biology,* 11:130.

Rajewsky, B. 1939. *Ztschr. Krebsforsch,* 49:315.

Saccomanno, G., V.E. Archer, O. Auerbach, M. Kuschner, R.P. Saunders and M.G. Klein. 1971. *Cancer.* 27:515.

*The Toxic Substance Control Act, Title III Indoor Radon Abatement.* Public Law 100-551, 100th Congress, October 28, 1988, Washington, DC.

The Union of Concerned Scientists. 1975. *The Nuclear Fuel Cycle.* Cambridge, MA: The MIT Press, 291 pp.

de Villiers, A.J. and J.P. Windish. 1964. *British Journal of Industrial Medicine.* 21:4.

Wagoner, J.K., V.E. Archer, B.E. Carroll, D.A. Holaday and P.A. Lawrence. 1964. *Journal of the National Cancer Institute.* 32:787.

*Environmental Radon: Occurrence, Control and Health Hazards.* Edited by S. K. Majumdar, R. F. Schmalz and E. Willard Miller © 1990, The Pennsylvania Academy of Science.

# *Chapter Two*

# INDOOR RADON IN THE UNITED STATES AND CANADA—1950-1984

**ROBERT F. SCHMALZ**
Department of Geosciences
The Pennsylvania State University
University Park, PA 16802

## INTRODUCTION

In his book, *De Re Metallica,* published in 1556, Georgius Agricola observed that miners from Joachimstahl and Schneeberg, Saxony (Bavaria), were subject to "a terrible consumption" which was almost invariably fatal. He attributed the disease to "corrosive dust" and "pestilential vapours" given off by the ores and inhaled by the miners at work (Agricola 1556). Miners in the Freiberg District continued to be plagued by the same affliction until well into the present century. The disease was identified as lung cancer in the 1920's, and by 1929, radon gas, known to be abundant in the mines of the Freiberg District, was regarded as the principal causative agent (see Brill, D.R., this volume).

Nuclear research and the development of the atomic bomb during World War II stimulated vastly expanded production of uranium from mines in Canada, along the margin of the Canadian Shield, and in the United States, mainly on the Colorado Plateau but in other areas (including Pennsylvania) as well. Radon-induced lung cancer was an acknowledged health hazard for mine and mill workers involved in this rapidly growing industry, and special precautions were taken to monitor the workers' exposure and to protect them against excessive radiation. Long-term epidemiological studies were instituted to provide essential data relating exposure levels to the incidence of lung cancer and other radiation-induced illnesses later in life (Fabrikant, et al. 1988). It was generally believed, however, that the health risk was largely confined to a rather small population directly involved in mining and processing the uranium, thorium and radium ores and the materials produced from them.

Beginning in about 1950, and continuing through the present day, a series of discoveries showed that the population "at risk" was very much larger, and that indoor radon in his own house was probably the principal source of the average citizen's exposure to ionizing radiation. Today, the National Research Council estimates that exposure to naturally occurring radon may be responsible for as many as 22,000 lung cancer deaths among non-smokers annually in the United States (Fabrikant, et al. 1988). The discoveries which led to this dramatic change of view can be divided into two broad categories: first, between 1950 and 1984, a number of investigations revealed that some segments of the non-mining population were exposed to excessive concentrations of radon released from certain building materials and as a result of unwise construction practices. Later, mainly since 1984, it was discovered that a vastly larger population, perhaps 25 to 50 million persons in the United States, are exposed regularly to unsafe levels of naturally-occurring radon in their homes and work places. These discoveries were widely discussed in the media, and stimulated public concern culminating in the enactment of the Indoor Radon Abatement Act (PL 100-551) which sets as a national goal that,

> ". . . the air within buildings in the United States should be as free of radon as the ambient air outside of the buildings."
>
> (15 UCC 2661)

The consequences and control of naturally occurring radon are discussed in detail elsewhere in this volume; it is the purpose of this chapter to review, briefly, the discoveries of the period between 1950 and 1984, in which radon exposure was found to result from a variety of human errors, particularly in housing construction in the United States and Canada.

## RADON SOURCES

Before examining the history of indoor radon occurrences in Canada and the United States since World War II, it is appropriate to summarize very briefly the origin of the gas itself.

Radon is a colorless, odorless, water-soluble gas, one of the "noble" gases (helium, neon, argon, xenon, and krypton) which form few chemical compounds. The isotopes of radon are produced by the radioactive decay of radium as shown in the two decay series in figures 1-a and 1-b. Only two radon isotopes, radon-220 and radon-222, play a significant role environmentally.

The uranium series (1-a) is the primary source of naturally occurring radon. Uranium-238 decays by a series of alpha and beta emissions to produce radon-222. This radon isotope has a half-life of 3.8 days which allows ample time for the gas to diffuse through the soil or to be carried in aqueous solution from its source into the human biosphere. The principle health hazard presented by radon is not due to alpha emission from the gas itself, but from the solid daughter isotopes which may be trapped or absorbed in the body: polonium-218 and polonium-214 deposited on the bronchial epithelium are believed to be the primary cause of radon-induced lung cancer, while lead-210 and polonium-210 appear to be the principal factors of radiation damage to bone and soft tissue elsewhere in the body (Fabrikant, et al. 1988).

The decay of thorium-232 (figure 1-b) also produces isotopes of radon (radon-220 and radon-219) but these have such short half-lives that their progeny are of relatively minor consequence—ordinarily their abundance is less than one-third that of the radon-222 daughters.

(1-a) The uranium series:

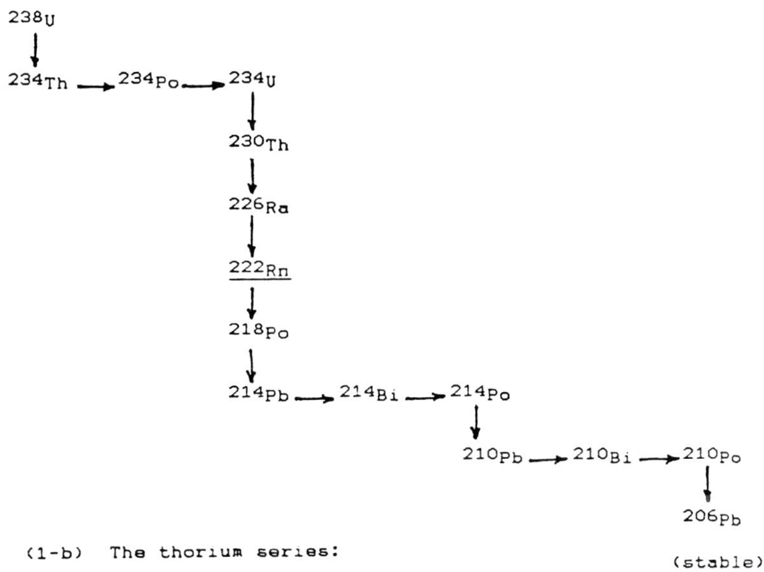


(1-b) The thorium series:

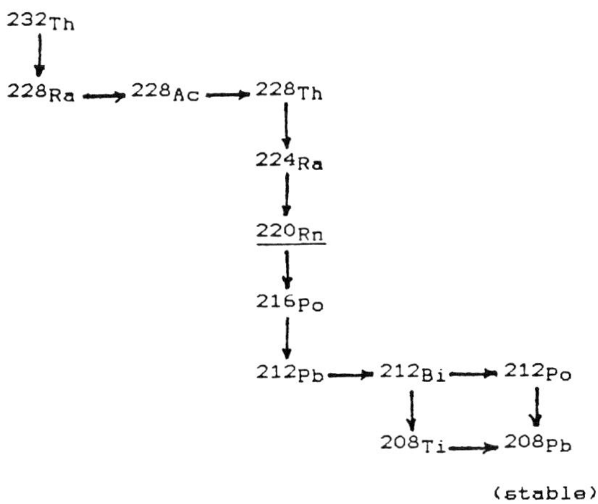


FIGURE 1. Principal nuclides in the radioactive decay series of uranium-238 and thorium-232. Vertical displacements represent alpha decay, horizontal displacement represent beta decay.

Radon is produced by the decay of radium in rock or soil, and its occurrence may therefore be related to the natural distribution of radium and its parent elements (notably uranium and thorium). Uranium and thorium are usually found in sialic rocks (eg: alkaline granite, granodiorite, granite-gneiss, pegmatite, and extrusive igneous rocks of similar composition); both elements are also found in hydrothermal veins usually as pitchblende. Uranium is soluble in oxidizing solutions, but precipitates when reduced. Consequently, uranium (and its progeny) may be concentrated in reducing sedimentary environments represented by black shales, organic-rich limestones or organic debris enclosed in sand- or siltstones.

Although radium rarely occurs except in close association with its progenitors, it is a divalent alkaline earth element and behaves geochemically like the other elements in the group (berylium, calcium, magnesium, strontium and barium); it has a particularly close geochemical affinity with barium (indeed, the two elements almost invariably occur together). The half-life of radium is short (1,620 years) and therefore the element is found in the absence of its progenitors only under unusual conditions and in geologically recent deposits (Choppin & Rydberg 1980).

Finally, as a chemically inert but water-soluble gas, radon may travel long distances from its immediate radium progenitor, in the gas phase by convection or diffusion, or in aqueous solution. Consequently, radon may be relatively abundant in areas far removed from any identifiable concentration of radium, thorium or uranium. This fact makes prediction of radon distribution particularly difficult, and has frustrated many efforts to use broad regional radiation surveys to define local areas of greater-than-average radon hazard (eg: Otten 1989; Peake 1989).

If radon is present in soil or bedrock in significant quantities, there are four primary pathways by which it may enter a building:

i) as a gas, by convection or diffusion from the soil or rock adjacent to the structure (usually through the foundation, slab or crawl-space),
ii) by decay of radium contained in the materials of which the structure is built or introduced into the structure for practical or decorative purposes,
iii) by degassing of water employed for domestic purposes, particularly as a result of heating, and
iv) as a component of commercial natural gas fuel.

Although there are exceptions it appears that in most cases today the first of these pathways is by far the most important. Building materials were significant contributors to indoor radon concentrations in the United States and Europe in the past, but the imposition of more restrictive building standards has eliminated the problem in most areas today.

## URANIUM MINING

The economics of uranium mining bear directly on the emergence of the radon problem in houses in Canada and Colorado. For this reason we must include a summary of that activity.

Uranium has been mined in the southern and central Rocky Mountain region of the United States since the early part of this century. The largest reserves are in New Mexico and Colorado; somewhat smaller quantities of ore are found in Wyoming, Utah and Montana. Both uranium and radium have been mined in Canada since 1933, at a number of localities extending in a broad arc from Blind River, Ontario, to Great Bear Lake, in the Northwest Territory. Because radon had been recognized as a carcinogenic agent since 1929, efforts were made to minimize the health hazard to mine and mill workers by monitoring radiation levels, regulating exposure time, and improving ventilation in mines and mills (Fabrikant, et al. 1988). Despite the fact that significant levels of radon had been detected in indoor air in buildings far removed from any recognized source (Schmid, 1934), radon was considered a serious health hazard only for those directly engaged in uranium and radium production.

Uranium production generates very large volumes of waste: even high grade ores rarely yield more than 2-4 pounds of $U_3O_8$ per ton. Open pit mining requires stripping large volumes of overlying rock ("overburden") to gain access to the uranium-bearing strata. In Wyoming, for example, nearly forty tons of rock must be moved for each ton of ore produced. The underground mines of Colorado are more efficient, but they produce at least one ton of waste for each ton of ore (US EPA, 1978-a). In the first step of beneficiation, the ore is milled to yield an ore concentrate containing approximately 75% $U_3O_8$. Thus, combined mine and mill waste generated at or close to the mine location may range between 1,500 and 60,000 tons for each ton of concentrate shipped. Ten years ago, the Environmental Protection Agency estimated that the quantity of uranium mill tailings in the United States was in excess of 110 million tons (US EPA 1978-a; US EPA 1979). Disposing of mine waste and of the substantially greater volume of mill tailings therefore presents a major problem for uranium producers.

Mine waste, essentially crushed and broken rock, is commonly used to backfill open pit mines. The volume of the waste usually exceeds the volume of the original rock, however, and the excess must be disposed of on the surface. Underground operations also generate a larger volume of waste than can be accommodated in disused shafts and tunnels below the surface. The vastly larger volume of mill tailings pose a more severe disposal problem. Firstly, because the quantity is great—1000 to 1500 times the tonnage of ore processed. Secondly, because the tailings have been finely pulverized and treated with chemicals which, although usually non-toxic, may nonetheless be inimical to the environment. Thirdly, although its significance was not fully appreciated at first, because the tailings contain most of the thorium and radium which was present in the original ore.

## DURANGO AND GRAND JUNCTION, COLORADO

Uranium mining, which was a small but established industry in the western United States prior to World War II, expanded rapidly, first with the Manhattan Project and after the war in response to the needs of a burgeoning nuclear power industry. The increased production from the mines was accompanied by a rapidly growing burden of mine waste and mill tailings. Beginning in 1953, and continuing for at least a decade, mine operators in the area of Durango and Grand Junction, Colorado, transformed the problem into an asset by utilizing the waste as construction material. The crushed rock of the mine waste made excellent landfill and concrete aggregate; the pulverized mill tailings were used to back-fill around foundations and to replace sand in mortar and in the manufacture of cement building blocks.

Because of the very limited penetration power of alpha particles (Choppin & Rydberg 1980) it was generally believed that the presence of alpha emitters such as uranium, thorium or radium embedded in concrete or plaster would pose little or no radiation hazard. Unfortunately, this concept overlooked the fact that as a gas, radon could escape from the containing material by diffusion, and its relatively long half-life would allow it to accumulate to dangerously high levels in the enclosed space of a house or other structure. By 1966, radon levels in scores of houses in both Durango and Grand Junction were found to exceed the level (4 pCi/L) deemed "Acceptable" by the Environmental Protection Agency. The practice of using mine and mill waste in construction was stopped, but a survey of 1,500 buildings identified at least 600 in which remedial action was indicated (Collé 1980). Management of uranium mine and mill waste is now strictly regulated, and the use of such materials in most ordinary construction is prohibited.

The Colorado experience clearly defined and directed attention toward a radiation hazard which had not been previously acknowledged: the accumulation of radon inside residential structures built of or upon rock or soil abnormally rich in natural alpha-emitters. The problem was clearly aggravated in structures built to withstand the rigors of a Colorado winter, and by the use of materials which had been artificially enriched in radio-nuclides.

## CANADIAN URANIUM MINES

The Canadian experience with indoor radon is very similar to that in Colorado. Radium was mined at several localities in Ontario, Saskatchewan and the Northwest Territory, beginning in the early 1930's. In the post-war years, many of the former radium mines produced uranium as well, and several new mines were opened primarily to supply uranium for the Canadian nuclear power industry. Notable among these uranium mining and processing localities were Port Hope, Elliot Lake and Bancroft, Ontario, and Uranium City, Saskatchewan. In all four communities, houses and other buildings were soon found to be contaminated by unacceptably high concentrations of radon (Findlay 1980).

The problem was recognized early at Port Hope, Ontario, where a major facility for the production of uranium oxide concentrate ("yellow cake") was built at the site of a forty-year old radium production plant. Houses (and other buildings) built after the War to accommodate the growing population of the community were erected on land which had previously been contaminated by waste from radium processing. Radon diffusing from the contaminated soil through foundation walls and floor slabs raised interior levels in many buildings above the 0.02 Working Level (approximately 2 pCi/L) limit set by the Atomic Energy Control Board of Canada. The problem was most severe during winter months when ventilation is restricted (to conserve heat) and when interior heating gives rise to a "stack" effect, drawing soil gas into the structure from beneath the frozen soil outside.

Radon control efforts at Port Hope focussed upon the removal of contaminated soil from around and beneath the affected structures, and the development of improved building methods for use in new construction (Eaton 1980).

At Elliot Lake, Bancroft and Uranium City, radon appears to be derived primarily from natural radium concentrations in bedrock and soil, although some mine and mill waste was used for landfilling on a limited scale in all three communities. The gas diffuses from its source to the buildings through the permeable soil and sedimentary deposits upon which the towns are built. It enters the buildings through "numerous unplanned openings" in the foundation walls or floor slab which result from "inadequate levels of workmanship common to normal construction practices" (Findlay 1980). The average cost of remediation (mainly sealing openings in the foundation) was found to exceed, usually by ten percent or more, the cost of preventive measures (foundation sealing, forced sub-foundation ventilation) in new construction (Findlay 1980).

To some degree the Canadian experience parallels that in Colorado: in both cases elevated indoor radon levels were observed in mining or ore processing communities where rocks and soil were known to be enriched in radio-nuclides; in both cases buildings were weatherized to withstand rigorous winter climates; and in both cases injudicious use was made of uranium mine and mill waste. There are, however, two significant differences. At Elliot Lake in particular, above normal radon levels were observed in structures built on rocks with average uranium/thorium/radium content, and in all the Canadian localities, radon entry could be traced to inadequacies inherent in ordinary construction practices.

## POLK COUNTY, FLORIDA

Central Florida has been the major U.S. producer of phosphate rock for many years. Primarily used for fertilizer, the rock serves also as an important feedstock for various chemical industries. The phosphate occurs as a mixed alluvial sand and nodular paleosol (?) deposit (land-pebble phosphate) of Pliocene age, which is recovered in large-scale, open pit "stripping" operations. The overburden is generally less than 40 feet thick, and rests upon a thin (3 foot) leached zone at the top of the phosphate stratum. The ore, composed mainly of calcium phosphate (apatite) with admixed calcium carbonate, silica sand and clay, contains up to 20 pCi/gm radium-226. The leached zone at the top of the phosphate bed is enriched in radium relative to the overburden above and the ore beneath (Guimond et al. 1978; Mills et al. 1977).

Industrial recovery and utilization of the Florida phosphate rock leads to three potentially hazardous radiation exposure pathways.

The first, of course, is the potential hazard to workers directly involved in the mining, processing or final application of the phosphate materials. The hazard is not inconsequential as the radium-226 content of phosphate fertilizer from Florida may reach 60 pCi/gm. This pathway does not relate directly to the present discussion of indoor radon, however.

The second pathway involves the peculiarities of the mining operation itself. As noted earlier, the phosphate rock lies beneath up to forty feet of overburden, including a leached zone, usually considered waste, of approximately 1-5 feet thickness. Strip mining is a continuous process which moves across the ore body in a systematic fashion. A drag-line is used to strip the overburden from the active part of the mine, depositing the waste in an area of the mine from which the ore has already been recovered. As soon as it is exposed, the ore is removed, and the mined-out area is ready to receive the waste overburden from the next section of the mine. In this process, the stratification of the overburden is inverted: the original surface layers are excavated first and are dumped at the base of the waste pile; the deeper layers then are dumped on top. Thus, the radium-enriched rock from the leached zone lies at or near the surface of the mined-out land after reclamation. Houses and other structures built on such reclaimed phosphate lands in Polk and Hillsborough Counties are found to have radon concentrations in excess of 0.05 WL (Johnson, W.C., 1979; Guimond, et al., 1978; Phillips, et al., 1980).

Here, as in Canada and in Colorado, the principal route by which the radon enters the structures appears to be by diffusion through openings in the foundation or floor slab. Where remediation is practical, the suggested corrective measures are similar to those proposed in Canada: sub-slab forced ventilation, sealing, and enhanced ventilation in the occupied portions of the structure.

The third pathway by which radon from the Florida phosphate industry reaches the human biosphere is through the use of by-product gypsum in the manufacture of building materials. The phosphate ore is subjected to a two-stage beneficiation process. The first stage is essentially mechanical concentration to separate the phosphorus-bearing apatite from carbonate minerals, clay and sand with which it is found. This process yields approximately two tons of sand, clay and carbonate "slime" for each ton of phosphate concentrate. Because of the geochemical affinity between calcium and radium, this mill waste is enriched in radium-226. Large quantities of mill tailings are used in back-filling the strip mines, and add to the evolution of radon from reclaimed lands.

In the second phase of beneficiation, most of the phosphate concentrate is treated with sulfuric acid to produce phosphoric acid and gypsum ($CaSO_4 \cdot 2H_2O$). Instead of discarding the by-product gypsum ("Phosphogypsum"), producers in Florida and Europe utilized it as raw material for the manufacture of plaster and plaster-based building products (mainly gypsum wall panels). Elevated indoor radon levels in houses throughout the southeastern United States, in England and in Europe have been traced to the use of radium-rich building materials manufactured from by-product phospho-gypsum (US EPA, 1978-b; UNSCEAR, 1977; O'Riordan, et al., 1972). Uranium mine and mill tailings, and phospho-gypsum products are not the only building materials which may contribute to elevated radon levels (see Table 1) but they are, perhaps the most conspicuous examples.

## THE ENERGY CRISIS OF 1973-74

The OPEC oil embargo of 1973-74 stimulated widespread efforts to conserve fuel in the United States, Canada and western Europe. Two aspects of that conservation effort are of particular relevance to any discussion of indoor radon.

TABLE 1

*Radium concentrations in natural and man-made building materials.*

| Material | $^{226}$Ra Bq/kg | $^{226}$Ra pCi/gm |
|---|---|---|
| Concrete | 10-80 | 0.3-2.1 |
| Clay brick | 20-200 | 0.5-5.5 |
| Portland cement | 10-50 | 0.3-1.4 |
| Granite | 100-200 | 3.0-5.5 |
| Volcanic tuff | 100-600 | 3.0-16.2 |
| Gypsum (natural) | 5-20 | 0.1-0.5 |
| Alum shale-based light concrete | 300-2500 | 8.1-67.5 |
| Fly ash | 50-300 | 1.4-8.1 |
| Ca-silicate slag | 1000-2000 | 27.0-54.0 |
| By-product gypsum | 500-2000 | 13.5-54.0 |

(modified from Stranden, 1988)

Firstly, the development of "energy-efficient" housing involved improved insulation and reduced air exchange rates in both new construction and, by retrofitting, in existing structures. Smaller, double- and triple-glazed windows, heavier insulation in walls and ceilings, careful caulking of joints and cracks, installation of vapor-barriers and other procedures were devised to reduce interior ventilation rates to values in the range of 0.2 to 0.4 air changes per hour (ach) (Haywood, et al., 1980; Fisk & Turiel, 1983). One consequence of the reduced ventilation in such "super-weatherized" houses was the accumulation of pollutants from a variety of sources - cigarette smoke, vapors exhaled from foam insulation and other plastics, cleaning solvents, carbon monoxide and carbon dioxide from space heaters, and even cooking vapors (Spengler & Sexton, 1983). Studies to determine the nature of the health hazard, if any, presented by these indoor polluants quickly identified radon among them, although usually the gas was present in concentrations which would have escaped notice with higher ventilation rates. Efforts to discover the source(s) of the radon identified domestic water supplies (where supplied from underground sources) (Hess, et al., 1983; Horton, 1985), and building materials as significant contributors in some locations. Diffusion of soil gas into the buildings through openings in the foundation or slab was found to be the most important pathway by which radon enters the buildings in most cases, however.

The second consequence of the oil embargo which bears on indoor radon concentrations was a growing interest in active and passive solar heating. Residential solar heating in colder climates requires a thermal reservoir in which heat can be stored during warm sunny days, and from which heat can be drawn at night or when the sun is obscured by clouds. Because of its high heat capacity, water is the most suitable storage medium, but its use requires a heat exchanger and large storage tanks. Rock-filled reservoirs are easier to design, construct and maintain, and are therefore widely used despite their reduced thermal efficiency. With the low air exchange rates required in a solar heated building, indoor air may remain in contact with granitic rocks in a thermal reservoir long enough to acquire a significant, even dangerous, radon concentration. Again, this effect would almost certainly be overlooked in a structure with ordinary air exchange rates.

The discovery of extremely high radon levels in the Stanley Watras house in 1984 stimulated widespread screening of houses in many parts of the United States. It appeared clear in the Watras case that the source of the gas was the soil beneath the house, and that it entered the building through cracks and openings in the substructure. This inference was confirmed by the results of the remediation program undertaken by the Philadelphia Electric Company and the Pennsylvania Department of Environmental Resources. Although the location of the Watras house suggested that the gas might be derived from sialic igneous and metamorphic rocks of the Reading Prong, measurements in neighboring houses revealed no systematic radon anomaly (see M. Reilly, elsewhere in this volume). Screening measurements in other parts of Pennsylvania, and elsewhere in the United States, soon showed that radon contamination was widespread, and occurred in houses far removed from areas where elevated radon levels might be anticipated (Cohen, 1986; Nero, et al., 1986).

## SUMMARY

Prior to the second World War it has been established that exposure to radon gas was linked to the incidence of lung cancer. Critical dose levels had not been established, however, and it was generally believed that any serious health risk was largely limited to those directly involved in the mining and processing of uranium, thorium and radium. Except in mines, the health hazards uniquely associated with these alpha-emitters were thought of mainly in terms of their inherent chemical toxicity or radioactivity; the hazard could therefore be reduced to acceptable levels by physical isolation or shielding or both.

With the expansion of uranium production during and after the War, however, it was apparent that the chemical and physical properties of radon gave it mobility in aqueous solution or in the gas phase which made conventional methods of containment or shielding ineffective. The consequences were especially apparent in localities where soils were naturally rich in the radon progenitors, or where human activity had concentrated those elements in and around residential or public buildings. The discovery that radon was concentrated inside structures under these circumstances greatly increased the size of the population considered to be at risk, but indoor radon contamination was still not regarded as a health hazard of general concern.

Screening studies stimulated by the discovery of elevated radon levels in houses in Pennsylvania's Reading Prong revealed for the first time how widespread the problem might be. Today, although we have acquired sufficient data to estimate the number of houses in which radon levels exceed the national average and/or the level considered "Acceptable" by the Environmental Protection Agency, we still are unsure at what concentration indoor radon becomes a significant health threat, and therefore how grave the problem it poses may be.

## REFERENCES

Agricola, G. *De Re Metallica.* Latin edition of 1556. Translated by H.C. Hoover and L.H. Hoover. New York, Dover Publications, Inc., 1950. 592 pp + 2 appendics, index.

Choppin, G.R. & R. Rydberg. *Nuclear Chemistry: Theory and Applications.* New York, Pergamon Press, 1980. 574 pp. + 14 appendices, indices.

Cohen, B.L. (1986). A national survey of $^{222}$Rn in U.S. homes and correlating factors. Health Physics, vol. 51, pp. 175-ff.

Collé, R. & P.E. McNall, Jr., eds. *Radon in Buildings: A Proceedings of a Roundtable Discussion of Radon in Buildings held at NBS, Gaithersburg, MD, June 15, 1979,* (SP-581). Washington, D.C., U.S. Department of Commerce, 1980. 77 pp.

Eaton, R.S. Radon Control in Housing in Canada. *In* Collé and McNall (1980), pp. 65-66.

Fabrikant, J.I., et al. *Health Risks of Radon and Other Internally Deposited Alpha-Emitters - BEIR IV.* Washington, D.C., National Academy Press, 1988. 395 pp. + 8 appendices, glossary, index.

Findlay, W.O. Application of Radon Standards to New and Existing Housing in Elliot Lake, Ontario. *In* Collé and McNall (1980), pp. 66-67.

Fisk, W.J. & I. Turiel. 1983. Residential air-to-air heat exchangers: Performance, energy savings and economics. Energy and Buildings, vol. 5, p. 197.

Guimond, R.J., W. Ellett, J. Fitzgerald, S.F. Windham & P. Cuny. *Indoor Radiation Exposure Due to Radium-226 in Florida Phosphate Lands.* (EPA 520/4-78-013). U.S. Environmental Protection Agency, Washington, D.C. 1978.

Haywood, F.F., A.R. Hawthorne and D.R. Stone. 1980. Measurements of Airborne Pollutant Concentrations Inside a Well-Insulated Structure with Low Ventilation Rate. *In* Collé and McNall, 1980. pp. 64-65.

Indoor Radon Abatement Act (PL 100-551). 1988. (15 USC 2661 et seq.)

Johnson, W.C. *Radon Daughter Concentrations in Structures in Polk and Hillsborough Counties, Florida,* 10th Annual National Conference on Radiation Control. HEW Publication (FDA)-79-8054, U.S. Department of Health, Education and Welfare, Washington, D.C. 1979. pp. 219-226.

Mills, W.A., R.J. Guimond & S.T. Windham. 1977. Radiation exposures in the Florida phosphate industry. Fourth International Congress of the International Radiation Protection Association, Paris, France.

Nazaroff, W.W. & A.V. Nero, eds. *Radon and Its Decay Products in Indoor Air.* New York, N.Y. John Wiley, 1988. 518 pp. + index.

Nero, A.V., M.B. Schwehr, W.W. Nazaroff & K.L. Revzan. 1986. Distribution of airborne radon-222 concentrations in U.S. homes. Science, vol. 234, pp. 992.

O'Riordan, M.C., M.J. Duggan, W.B. Rose & G.F. Bradford. 1972. *The Radiological Implications of Using By-Product Gypsum as a Building Material.* National Radiological Protection Board, Report 7. Harwell, Didcot, U.K.

Otton, J.K. 1989. Mapping of Radon Potential of Rocks and Soils. Professional Geologist, vol. 26, #5, p. 8, (Abstract).

Peake, R.T. 1989. Assessing Radon Potential of the United States. Professional Geologist, vol. 26, #5, p. 8, (Abstract).

Phillips, C.R., S.T. Windham & J.A. Broadway. Radon and Radon Daughters in Buildings: A Survey of Past Experience. *In* Collé and McNall (1980), pp. 37-44.

Schmid, von E. 1932. Messung des Radium-Emanationsgehaltes von Kellerluft. Mitteilung aus dem Physikalischen Institut der Universität Graz, No. 82, pp. 233-242.

Spengler, J.D. & K. Sexton. 1983. Indoor air pollution: A public health perspective. Science, vol. 221, p. 9-ff.

Stranden, E. Building Materials as a Source of Indoor Radon. *In* Nazaroff and Nero, 1988. pp. 113-130.

UNSCEAR 1977. *Sources and Effects of Ionizing Radiation.* Report to the General Assembly of the United Nations Scientific Committee on the Effects of Atomic Radiation (UNSCEAR). United Nations, New York. 1977.

U.S. EPA. 1978-a. *Radiation Protection Activities, Annual Report.* Office of Radiation Programs, EPA 520/4-78-003. Washington, D.C. 1978.

U.S. EPA. 1978-b. *Conference of Radiation Control Program Directors, Inc., Natural Radioactivity Contamination Problems.* Report: EPA 520-4/77-015. U.S. Environmental Protection Agency, Washington, D.C. 1978.

U.S. EPA. 1979. *Radiological Quality of the Environment in the United States: 1977.* (EPA 520/1-77-009). U.S. Environmental Protection Agency, Washington, D.C., 1979.

*Environmental Radon: Occurrence, Control and Health Hazards.* Edited by S. K. Majumdar, R. F. Schmalz and E. Willard Miller 

# *Chapter Three*

# THE INDEX HOUSE: PENNSYLVANIA RADON RESEARCH AND DEMONSTRATION PROJECT POTTSTOWN, PENNSYLVANIA, 1986-1988

**MARGARET A. REILLY**
Division of Environmental Radiation
Pennsylvania Department of Environmental Resources
Bureau of Radiation Protection
Harrisburg, PA 17120

## INTRODUCTION

During the closing days of December, 1984, there was brought to the attention of the Pennsylvania Department of Environmental Resources (DER), Bureau of Radiation Protection, a sequence of events which would have enormous impact on radiation protection programs at the state and national levels in the following years.

Until that time, the attention of the Bureau had been focused upon the accident at Three Mile Island and upon the many changes of policy and procedure required to respond to lessons learned from that event. In those years and in the twenty or so years of organizational life of the Bureau prior to that event, the business of the agency had dealt with a specific family of institutional and technical matters. The technical matters included interest in radiation-producing equipment such as X-ray generators and accelerators, discrete sources of radioactive materials and nuclear power stations. The institutional context of these radiation sources placed them, for the most part, in the hands of trained professionals with the knowledge to deal responsibly with the associated radiation hazard within an established structure of regulation, registration and licensure. The radiation protection community was well settled into a pattern suited to deal with controlled radiation sources used in a known setting. Procedures and standards had been worked out, authorities and responsibilities were established and the system was stable. Within this framework, natural radioactivity was viewed as a source of potential error in sensitive radiation measurements, and as a datum to which other radiation measurements might be referred (eg: "twice background"). It was a kind of neutral-to-benign environmental characteristic, ubiquitous and more or less constant. As a natural feature of the environment, "background" radiation was assumed to be harmless and beyond regulatory concern.

Under these circumstances, it is not surprising that in 1984 radiation protection agencies like the DER were unprepared to deal with natural radiation problems like indoor radon contamination. There was no existing system to identify radon sources, comparable to the registration and licensing of medical and industrial radiation sources. Neither was there any identifiable "user" who could be considered responsible and liable for remediation costs. The kinds of radiation-producing equipment and sources of radioactive materials then subject to regulation employed a common language and a familiar set of units; the language used in radon measurements, units, mechanisms and controls was to a large extent alien. In short, no radiation protection infrastructure existed to deal with the problem of radon contamination, and no one knew what to do about radon contamination once it was identified.

All of this was about to change.

## DISCOVERY, DECEMBER 19, 1984

About noon of December 19, 1984, the Senior Health Physicist at the Philadelphia Electric Company's Limerick (Nuclear) Generating Station called the Department of Environmental Resources with a question which soon proved to be a profound, world-class problem. A contract engineer working at the facility had been tripping the newly installed exit portal monitor every time he had reason to leave the site. Because tripping the portal monitor for any reason caused an administrative burden to the facility and delays and inconvenience to the engineer, a thorough investigation was immediately undertaken. Inasmuch as the reactor had not yet achieved criticality, fission products could not be responsible for the contamination the engineer was carrying past the exist portal monitor. This conclusion was quickly confirmed by gamma spectrometric analysis of the engineer's clothing, which was found to be contaminated by beta- and gamma-emitting decay products of radon.

The engineer had already begun to suspect that he was bringing the contamination from his home. Earlier that week he had used the portal monitor on his way into work, at which time the contamination on his clothing and his person set off all the detectors in the array. (It is useful to point out that the detectors used were large-area thin window proportional counters which are exquisitely sensitive to beta radiation as well as gamma emission. This makes them good detectors for reactor-produced fission products, all of which are beta emitters. It also makes them excellent detectors for radon decay products, several of which are beta emitters as well.) No one could dispute the engineer's conclusion that the source of the contamination was his own home; he asked the utility to check it.

## INITIAL INVESTIGATIONS

The utility arranged for its consultant to carry out 'radon' measurements in the engineer's house. The consultant employed a modified Kusnetz technique which uses airborne particulate alpha activity, with corrections, to determine the concentration of airborne radon decay products (Kusnetz, 1956). The unit of measurement is the "Working Level" (WL), which finds its origins in uranium mining. At first glance, a "Working Level" sounds innocuous, but in actuality continuous exposure to one Working Level delivers about 83 millirem per hour or 730 Rem per year to the bronchial epithelium.

The results of that initial survey, presented in Figure 1, were disturbing even with our limited level of understanding at the time. Radon decay product concentrations well in excess of 10 Working Levels were found in the house. Duplicate measurements yielded equal or even higher results, in some cases approaching 20 Working Levels. Due to the limitations of the sampling and counting arrangements which had been assembled on short notice, some uncertainty could be assigned to the absolute values observed, but the order of magnitude was certainly correct.

Because the problem appeared to be serious but unrelated to the operation of the Limerick Generating Station, the Senior Health Physicist asked the Bureau of Radiation Protection to take up the problem and talk with the engineer. The Bureau suggested a one day delay to allow its staff to become familiar with the language of radon, to gain some understanding of the risks presented by the radon concentrations involved, and to determine what steps the agency might take to help correct the situation.

## PREPARATION FOR THE FIRST VISIT BY DER/BRP

During the next twenty-four hours (December 20) the Bureau staff studied NCRP reports 77 and 78 (NCRP 84, 84-a), called federal agencies concerned with the issue, notified the DER management, estimated radiation doses due to radon exposure, and considered preliminary responses to the property owner's concerns.

On the following day, December 21, the problem was discussed with the property owner by telephone. He was justifiably concerned and asked whether

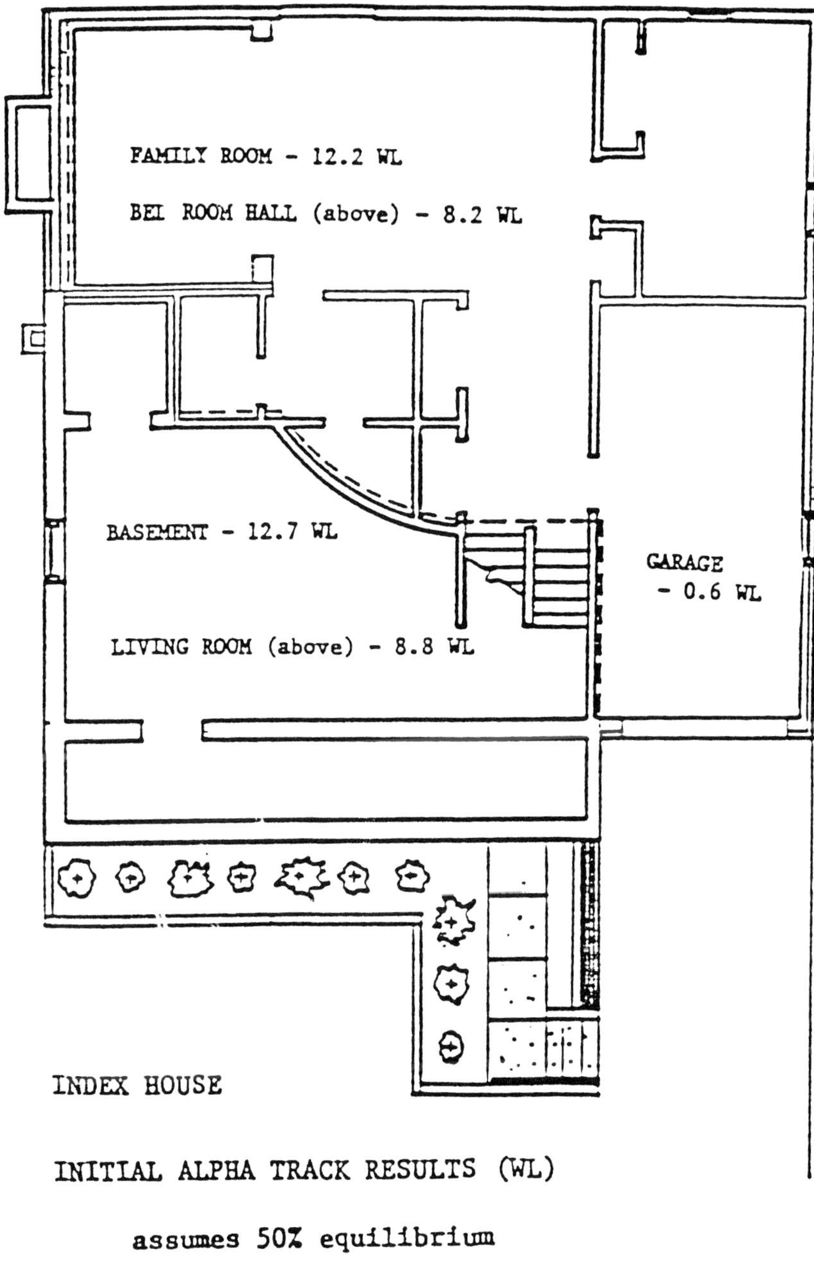


FIGURE 1. Floor plan of the Index House, showing initial alpha-track radon measurements expressed in Working Levels, assuming 50% equilibration.

he should get his wife and two small children out of the house. Since the dose equivalent to the bronchial epithelium from even 10 WL is about 0.8 Rem per hour, he was encouraged to do so. Arrangements were made for the Bureau staff to gain access to the house to make independent measurements on December 26th. The family spent the holidays with relatives in a neighboring state.

During the two business days between the initial contact and the first visit to the property, the focus of the Bureau's effort was on gathering and evaluating the available instruments, devices and methods with which to approach the problem. Among the radon sensors selected were alpha track detectors. These are passive film detectors which are exposed for several days (or longer), then developed chemically to make damage tracks in the plastic caused by the passage of alpha particles visible under the microscope. The number of alpha tracks, adjusted for the exposure time interval, is related to radon concentration in the air. The results are expressed in picocuries (of radon) per liter of air (pCi/L). Based on the sensitivity of the detectors used, it was expected that an exposure time of at least one week would be required. In an attempt to acquire earlier results, air samples were also taken using activated charcoal. This method depends on the adsorption of radon gas by the charcoal, with subsequent counting of ingrown decay products by gamma spectroscopy in the Radiation Measurements Laboratory in the DER laboratory at Harrisburg. Additional observations were made in the house using microR meters for measurement of low intensity gamma radiation, and a thin window alpha scintillation survey meter. Finally, thermoluminescent dosimeters (TLD's) for passive longer-term measurement of gamma radiation were annealed and packed.

During the same two days, geological aspects of the problem were investigated in an effort to identify possible source(s) of the radon gas. The house is situated on an edge of the Reading Prong, a physiographic province extending from near Reading, Pennsylvania, northeastward past Allentown, Bethlehem and Easton, into New Jersey and New York. The area occupied by the Prong in Pennsylvania is about 300 square miles, underlain by Pre-cambrian metamorphic rocks (mainly granitic gneiss) which contain moderate concentrations of uranium and thorium. The landscape is physiographically distinct, being somewhat hilly compared with the surrounding countryside, and is easily distinguished from an aircraft.

On December 26th, several of the BRP staff met the engineer and his wife, and visited the house. The house is a split-level style. The ground level garage with kitchen above constitutes one portion, and the balance of the house, basement and family room below and bedrooms, living room and bath above, the other. The house is heated by oil-fired hot water; domestic water is drawn from a private well; sewage disposal is by an on-lot septic system. The house is located in a neighborhood which until relatively recently had been an orchard. The site upon which the house is situated slopes steeply downward from the road to the north edge of the property. In 1984 the house was seven years old, and had had two previous owner-occupants. The engineer and his family had lived there for just under one year.

## FIRST VISIT, DECEMBER 26, 1984

During the visit of December 26th, duplicate alpha-track detectors and duplicate TLD's were posted in the on-grade garage, in the below-grade family room, in the living room, and in the bedroom hall. Since the couple expressed some concern that the source of the radon might be the structural materials of the house, a single alpha-track detector was placed in a standpipe serving the septic system. The rationale for the standpipe alpha detector was that if a detector exposed to soil gas but outside the house showed high readings, the source of the radon was probably soil gas rather than building materials. Both TLD's and alpha-track detectors were to be exposed in place for seven days.

Air samples collected on activated charcoal were analysed by gamma spectroscopy using an intrinsic germanium detector. Since values for collection efficiency of radon gas were not available from the manufacturer, the results could not be used for direct measurement of radon gas concentration. The gamma spectrum analyses did show, however, that if thoron was present, its concentration was vanishingly small compared to radon. If thoron ($^{220}$Ra) were present, it would seriously complicate subsequent measurements and dose estimates. Several gamma measurements were made using the microR meter. The highest measurement was 140 microR per hour in the service area of the basement. The gamma exposure rate in the house was a combination of gamma 'shine' from the airborne radon decay products plus gamma 'shine' from uranium and its decay products in the bedrock beneath the floor. Measurements in the back yard ranged from 40 microR per hour, presumably all from ground 'shine' from radon decay products in the native soil. The alpha survey meter was not directly useful in making measurements in the house since the plateout of alpha-emitting radon decay products on the window of the detector yielded a background which ingrew significantly with time. (It should be noted that radon decay products produced by alpha decay, particularly polonium-218, are born with a net positive static charge due to the stripping off of orbital electrons from the atom during decay recoil. This charge drives the daughter atom to attach to any surface which satisfies the charge requirement. Common attachment surfaces include airborne dust particles, clothing, detector windows, lung epithelium and television screens.)

Arrangements were made to return to the house on January 2, 1985, to collect the passive detectors and to take samples of soil, well water, and other materials. During the drive back to the rendezvous point a microR measurement indicated a gamma intensity in the vehicle of 70 microR per hour. At the rendezvous point, up to 100 microR per hour was recorded on staff clothing. Normal background in the area is 10-12 microR per hour. Staff exposure time in the house had been less than two hours. It was now quite clear that the radon decay products adhering to the engineer's body and clothing would have been sufficient to activate the portal monitors at the Limerick plant.

## SECOND VISIT, JANUARY 2, 1985

On January 2nd staff members from the Bureau of Radiation Protection returned to the house to collect the passive radon samplers set out inside the house on the visit of December 26th. Well water and outside soil samples were collected at the same time. Alpha-track detectors and TLD's placed in the house during the earlier visit were collected for analysis. The observed radon concentrations and TLD results are presented in Tables 1 and 2, and Figure 1, respectively. The same data, normalized to one year, are shown in Figure 2 by location in the house. Analysis of the well water sample by gamma spectroscopy showed a radon concentration of 15,000 picocuries per liter. Although high, this value is not disproportionately high when compared to well water samples collected from other households in the region. The soil samples were analysed for radium by gamma spectroscopy and found to range from 1.7 to 16.9 picocuries (radium) per gram. These values appear to be significantly higher than the national average (approximately 1 picocurie per gram).

During the visit on January 2nd, the staff observed that the family had returned to live in the house. This was not unexpected, since they had so simple alternative available, but it appeared inadvisable. On January 5th a letter from the Secretary of DER was hand-delivered to the family recommending their vacating the building. They moved out shortly thereafter.

TABLE 1

*Index House, Radon Concentrations, Expressed as Radon Concentration, and Working Levels (Average of Duplicates)**

| Location | Radon Concentration pCi/L | Working Levels** |
|---|---|---|
| Garage | 114.2 ± 10.2% | 0.57 |
| Basement | 2535.2 ± 4.3% | 12.68 |
| Family Room | 2441.3 ± 4.4% | 12.21 |
| Living Room | 1753.1 ± 5.2% | 8.76 |
| Hall - Bedroom | 1642.7 ± 5.4% | 8.21 |
| Septic Standpipe | 2666.8 ± 4.2% | 13.33 |

* Exposure time: 7 days.

** Assumes 50% equilibrium with Rn concentration divided by 200 to yield radon decay product concentration in Working Levels.

## JANUARY-MARCH, 1985

Following the January visit, efforts of the BRP staff were concentrated on the many problems associated with devising a remediation plan. The issue was extremely complicated. No existing government agency was authorized to carry out remediation of the property, even if funds were available to defray the cost. Only two engineering firms in North America had any experience in dealing with indoor radon contamination, and that experience was confined to correction of the problems resulting from the use of uranium mine and mill tailings in house construction. There was some question whether that experience could be applied directly to the present problem.

## APRIL-JUNE 1985

In early April, the Philadelphia Electric Company, owner of the Limerick Generating Station where the problem was first discovered, decided to undertake the remediation of the house as a research project to determine whether a house with such a profound indoor radon problem could be corrected at reasonable cost, and to evaluate the effectiveness of various remediation procedures. The Company contracted with ARIX, of Grand Junction, Colorado, one of the two North American firms experienced in radon remediation, to conduct the demonstration project. The project was conceived and planned as a multi-phase effort.

*Phase One:* The first phase consisted of excavating around the three below-grade foundation walls to the footing. The house is set into the hillside in such a way that the fourth foundation wall is wholly above grade. After sealing all foundation cracks with hydraulic cement, a sealant membrane of TROCAL (registered trade mark) was fastened to the outside of the wall using sheet metal flashing at the top edge. A four inch perforated drain tile was installed on compacted roadbase crushed stone at the level of the footing, draining downhill

TABLE 2

*Index House, TLD Results (Average of Duplicates)**

| Location | mrad/std. month** | mrad/year |
|---|---|---|
| Garage | 42.4 | 509 |
| Basement | 82.4 | 989 |
| Family Room | 74.1 | 890 |
| Living Room | 71.6 | 859 |
| Hall - Bedroom | 63.0 | 757 |

* Exposure time: 7 days.

** Standard month: 30.4 days.

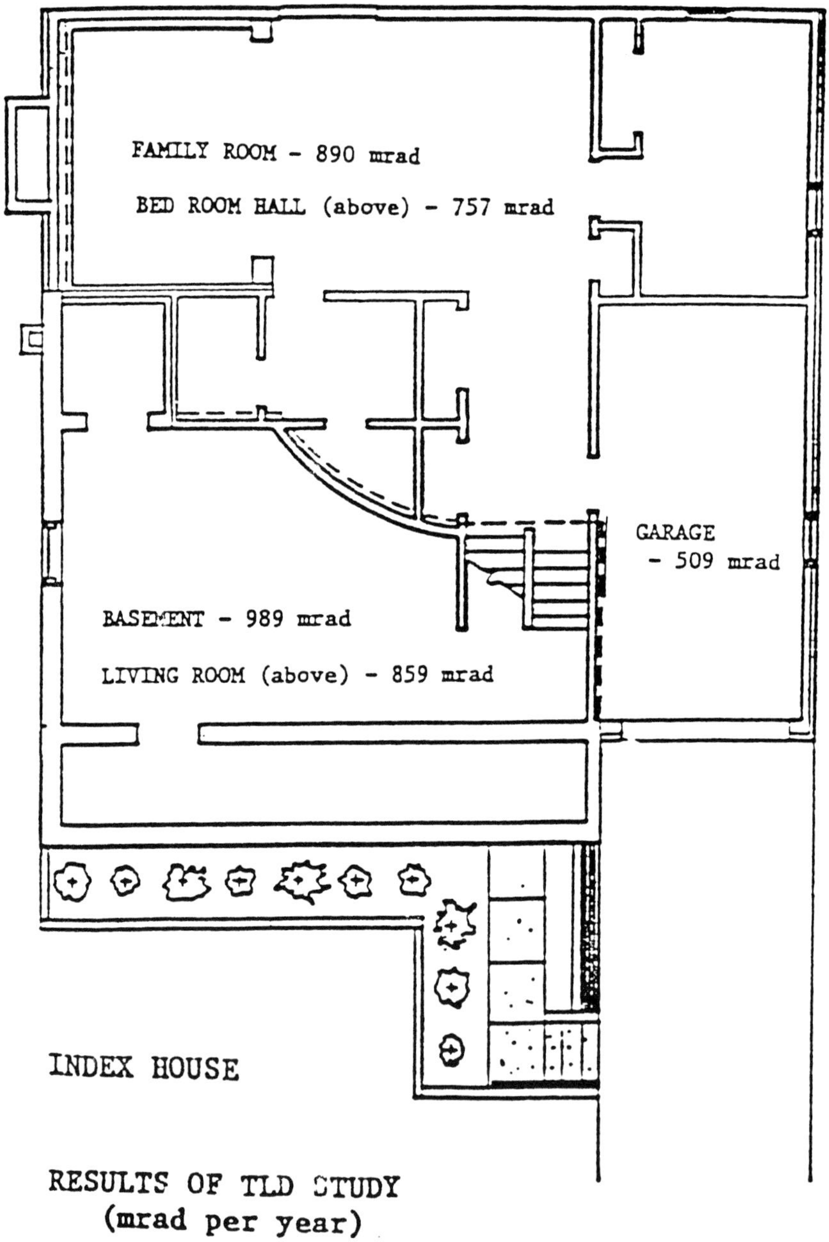


FIGURE 2. Floor plan of the Index House, showing TLD results, normalized to one year exposure in millirads.

to daylight at existing grade at the rear of the house. The purpose of the TROCAL membrane was to prevent the infiltration of radon-bearing ground water through the foundation walls as well as to act as a barrier against radon gas itself. (Note: Working drawings for remediation procedures developed in the course of the demonstration project are included in the DER/BRP publication, "General Remedial Action Details for Radon Gas Mitigation" which had been reprinted (with permission) in the Appendix of this volume. See also: Henschel, 1988).

The drain tile was covered with drainage fabric to inhibit future clogging of the tile. The excavation was then back-filled with washed limestone gravel to within a foot of existing grade. The original plan to backfill with the native soil which had been excavated was abandoned because it was feared that angular fragments in the excavated material might puncture the TROCAL membrane. The limestone fill was covered with drainage fabric, again to prevent clogging, and the excavation filled to grade with topsoil and relandscaped.

The net effect of this phase was a 39% reduction in the concentration of radon decay products inside the house.

*Phases Two and Three*: The second and third phases of the project sought to attack the problem from inside the house by sealing recognized existing penetrations.

In Phase two, an interior stub foundation wall separating the basement from the family room was sealed with two coats of epoxy paint, fitted with a soil-gas accumulator of perforated pipe, and exhausted to a roof vent.

In Phase three, all openings in the basement floor slab and all joints and cracks in the floor slab and foundation walls were sealed. The unused penetration for a sump was filled with soil and capped with concrete, and the joints between the concrete and the floor slab were sealed with silicone sealant. A ¾″ French drain at the joint between the floor slab and the foundation wall was sealed with siliconc, and all existing cracks and joints were dressed by power hammer, cleaned and filled with flowable silicone.

The combined effect of phases one, two and three was a reduction in indoor radon decay products of 78%.

*Phase Four:* The final remediation phase was originally conceived as the installation of a relatively simple sub-slab passive ventilation system. However, problems encountered in attempting to dress and seal floor cracks in Phase three drove the final phase of the project to a treatment which was more drastic than would be practicable in most other homes with a profound radon problem. The basement slabs in the Index House had apparently been poured directly on the native rock substrate, with no gravel ballast beneath. Further, the slab thickness ranged from two to six inches. This condition combined with the absence of ballast would foster crack development in the future. The original plan to cut a two-foot wide perimeter drain inside the basement wall and to install a perforated tile sub-slab ventilation system in an assumed gravel ballast layer, was obviously impossible.

Phase four, then, required the complete removal of floor slabs on the basement and family room levels to provide space for the actual sub-slab system. About eight inches of the underlying rock had to be hammered out. Because the bedrock beneath the house was, in effect, a low grade uranium "ore", it was considered essential to provide some form of gamma shielding against bedrock ground "shine". This shielding requirement was addressed by substituting diabase gravel for the limestone gravel originally specified as sub-slab ballast. The radiologically inert diabase, with a 15% iron content, has a higher effective atomic number than does limestone, and consequently offers greater gamma protection. Suitable crushed diabase was available locally.

The eight inch deep excavation under the entire house was filled with crushed diabase. A four inch perforated drain tile line, imbedded in the gravel, was installed around the perimeter on both levels of the ground floor. Each of these perimeter systems was attached to a solid tile riser which extended up through the house to the roof. A slightly negative pressure, on the order of a few Pascals, was maintained in the system by a combination of convection and suction from wind-driven turbines on the top of each riser. The gravel was covered by two layers of 30-pound felt, a sealed TROCAL membrane, and two inches of sand. The felt and sand were provided to protect the TROCAL membrane against accidental puncture during or after construction. Finally, a new four inch, wire-reinforced slab was poured over the entire floor on each level.

The combined effect of the entire effort (Phases one through four) reduced radon decay products in the basement level from 12.8 WL to 0.003 WL, or 99.9%. The residual gamma intensity was 17 microR per hour, only slightly greater than regional background. These results are summarized in Table 3.

*Waste Materials:* During the course of the project, substantial quantities of waste rock and soil were generated by the excavation of the below-grade foundation

TABLE 3

*Radon concentrations and radon progeny concentrations at four locations in the Index House, at the time of discovery, and following each phase of remediation.*

| Location | Initial* | | Post-phase I | | Post-phases II & III | | Post-phase IV | |
|---|---|---|---|---|---|---|---|---|
| | pCi/L | WL | pCi/L | WL | pCi/L | WL | pCi/L | WL |
| Basement | 1765 | 12.8 | 1360 | 9.2 | 365 | 2.8 | 0.41 | 0.003 |
| Family Room | 2026 | 15.2 | 1579 | 10.3 | 532 | 3.1 | 0.88 | 0.004 |
| Living Room | 1701 | 14.0 | 943 | 7.2 | 361 | 2.5 | 0.38 | 0.003 |
| Bed Room | 1453 | 12.0 | 1058 | 6.1 | 476 | 2.4 | 0.52 | 0.004 |

* These values differ slightly from those of Table 1 because they were made at a different time, and under slightly different conditions.

(Data from McKelvey, 1989)

walls (Phase one), the original basement floor slabs and the rock from under the old basement floor (Phase four). The original plan called for disposal in a commercial landfill. Concerns arose about possible future controversy, however misled, about hazards posed by these materials. To avoid any possible problem, plans were changed and the debris was deposited in the backyard of the property, which fortunately was large enough to accommodate it. The material was covered with topsoil and landscaped.

The project was completed in late June, 1985. The family returned to their home in time to celebrate the fourth of July.

## CONCLUSION

In retrospect, it is reasonable to speculate that the whole national interest in the hazards of indoor radon would not have materialized had not several things happened at the right time and in the right sequence. Firstly, it was required that this house with extremely high radon concentrations be purchased and occupied by someone who worked at a nuclear power station. Second, it was necessary that the facility have beta-sensitive portal monitors. (This type of equipment was not in universal use at the time.) Thirdly, it was required that the engineer have reason to enter the restricted area of the facility from which he could exit only through a portal monitor. (Many contractor engineers have no reason to go on-site). Finally, it was essential that the portal monitors be installed and in operation before the engineer's transfer to another site. (The engineer was originally scheduled for transfer to another state about the time the radon problem was found.)

In the years since the discovery of the Index House an infrastructure to address the problem of indoor radon has evolved. Exposure criteria have been set, assessment techniques and remediation methods have been developed and evaluated. Specialists within the private sector have been trained and have subsequently gained experience and insight. The regulatory structure to control the quality of testing and remediation services is in place. Public concerns about the problem have driven the establishment of some sound commercial practices which are very effective in mitigating the problem. These include radon screening for real estate transactions and the adopting of new construction techniques which protect against radon.

## REFERENCES

Anon. *General Remedial Action Details for Radon Gas Mitigation.* Pennsylvania Department of Environmental Resources, Bureau of Radiation Protection, Harrisburg, PA. May, 1985. 36 pp.

Henschel, D.B. 1988. Radon Reduction Techniques for Detached Houses: Technical Guidance. (2nd Edition, Revised January, 1988), (EPA/625/5-87/019). United States Environmental Protection Agency, Air and Energy Engineering Research Laboratory, Research Triangle Park, N.C. 192 pp. + 2 appendices.

Kusnetz, H.L. 1956. Radon Daughters in Mine Atmospheres—A Field Method for Determining Concentrations. American Industrial Hygiene Association Quarterly. Volume 17, p. 85.

McKelvey, W.F. 1989. *Close Out Evaluation and Report of the Remedial Action Demonstration.* Performed for Philadelphia Electric Company by VERSAR, Inc. March 1989.

National Council on Radiation Protection and Measurements (NCRP). 1984. Exposures from the Uranium Series with Emphasis on Radon and its Daughters. NCRP Report 77, Washington, D.C.

National Council on Radiation Protection and Measurements (NCRP). 1984a. Evaluation of Occupational and Environmental Exposures to Radon and Radon Daughters in the United States. NCRP Report 78, Washington, D.C.

*Environmental Radon: Occurrence, Control and Health Hazards.* Edited by S. K. Majumdar, R. F. Schmalz and E. Willard Miller 

# *Chapter Four*

# RADON IN RESIDENCES: A CARTOGRAPHIC ANALYSIS

**GEORGE A. SCHNELL[1] and MARK MONMONIER[2]**

[1]Department of Geography
The College at New Paltz
State University of New York
New Paltz, NY 12561
and
[2]Department of Geography
Syracuse University
Syracuse, NY 13210

## INTRODUCTION

Public concern with radon had an early focus on a specific physiographic feature—the Reading Prong in eastern Pennsylvania. Yet, evidence accumulated from across the country indicates that the radon threat is by no means limited to southeastern Pennsylvania. Three factors seem to control the very existence of the risk as well as its magnitude, namely, the geologic environment, the architectural style and condition of the dwelling, and the lifestyle of the occupants. Homes with high levels of radon tend to be located where the parent material is uranium bearing and the soil or water have been contaminated. Contaminated dwellings also tend to have foundations that admit the gas through cracks, holes, or sumps, and have occupants whose daily activities lower the air pressure in the building and draw in the gas.

This chapter addresses the geographic variation in the presence of radon at relatively high levels. Its focus is the Commonwealth of Pennsylvania but it considers the incidence of residential radon in adjoining counties in contiguous states, and by state throughout the nation. Cartographic analysis provides a robust assessment of the broad impact of physiography, the local effects of housing and lifestyle, and the quality of the best available spatial data. By promoting a fuller understanding of the pattern and magnitude of the risk, radon maps constitute a basis for a more effective and efficient prophylaxis.

Further, county-unit maps of age-adjusted mortality rates for successive decades demonstrate inconsistent and puzzling linkages between the geographies of radon and cancer.

## DATA

Radon maps for this study are based upon data compiled by the University of Pittsburgh Radon Project, directed by Bernard L. Cohen. These radon-level measurements, taken between February 1986 and April 1988, reflect tests using charcoal canisters collected from 38,872 living areas and 28,704 basements throughout the United States and reported at state, county, and Zip Code area levels of aggregation. Living areas include living rooms, dining rooms, family rooms, bedrooms, kitchens, and hallways. These samples include data from 5,790 living areas and 5,950 basements in Pennsylvania. The data largely reflect home testing by residents using a kit obtained from the Pittsburgh Radon Project and returned for analysis. In addition to seasonally adjusting their data to the national average, the University of Pittsburgh researchers excluded measurements for which the client reported that the house was either tested previously or within five miles of a house believed to have a high radon level. These exclusions would tend to dampen the effects of a rush to test in the vicinity of the Watras house and other notorious or prominent sites. In addition, county-unit summaries for basements exclude counties with fewer than four basement readings, and counties with fewer than four living-area tests are likewise omitted. Aggregated results for the five-digit Zip Code areas are sometimes based on even smaller samples, as few as two basement or living-area measurements.

The Radon Project tabulations include four areally aggregated summary measures: the geometric mean, the arithmetic mean, the percentage of tests between 4 and 20 pCi/L, and the percentage of measurements over 20 pCi/L. Radon maps in this chapter are based on three basement measurements and one living-area measurement. For basements, we mapped the geometric mean, which is less affected than the arithmetic mean by a few extremely high measurements. For basements and living areas separately, we added the two Radon Project percentages to obtain a cumulative percentage of more than 4 pCi/L, which was mapped. According to Environmental Protection Agency (EPA) and Centers for Disease Control (CDC) guidelines, 4 pCi/L is the "recommended action level" at which the homeowner should attempt to lower the radon level. EPA/CDC comparable-risk data regard 4 pCi/L as more serious than two hundred chest X-rays per year, and 20 pCi/L as more serious than smoking a pack of cigarettes a day.[1] We also mapped the percentage of basements tested with readings over 20 pCi/L.

## COUNTY-UNIT RADON MAPS

Four county-level maps show spatial variation for selected radon measurements. These maps represent the 67 counties of Pennsylvania and 38 contiguous counties in New York, New Jersey, Delaware, Maryland, West Virginia, and Ohio. Class breaks reflect the statewide averages for Pennsylvania of 50 percent for basements tested at more than 4 pCi/L, 10 percent of basements tested at more than 20 pCi/L, and 31 percent of living areas tested at more than 4 pCi/L. Moreover, the state's geometric mean was 4 pCi/L. For all four maps, half the state average served as the lower break, but the upper breaks for these four-category maps varied.

The absence of an area symbol indicates counties with fewer than four measurements. Seven of the 38 counties surrounding Pennsylvania lacked some or all structural radon data: three lacked only basement measurements, two lacked living-area data, and two lacked both. Within the Commonwealth, seven counties (Clinton, Jefferson, McKean, Snyder, Sullivan, Union, and Wyoming) lacked basement data, four counties (Cameron, Fulton, Mifflin, and Montour) lacked living-area data, and four counties (Forest, Juniata, Potter, and Tioga) lacked both measurements.

The areal pattern of above-average geometric means for basements, as indicated by the two highest categories in Figure 1, shows a concentration of radon-contaminated residences in the southeastern quarter of Pennsylvania. This pattern extends beyond the state line into central Maryland and western New Jersey, and includes the Reading Prong's southern physiographic counterpart, the Blue Ridge Province in south-central Pennsylvania and Maryland. But this region of high readings is broken by below-average values in Adams County and parts of the Philadelphia metropolitan area. Although the high-rate area includes a number of comparatively populous counties such as Cumberland and Lancaster, with geometric means of more than twice the state figure, a number of similarly ranked counties are located in other parts of the state.

The distribution of homes tested with basement readings over 4 pCi/L has a pattern generally similar to that of the geometric mean. Most of the highly ranked counties in Figure 2 occur in the southeastern quarter of Pennsylvania, as in Figure 1, and adjacent parts of Maryland and New Jersey. Counties with rates lower than the 50-percent Commonwealth average include Adams, Bucks, Delaware, Montgomery, and Philadelphia. Moreover, some higher-rate counties fall outside the southeast, most notably, Centre County, Pennsylvania, and Sussex County, New Jersey.

Figure 3, which shows the proportion of houses with extremely high basement readings, again features southeastern Pennsylvania and a few neighboring counties. And, as before, higher-than-average rates are scattered elsewhere throughout the area mapped, and Adams County and most of the Philadelphia

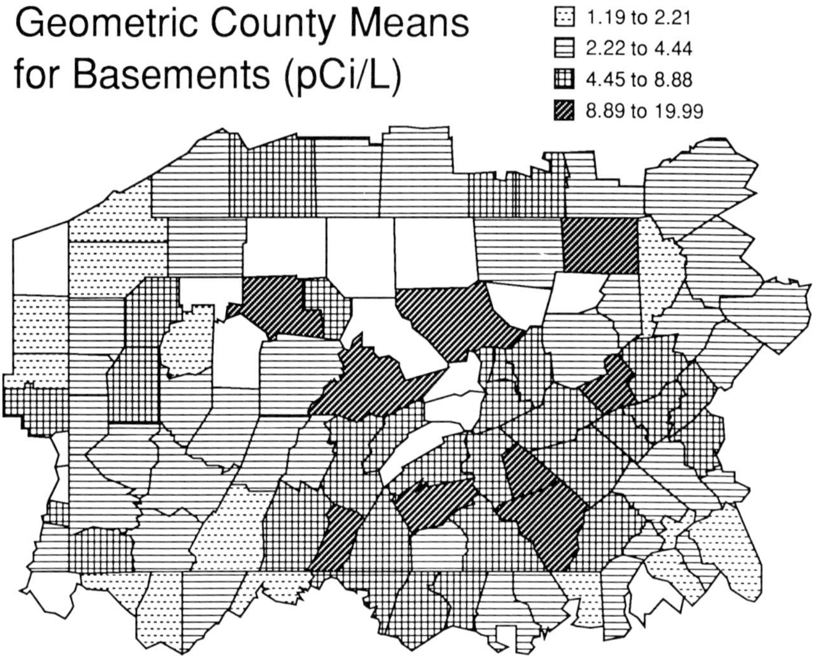


FIGURE 1. Geometric means for county-unit basement radon levels.

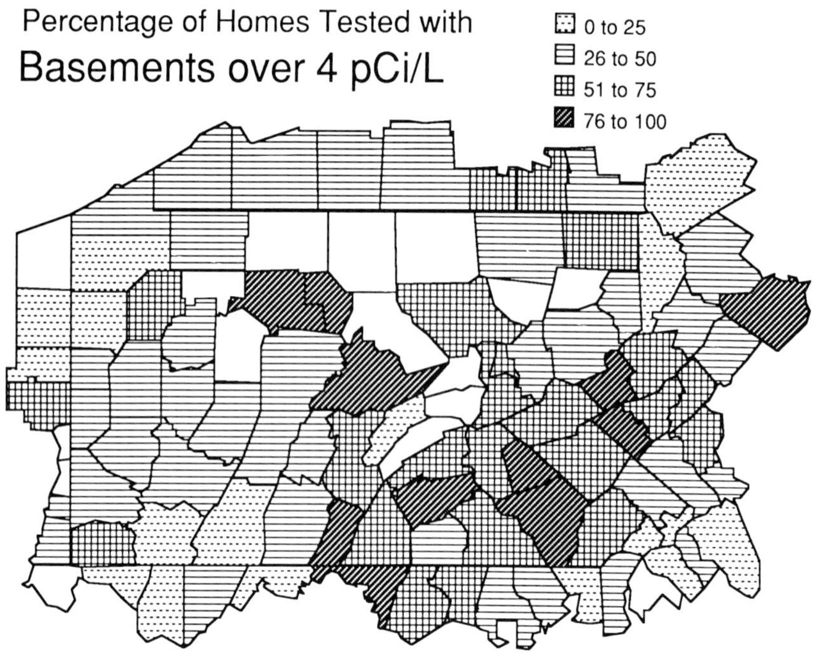


FIGURE 2. Percentage of homes tested with basement radon level above 4 pCi/L.

metropolitan area are below average.

Figure 4, which represents radon levels in living areas, again portrays the southeast as the largest cluster of high-risk counties. Counties with high readings are located elsewhere, some of which are repetitious of the earlier maps. Nonetheless, a few counties, such as Allegany County, New York, and Bedford County, Pennsylvania, have markedly high rates for living areas but less significant levels in basements. These county-level maps, Figures 1 through 4, thus suggest that the radon hazard is neither confined to, nor consistently high within, the Reading Prong and Blue Ridge areas.

Figure 5 suggests that Pennsylvania's radon problem might well be less severe than the risk in several mid-continent states. Figure 5, like Figure 4, portrays living-area radon levels and both have the same class breaks. On the national map, a group of six contiguous states has rates higher than Pennsylvania's average of 31 percent—New Mexico, Colorado, Wyoming, North Dakota, South Dakota, and Iowa. Others with more than 15 percent of the living-area readings exceeding 4 pCi/L are either contiguous to this mid-continent belt or part of a contiguous eastern cluster that includes Pennsylvania. Despite the coarseness of state-level rates, these clusters suggest a strong correlation with geology.[2]

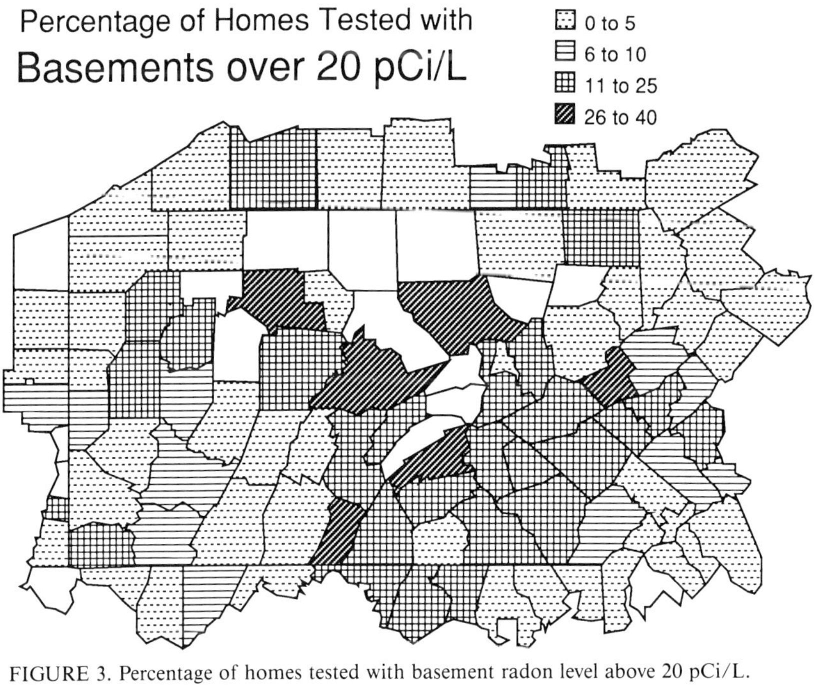


FIGURE 3. Percentage of homes tested with basement radon level above 20 pCi/L.

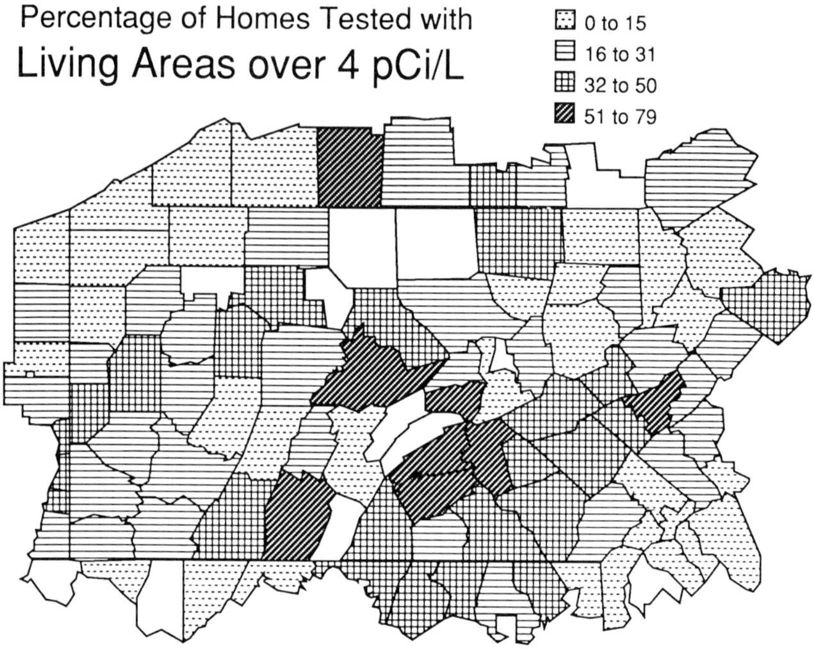


FIGURE 4. Percentage of homes tested with living-area radon level above 4pCi/L.

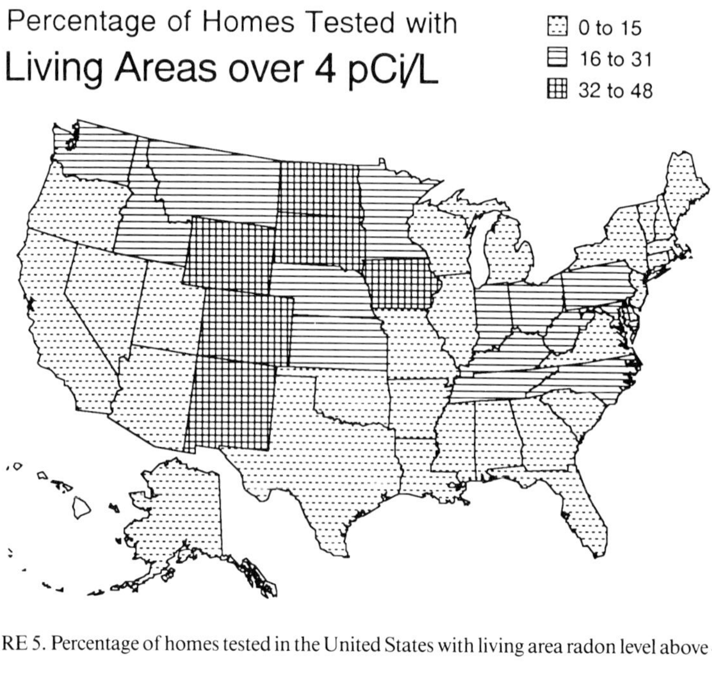


FIGURE 5. Percentage of homes tested in the United States with living area radon level above 4 pCi/L.

## ZIP CODE AREA RADON MAPS

Three additional maps, based on smaller and inherently more homogeneous areal units, represent an attempt to refine this areal description of the radon threat. These maps use point symbols to locate 5-digit Zip Code areas in which significantly more than one-fifth of the homes tested have radon levels above 4 pCi/L or 20 pCi/L, with the former level applied to basements and living areas, and the latter to basements only.

Binomial probabilities were used to assess significance at the 95-percent level. Because of chance sampling error, an area with 40 percent of its homes above either 4 or 20 pCi/L, for example, would not have significantly more than a fifth of its residences at this level if only 12 homes were tested. If samples of size 12 are drawn at random from a population in which just less than 20 percent of all objects have property "A," 5 percent of these samples will have rates for "A" above 41.6 percent. Thus, if the effect of sampling error is not to exceed 5 percent, a Zip Code area with only 12 readings would not be considered to have a rate significantly higher than one-fifth unless more than 41.6 percent of homes had readings at more than 4 or 20 pCi/L. As in other examples, the binomial distribution would account for sample size by requiring a rate above 50 percent for a Zip Code area with 6 tests, a rate above 33.33 percent for an area with 24 tests, and a rate exceeding 29.7 percent for an area with 48 tests.[3]

Figure 6, portraying 201 Zip Code areas with significantly more than one-fifth of the basements tested exceeding 4 pCi/L demonstrates not only a strong

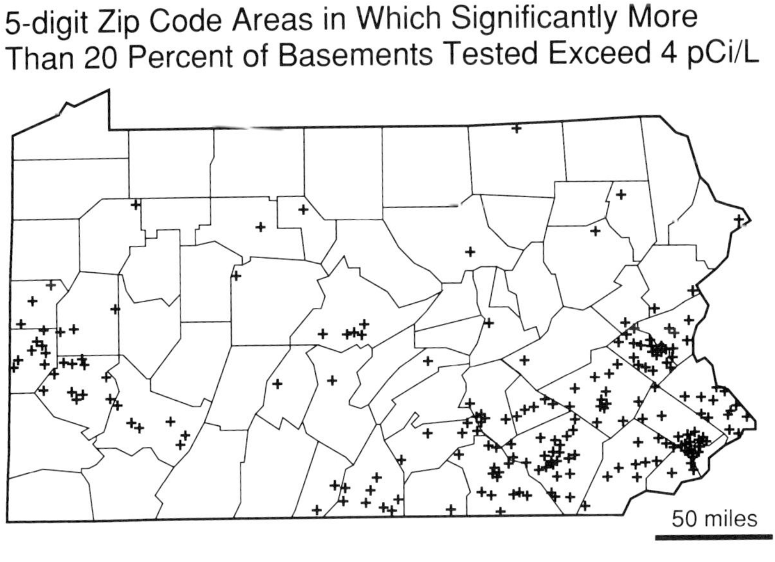


FIGURE 6. Five-digit Zip Code areas in which significantly more than 20 percent of basements tested exceed 4 pCi/L.

concentration in southeastern Pennsylvania but also a secondary cluster in western Pennsylvania. The southeastern cluster sweeps broadly across the Reading Prong, Blue Ridge, and Piedmont provinces, and includes a large number of suburban sites in the Philadelphia metropolitan area. The western cluster is a belt of 5 counties extending from Westmoreland County across Allegheny to Beaver, Lawrence, and Butler. Over a dozen isolated sites are scattered throughout the state, and a minor cluster of four Zip Code areas occurs in southern Centre County (in State College and vicinity).

A similar but less dense pattern exists for the 93 Zip Code areas for which significantly more than 20 percent of the living areas in homes tested exceeded 4 pCi/L (Figure 7). The southeast, however, now consists of a broad belt with a few dense clusters and substantial empty areas in Adams County and the Philadelphia region. The western Pennsylvania area is still evident with a notable core in the populous Allegheny County. Figure 7 also shows five significantly high living-area sites in Centre County and few such sites elsewhere.

The 12 Zip Code areas shown on Figure 8 approach more closely the concept of a radon "hot spot." Significantly more than one-fifth of the basements tested in these sites had radon concentrations above 20 pCi/L. Except for one Zip Code area in State College, all are in southeastern Pennsylvania. Half of these high-concentration sites are in Lehigh and Northampton counties, but these six are north of the Reading Prong in the Great Valley Province underlain principally by Ordovician limestone. High-concentration sites in Lebanon and Cumberland counties are also in the Great Valley, and the three Piedmont sites

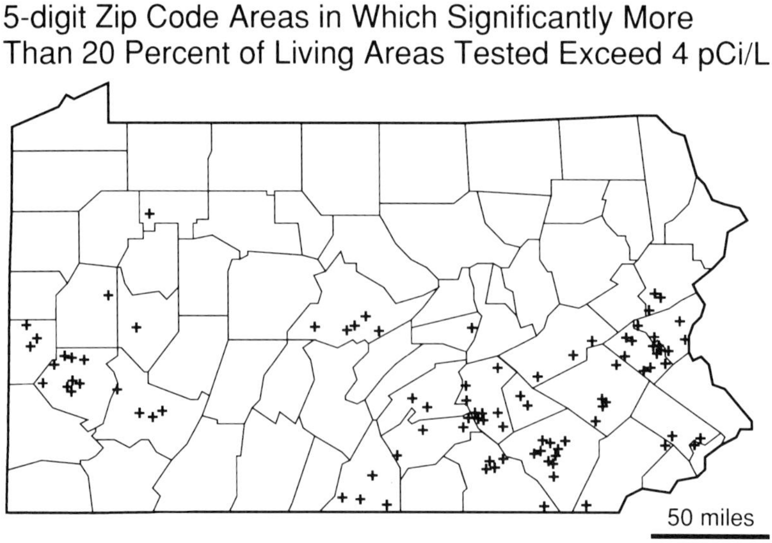


FIGURE 7. Five-digit Zip Code areas in which significantly more than 20 percent of living areas tested exceed 4 pCi/L.

in Lancaster County are also in areas atop Ordovician or Cambrian sedimentary and metamorphosed sedimentary rocks.

## INTENSITY OF TESTING

Because our binomial probability criterion for significance tends to suppress sites with few participants in the Radon Project's testing program, the three Zip Code maps might underrepresent the extent of the radon hazard. Moreover, inspection of the data suggests substantial geographic variation in the intensity of testing. To explore these geographic variations, we computed location quotients for counties. The location quotient is a standard index for showing relative concentration in the parts of a region. A ratio of two proportions, the location quotient was calculated here as the ratio of radon tests in a given county to the number of owner-occupied detached housing units in that county divided by the ratio of tests in Pennsylvania to the number of owner-occupied detached housing units in the state.[4] Thus, a county with a location quotient of 1.0 would have a testing rate similar to the state as a whole, whereas a location quotient above 1.0 would indicate more intensive testing and less than 1.0 would indicate less intensive testing. To explore possible geographic variations in test site within the home, we carried out separate tabulations for basement testing, living-area testing, and both types of testing combined.

These calculations, which assume that renters are less likely than homeowners

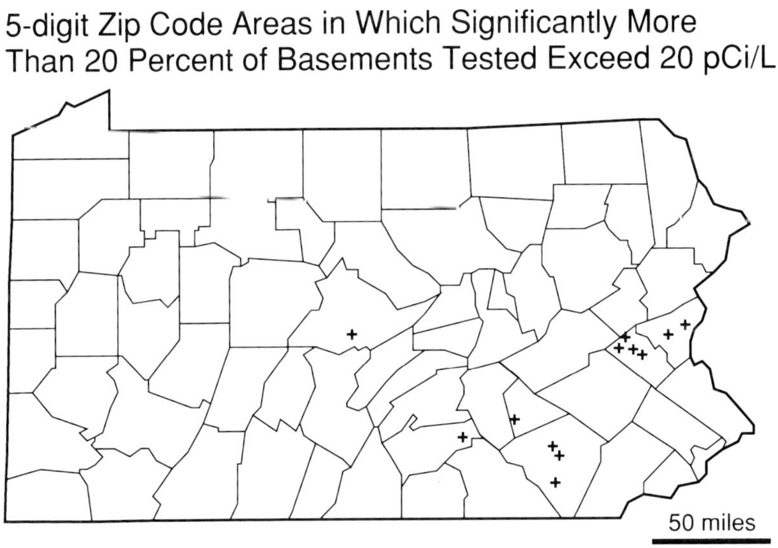


FIGURE 8. Five-digit Zip Code areas in which significantly more than 20 percent of basements tested exceed 20 pCi/L.

to participate in radon testing, might discriminate against counties such as Philadelphia[5] and Allegheny with large numbers of renters in multi-family housing. Yet this assumption seems reasonable for tenants of high-rise structures and other apartments located well above the first floor and with little concern for the incidence of radon at dangerous levels. Moreover, homeowners have a vested interest given the expense of mitigation as well as the possible difficulty of selling a house that might have a radon problem.

The geographic patterns of location quotients (Figures 9-11) are markedly similar. Four counties (Forest, Juniata, Potter, and Tioga) had only three or fewer tests and are therefore excluded. Higher values are largely in the southeast. Lehigh and Pike counties have location quotients over 2.0 for both basement and living-area testing; Bucks, Cumberland, and Franklin counties are above 2.0 for basement testing; and Lancaster and Monroe counties have living-area testing rates more than double the state average. Southeastern counties with location quotients less than 0.5 include Adams, Delaware, and Philadelphia. Location quotients above 1.0 occur outside the southeast in Centre County and in Allegheny County and some of its neighbors. The high geographic similarity between the location quotients portrayed in Figures 9, 10, and 11 and the county-unit and Zip Code area radon rates suggest that homeowners outside the high-radon zones seem to have a limited interest in testing. Although such testing might well reveal other "hot spots," it seems doubtful that a concentration as extensive as the one in southeastern Pennsylvania would be identified.

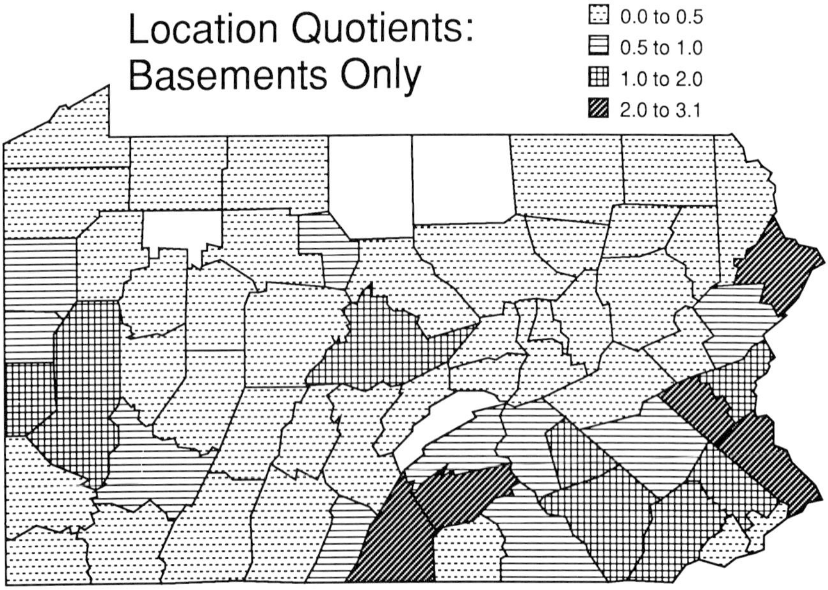


FIGURE 9. County-unit location quotients for basement tests.

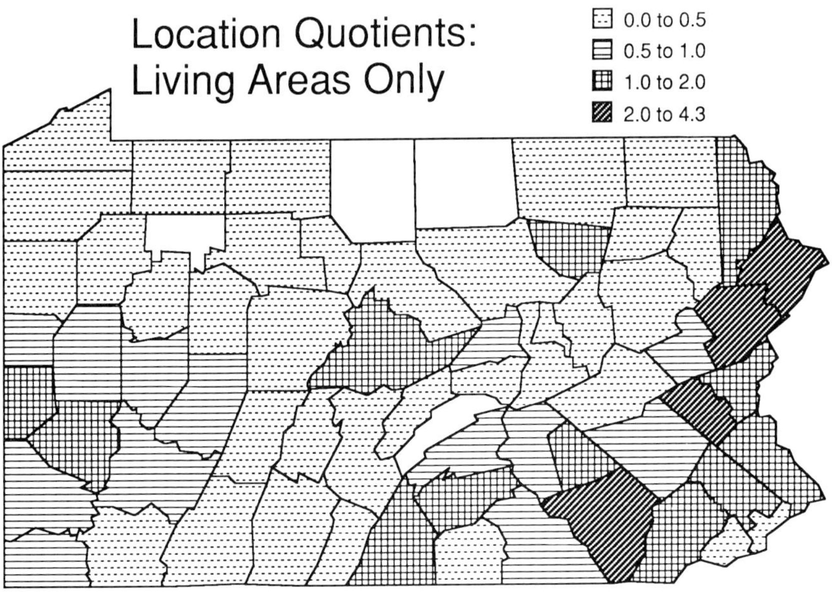


FIGURE 10. County-unit location quotients for living-area tests.

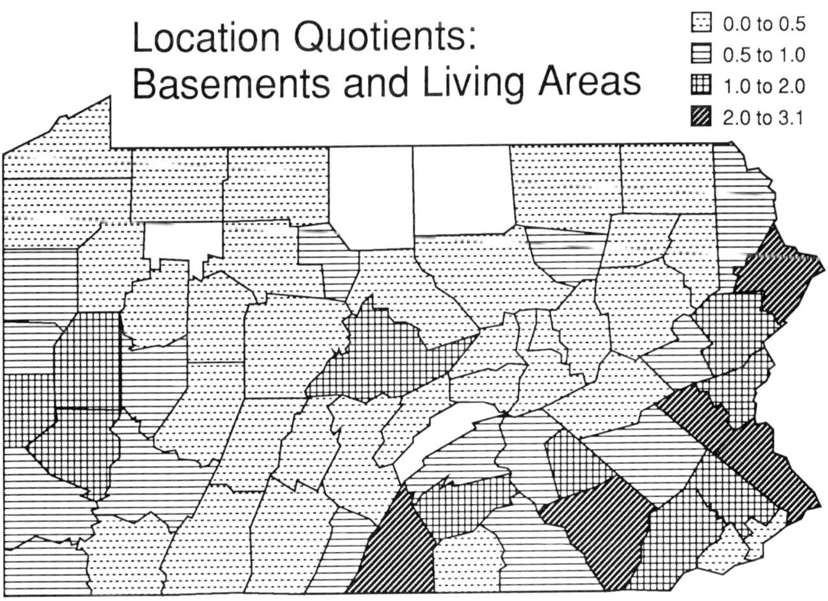


FIGURE 11. County-unit location quotients for basement and living-area tests combined.

Moreover, the relationship between the intensity of testing and the concentration of radon, as revealed in the Radon Project data, would seem to reflect a justifiable anxiety in, and public awareness of, broad areas identified as having high readings.

## RADON AND LUNG AND OTHER RESPIRATORY CANCER: AN ENIGMA

A final set of maps addresses the geographic correlation of radon and cancer mortality. As a serious health hazard, radon might be expected to exhibit a geographic pattern similar to that of cancer mortality. Figures 12 through 14 display county-level trends in the age-adjusted mortality rates for the 1950s, 1960s, and 1970s, respectively.[6] These maps portray disease-specific rates for lung and other respiratory cancers. To promote cultural and socioeconomic homogeneity and to minimize the effects of occupational exposure to carcinogenic effects outside the home, these rates are for white females only. Similar class breaks for all three maps reflect age-adjusted statewide means of 5.2, 7.5, and 14.1 deaths per 100,000 persons for the 1950s, 1960s, and 1970s, respectively.

The series of cancer-mortality maps reveals three salient patterns—a progressive increase from decade to decade in respiratory cancer, substantial

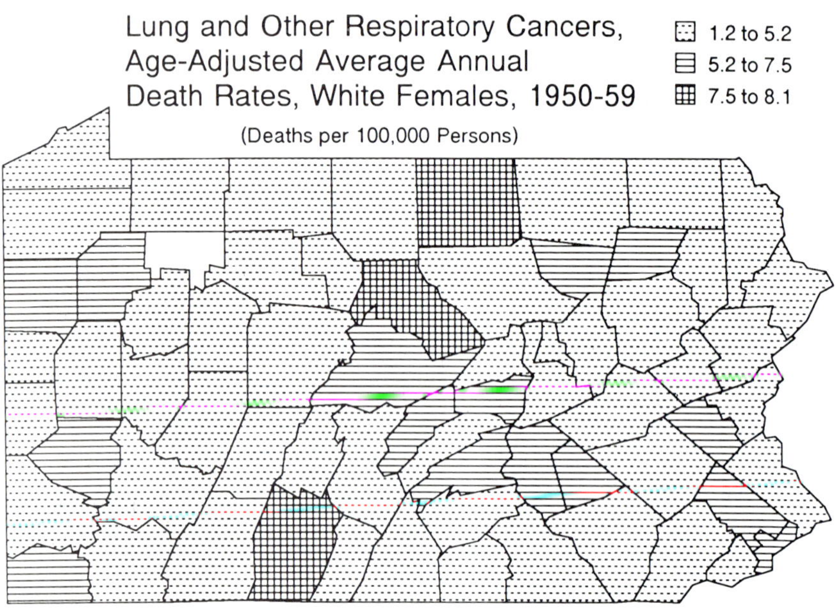


FIGURE 12. Age-adjusted mortality rates for white females, for lung cancer and other respiratory cancers, 1950-59.

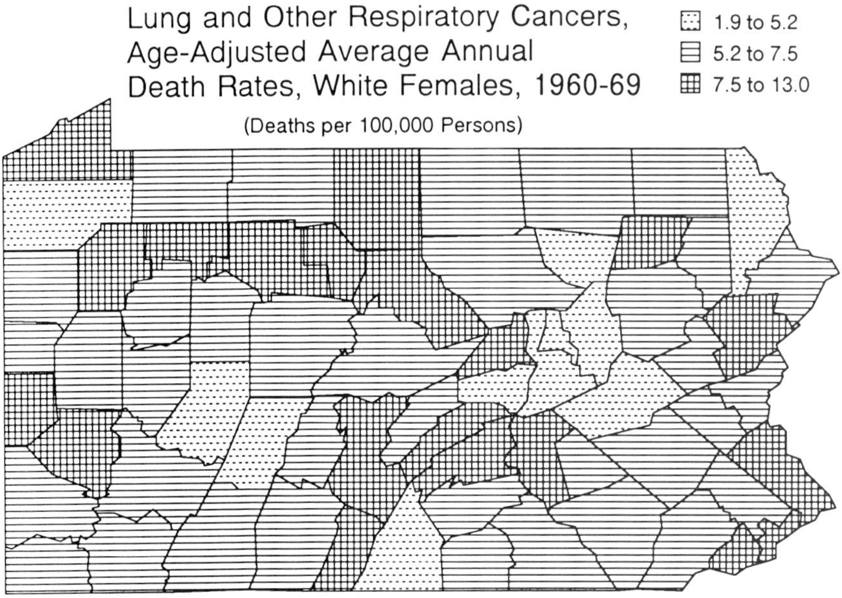


FIGURE 13. Age-adjusted mortality rates for white females, for lung cancer and other respiratory cancers, 1960-69.

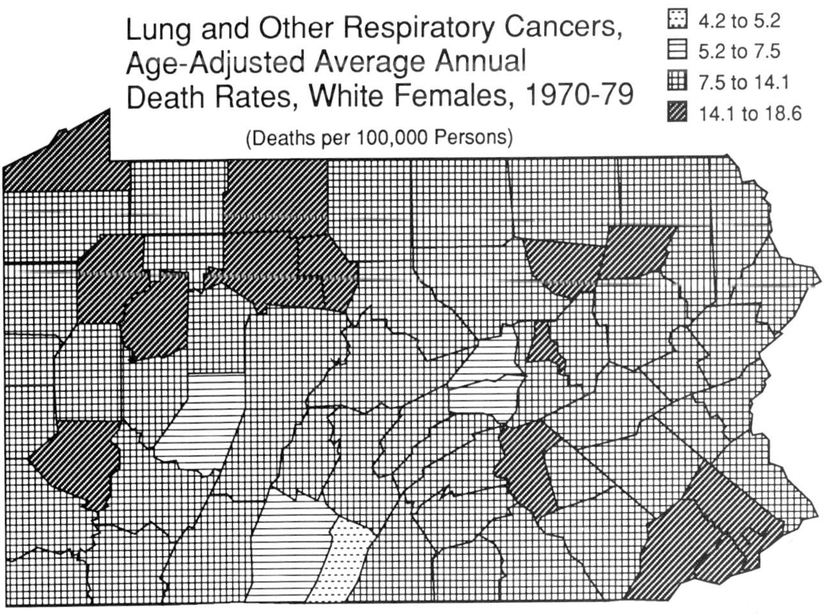


FIGURE 14. Age adjusted mortality rates for white females, for lung cancer and other respiratory cancers, 1970-79.

fluctuation from decade to decade in the location of high and low cancer-mortality rates, and no apparent linkage to the degree of radon contamination in homes. If anything, in fact, the correlation between cancer and radon is negative. High lung cancer rates in the 1960s and 1970s, at least, are common for heavily urban industrial areas such as Allegheny County, Dauphin County, Erie County, and most of the counties of metropolitan Philadelphia. Counties with high measured concentrations of radon in structures, such as Lancaster County, Lehigh County, Northampton County, and Centre County, had age-adjusted cancer rates for the two most recent decades below the state average. Relatively high cancer mortality is common in clusters and scattered patches outside southeastern Pennsylvania, Centre County, and the western Pennsylvania radon belt. And the highest rates within southeastern Pennsylvania occur in counties with comparatively low incidences of radon in dwellings. (For further discussion of the apparent negative correlation, see Chapter 20 by David Bodansky and Chapter 24 by Bernard L. Cohen.)

Respiratory cancer is a complex disease involving myriad causal factors, the most significant of which is smoking. Any attempt to account for a relationship between cancer mortality and radon would have to consider first the demonstrated causal link with tobacco smoking. Moreover, the high local variability of radon contamination, which might affect one structure but not its neighbors, as well as the even greater variation among household occupants as to their smoking habits render problematic any geographic analysis based on county data.

## CONCLUDING REMARKS

The distribution of radon levels in residences, a prerequisite for any government effort to mount an effective, broadly based program to identify and mitigate contamination, deserves a greater research effort. Despite the identification of four major factors in the determination of how much radon can get into a house—the presence of radium in soil, the facility of movement of radon through soil, entry into the house through openings or porous building materials below the surface, and reduced air pressure in the house[7]—the fact is that further, more intensive testing is needed. Indeed, a perusal of the NURE map's notes cautions against using such a map (of potential) as a sole predictor of high radon in specific communities or homes. It is obvious, therefore, that more testing on a broad scale is necessary for more nearly accurate studies of the distribution of radon levels in residences, at least in areas where prior testing revealed the need for further study. Perhaps then the radon threat can be more fully assessed and its connections to health risks associated with cancer can be better understood.

## ACKNOWLEDGMENTS

The authors gratefully acknowledge Dr. Bernard L. Cohen, past Director of the University of Pittsburgh Radon Project, for providing radon test results from the Radon Project which formed the basis of this study.

## REFERENCES AND NOTES

1. U.S. Environmental Protection Agency. 1986. *A Citizen's Guide to Radon.* Document OPA-86-004, 10-11.
2. The National Uranium Resource Evaluation (NURE) project measured radiation from radon decay in soil by air and the EPA released a map, "Areas with Potentially High Radon Levels," which was published in August 1987. The NURE Map, which covers much of the 48 states and Alaska, focuses on geological factors underlying the potential radon hazard. The NURE Map shows substantial areas outside the higher-rate clusters in Figure 5, and illustrates that the high-potential regions within the clusters on Figure 5 occupy less than half the area of the states so classified.
3. Ipsen, J. and P. Feigl. 1970. *Bancroft's Introduction to Biostatistics*, 2nd ed. Harper and Row, New York, NY 65-69.
4. Number of owner-occupied, detached housing units was calculated from 1980 Census of Population and Housing data (Summary Tape File 3A) provided by The Institute of State and Regional Affairs, Pennsylvania State Data Center, The Pennsylvania State University, Harrisburg, PA.
5. Since Philadelphia is a city (county) of row houses, owner-occupied *attached* housing units were added to the detached number in that one case—otherwise, the location quotient would have been badly distorted.
6. Riggan, W.B. et al. 1983. *U.S. Cancer Mortality Rates and Trends, 1950-1979.* U.S. Government Printing Office for the Environmental Protection Agency and the National Cancer Institute. Document no. EPA-600/1-83-015b, vol. II, 173.
7. Tanner, A.B. 1986. *Indoor Radon and Its Sources in the Ground.* U.S. Department of the Interior, Geological Survey. Open-file report 86-222. 1-2.

*Environmental Radon: Occurrence, Control and Health Hazards.* Edited by S. K. Majumdar, R. F. Schmalz and E. Willard Miller © 1990, The Pennsylvania Academy of Science.

# *Chapter Five*

# RADON PROPERTIES*

**C. RICHARD COTHERN**

Executive Secretary
Science Advisory Board
A-101F
U.S. Environmental Protection Agency
Washington, D.C. 20460

## INTRODUCTION

This chapter describes the characteristics or properties of radon and its progeny that are necessary to understanding its importance as an indoor air contaminant. The first section describes some basic and general nuclear properties including structure, nomenclature, half life and types of radiations. The second section describes the characteristics of the naturally occurring radioactive series of which radon is a part. Finally the units used to quantify the decay properties of radon and its progeny are delineated. Both historical and international (SI) units are discussed.

This chapter is necessarily brief and for more details concerning units, their interconversion as well as other aspects of environmental radon the reader is referred to a supplemental textbook on this subject.[1]

## NUCLEAR PROPERTIES

Radon is a colorless, odorless naturally occurring radioactive gas. It is part of the column in the periodic table for noble gases and as such is almost chemically inert. It does form compounds such as clathrates and complex fluorides.[2]

The nuclear properties of atoms are associated with the nucleus of which the primary constituents are neutrons and protons. The number of protons in the nucleus determines the element and its atomic number. A given element

*The thoughts and ideas expressed in this chapter are those of the author and are not necessarily those of the U.S. Environmental Protection Agency.

can have a wide variety of different numbers of neutrons. Variation in the number of neutrons does not change the element's chemical properties since the number of electrons is determined by the number of protons, but it does produce large differences in the stability of the element to radioactive decay. Atoms with the same number of protons but differing numbers of neutrons are called isotopes. Generally those isotopes with the highest and lowest numbers of protons are the most instable.

If an atom has 86 protons, it is radon. There are three well-known isotopes of radon, containing 133, 134 and 136 neutrons, respectively. The atomic mass number is the sum of the number of protons and the number of neutrons in the nucleus, and this sum identifies the different isotopes. The three isotopes of radon have atomic mass numbers of $86 + 133 = 219$, $86 + 134 = 220$ and $86 + 136 = 222$, respectively. Symbolically, these are written as $^{219}Rn$, $^{220}Rn$ and $^{222}Rn$. If the use of superscripts is awkward, then the isotopes can be written as Rn-219, Rn-220 and Rn-222, respectively.

There are 27 known isotopes of radon ranging from $^{200}Rn$ to $^{226}Rn$. The half-lives (the time for one-half of the radioactive isotopes to decay) of these isotopes are less than one hour, except for $^{222}Rn$ (3.8 days), $^{210}Rn$ (2.5 hours) and $^{211}Rn$ (14.7 hours).

The atomic mass numbers are not the exact masses of the atom. They only reflect the total number of neutrons and protons. They are, however, rough approximations of the actual masses. The energy releases in radioactive decay comes from the differences in the actual masses of the parent and progeny atoms, determined through Einstein's well known equation, $E = mc^2$. In this equation, E is the energy, m, the mass, and c, a constant, viz., the speed of light.

Generally speaking, in nature a system will gravitate toward the lowest energy state of the most stable situation possible; e.g. water runs downhill, unlike charges attract each other causing an electron to "fall" into the orbit closest to the nucleus, and snow falls to the ground. In this same way, if a nucleus can move to a lower energy state (a state in which the protons and neutrons are more tightly bound in the nucleus) by emitting radiation, it will. Such a nucleus is radioactive or unstable to decay. Stable nuclei are unable to lose energy by emitting radiation.

A nucleus could emit all kinds and combinations of radiation. However, because systems tend to seek stability and because of the nature of the nuclear force, the most likely (or most stable) radiations to be emitted are a helium nucleus (two protons plus two neutrons), called an alpha particle; an electron, called a beta particle; and a type of high-energy electromagnetic radiation called a gamma ray.

The beta particle ejected from a nucleus is created by a neutron decaying into a proton (which remains in the nucleus). As a result of this process, the resulting atom after beta decay has one more proton in its nucleus and thus has become the atom of a different element with an atomic number one greater than the parent atom. The nucleus can emit energy in the form of a gamma ray without emitting a particle that has rest mass, like the alpha or beta particle. The gamma ray is part of a family of electromagnetic radiations including light, radio waves, infrared radiation, ultraviolet radiation and X rays.

The process of alpha and beta radioactive decay leads to the creation of a different element, whereas gamma decay does not. The isotope that decays is called the parent and the resulting isotope, if different, is called the progeny. For example, $^{222}Rn$ decays to the progeny $^{218}Po$ by emitting an alpha particle. This decay sequence is written symbolically:

$$^{222}\text{Rn} \text{ ------> } ^{218}\text{Po} + {}^{4}\text{He}$$

An example of beta decay is

$$^{228}\text{Ra} \text{ ------> } ^{228}\text{Ac} + \text{beta particle}$$

Note that in beta decay the atomic mass number remains unchanged while the element changes.

The radiations of alpha and beta particles and gamma rays come in a variety of energies and as such can cause many different effects when they interact with matter. In general most of the radiations are capable of ejecting electrons from atomic orbits, a process called ionization. Because the resulting ion is so reactive, it is a way of detecting radiation.

The early attempts to measure the output of X rays or emanations of radioactive particles involved measuring the materials heat output with an air thermometer or a bolometer. It was not until 1905 that ionization was first used to measure radioactive decay. The physical quantity measured was the magnitude of the charge created by the interaction of ionizing radiation with matter. The detection schemes used ranged from that of discharging an electroscope to chambers across which voltages were applied for collecting the charge.

## NATURAL RADIOACTIVE SERIES

Numerically there are four possible radioactive series or chains of naturally occurring radionuclides. In general these can be characterized by the atomic mass numbers (total number of protons and neutrons) and its constituents by the simple expression:

$$A = 4n + m$$

where A is the mass number and n is the largest whole number integer that can be divided into A and give four with a remainder of m. Thus, the four possible series are 4n, 4n + 1, 4n + 2 and 4n +3.

Of the four possible series three are found in nature and account for most of the natural radioactivity of terrestrial origin. They are the 4n series headed by $^{232}$Th, which is known as the thorium series; the 4n + 1 series, headed by $^{238}$U, which is known as the uranium series; and the 4n + 3 series, headed by $^{235}$U and known as the actinium series. The radionuclide heading each of these series has a half-life that is long relative to the age of the earth. The fourth series, the 4n + 1 or the neptunium series, has not been found in nature with the exception of its essentially stable end product $^{209}$Bi. The major isotopic decay steps for the three naturally radioactive series are shown in Figures 1, 2 and 3. These figures show the higher mass nuclei at the top and the higher atomic number nuclei to the right. The decay then is to the left.

**THE URANIUM SERIES**

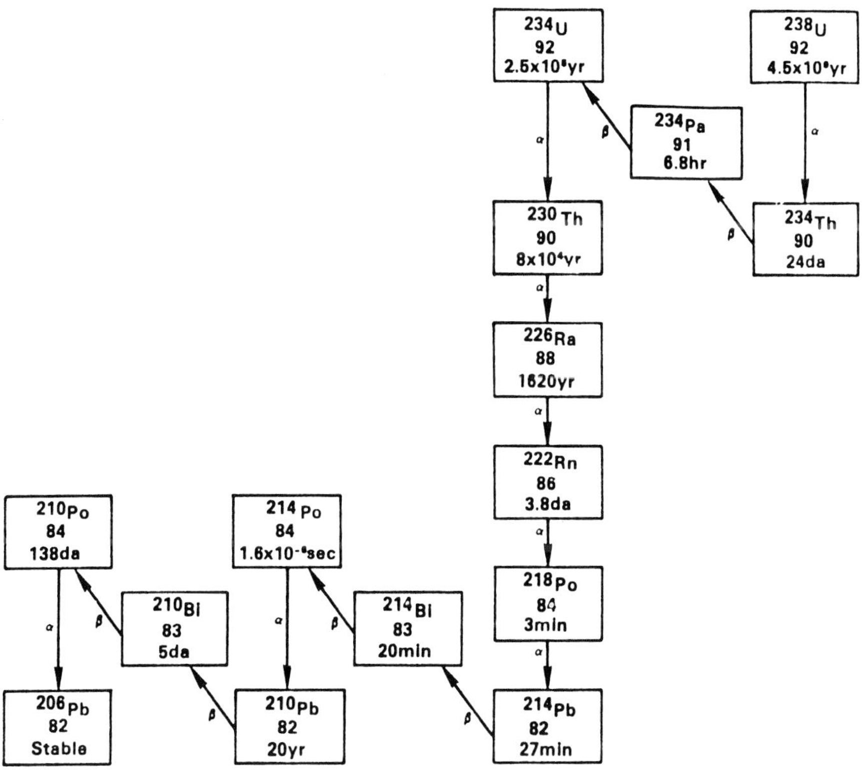


FIGURE 1. Schematic diagram of the major decay sequence for the naturally occurring uranium series.

## UNITS FOR RADON AND RADON PROGENY

In order to develop an expertise in an area of science one specific requirement is a study of the units of measure. Because units develop historically, they are often cumbersome, complex and seemingly illogical—units for radon and radon progeny are no exception. Only a few of the units used to describe radon and radon progeny are discussed here. For a more complete discussion of units and other aspects of radon the reader is directed to a supplemental textbook.[1]

The units necessary for the description of the decay of radon and radon progeny include the rate of radioactive decay, the amount of energy released

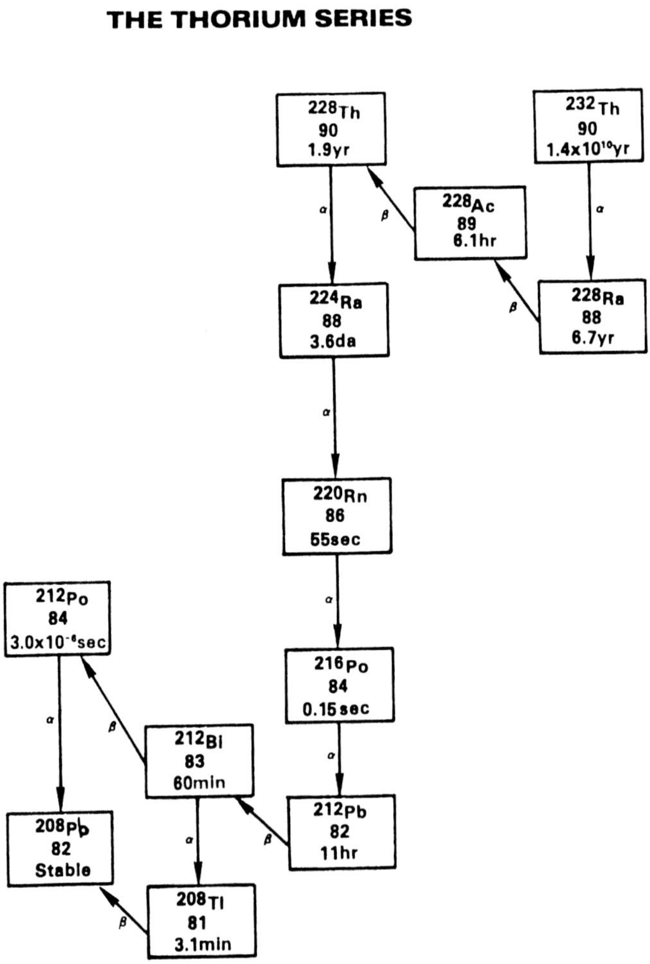


FIGURE 2. Schematic diagram of the major decay sequence for the naturally occurring thorium series.

and the kinds of particles emitted. Units that relate to the energy absorbed by various humans organs and its effect on these organs are also discussed here. It is useful and important to separate mentally the units that describe the rate of decay and the amount of energy emitted from those that relate to body organs and associated human health effects. The former are quantities that are actually measured, and the latter are health effect related information that is derived from a combination of the former and radiobiological or epidemiological data.

There are, generally speaking, three systems of units currently in use; two

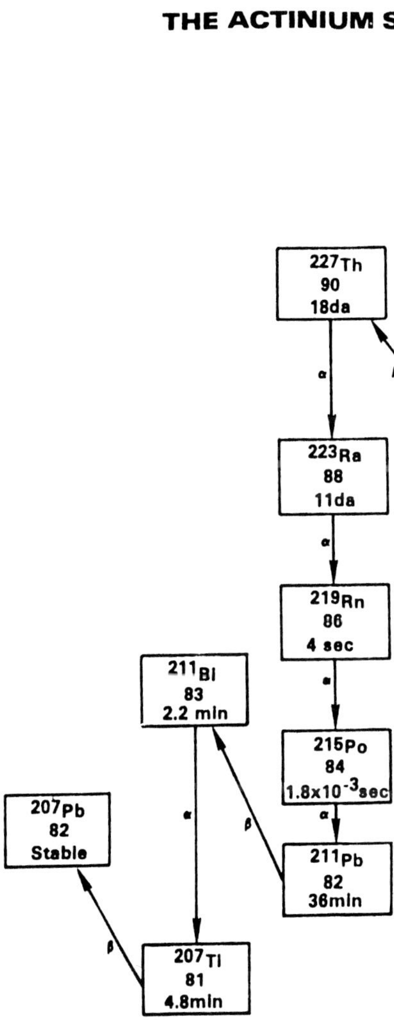


FIGURE 3. Schematic diagram of the major decay sequence for the naturally occurring actinium series.

of these systems are of historical importance, and the third is becoming more universally used. What is here called the conventional system arose historically to describe the amount of radioactivity in an environmental medium and dose imparted by ionizing radiation. The unit activity expressed in curies was designed to describe the physical properties of sources in the environment, whereas other units such as rad and rem were developed to characterize the risk imposed on a population by some amount of radiation. Some units, such as the working level month, are a hybrid combination, designed to produce a measure of the physical properties of the environment, but they are also approximately valid as indices of risk. Important units within the conventional system are the curie (Ci), the rad and the rem. During the 1950's, in order to introduce safety standards for radon levels in uranium mines, the additional conventional unit of the working level was created. Finally, during the 1970's the International System of Units (SI) was developed, and it is becoming widely used. The international system units corresponding to those mentioned above for the conventional system are becquerels, grays and sieverts and these are described below.

### A. *Activity*

The activity of a radioactive sample is a measurement of the rate of decay of the sample. The conventional unit for activity is the curie. It was named after the Curies, a wife-and-husband team who discovered radium and contributed extensively to the development of the field of radioactivity. The magnitude of the curie (unit symbol, Ci) was historically based on the activity of radium and is equal approximately to the rate of decay of one gram of $^{226}$Ra. For general use the value of the curie has been set at $3.7 \times 10^{10}$ disintegrations per second. The SI unit of activity is called the becquerel and the unit symbol is Bq. This unit was named for Henri Becquerel, who shared the Nobel prize with Marie and Pierre Curie in 1903 for discovering radioactivity. One becquerel equals one disintegration per second.

### B. *The Working Level*

The historical unit widely used to describe potential alpha energy concentration (the energy which could be released in the lung if an atom underwent the entire decay series after depositing in the lung) is the working level (WL). The WL was first suggested in 1955 by representatives of several uranium-mining states who met to determine what a safe exposure level of radon was. They concluded that since the damage to the lungs was due to the progeny involved and since the alpha particles from the progeny with short half-lives were the most significant part of the radiation, any standard should be based on the first four short lived progeny of radon. It was decided that a standard should be based on a single unit, rather than a specification of the individual concentrations for each progeny. They proposed that a safe level at which to work (a working level) would be 100 pCi/L of each of the progeny $^{218}Po$, $^{214}Pb$, $^{214}Bi$ and $^{214}Po$. The group felt that this level would be not more than one-tenth of that in the German and Czechoslovakian mines and that "over his working lifetime a worker exposed to this level for a period of forty hours a week would not suffer biological damage". The representatives observed that this standard was more precisely stated as 1.3 x $10^5$ MeV of potential alpha energy per liter of air. This numerical value is derived from the alpha energy released by the total decay of the short-lived radon progeny products in radioactive equilibrium with 100 pCi/L of $^{222}Rn$ per liter air as shown in Table 1.

The working level is a measure of exposure rate. It describes how much energy is found in the alpha particles in a liter of air. In order to relate this energy to biological deposition and absorbed dose, many additional factors must be considered. These include the breathing rate, the volume of air per breath, the time the radon progeny remain in the body and the half-lives of the progeny.

Exposure to one working level for one working month (approximately 170 hours) is called one working level month (WLM). This unit is commonly used for occupational exposure because it includes time.

## C. *Dose*

The absorbed dose, D, is a measure of the average density of energy absorbed by a mass of cells. The unit of absorbed dose in the SI system is joules per kilogram. The name given this unit is the gray, for which the unit symbol is Gy. In the conventional system, the unit for absorbed dose is based on the centimeter-gram-second (CGS) system and is defined as 100 ergs/gram and is called the rad (unit symbol, rad). Thus,

TABLE 1

*Decay Characteristics Relating to the Determination of the Definition of a Working Level*

| Radionuclide | Half Life | Number of atoms per 100 pCi | Alpha particle energy for complete decay per atom (MeV) | Total alpha particle energy for complete decay (MeV) |
|---|---|---|---|---|
| Rn-222 | 3.8 day | 1,760,000 | | |
| Po-218 | 3.05 min | 977 | 6.00 + 7.69 | 0.134 x $10^5$ |
| Pb-214 | 26.8 min | 8590 | 7.69 | 0.661 x $10^5$ |
| Bi-214 | 19.7 min | 6310 | 7.69 | 0.485 x $10^5$ |
| Po-214 | 164 microsec | 0.0009 | 7.69 | 0.000 x $10^5$ |
| | | | A Total of | 1.3 x $10^5$ |

one joule/kilogram = 10,000 ergs/gram = 100 rad

and

one Gy = 100 rad.

The quantity dose equivalent, H, is the product of the absorbed dose, D, the quality factor, Q, and the factor, N, which is the product of all other modifying factors. Dose equivalent is a measure both of the energy absorbed by a mass and the radiobiological effectiveness of the absorbed energy. The conventional name of this unit is the rem and the SI unit is the sievert. To translate between the units use the relationship;

1 sievert = 100 rem = 1 joule/kilogram

The dose equivalent to an organ may be related to a risk by multiplying H by a risk factor in units of risk per unit dose equivalent. This factor is different for the different organs of the body. In order to make reporting easier, the International Commission on Radiation Protection (ICRP) introduced the idea of the effective dose equivalent.[3] In this approach, the dose equivalent to an organ is multiplied by a weighting factor which accounts for the radiosensitivity of that organ. The weighting factor is equal to the risk incurred by an individual and the next generation from irradiation of a given organ to a value of H in the individual divided by the risk which would have been incurred if the whole body of the individual had received a dose equivalent of H. As a result, the actual risk imposed by a specific value of the effective dose equivalent is constant regardless of the organ irradiated. The total risk to an individual (and the next generation of children resulting from that individual) is equal to the sum of the effective dose equivalents for all organs of the body.

Using the weighting factors, the contributions to the total effect on the human body can be calculated by summing the annual dose equivalents to different organs ($H_T$) after weighting them with the corresponding weighting factors ($W_T$) as follows:

$H_E$ = summation over T of ($W_T H_T$),

where $H_E$ is the annual effective dose equivalent. To get some feel for the size of this unit the annual effective dose equivalent for natural background is listed in Table 2 and is approximately 200 mrem/year of approximately 2 mSv/year. About one-half of the dose equivalent arises from inhalation of $^{222}Rn$ and its decay products. It is unfortunate that both dose equivalent and effective dose equivalent are expressed with the same unit since this can lead to some confusion.

Some helpful conversion factors between different systems for radon and radon progeny are shown in Table 3.

TABLE 2

*Calculation of the Annual Effective Dose Equivalent to Humans from Natural Background Using Data from the UNSCEAR report*[4]

| Organ | Weighting factor | Annual dose equivalent (mSv/year) | Annual effective dose equivalent (mSv/year) |
|---|---|---|---|
| Gonads | 0.25 | 0.97 | 0.24 |
| Breast | 0.15 | 0.95 | 0.14 |
| Lung | | | |
| Mean Dose | 0.12 | 0.96 | — |
| Tracheal/bronchial | 0.06 | 14.0 | — |
| Pulmonary | 0.06 | 1.8 | — |
| Total | — | — | 1.0 |
| Red bone marrow | 0.12 | 1.1 | 0.13 |
| Bone Surfaces | 0.03 | 1.9 | 0.057 |
| Thyroid | 0.03 | 0.88 | 0.026 |
| Other | 0.30 | 0.97 | 0.29 |
| | | Total 1.9 mSv/year or approximately 200 mrem/year | |

TABLE 3

*Some Conversion Factors Between Systems For Radon and Radon Progeny*

1 Bq = 27 pCi
1 pCi/L = 37 Bq/m$^3$
1 joule-hour/m$^3$ = 285 WLM
1 WLM = 0.0035 joule-hour/m$^3$
1 WL = 2.08 x 10$^{-5}$ joule/m$^3$
1 WL = 51.6 WLM/year
1 WLM/year = 4.30 x 10$^{-7}$ joules/m$^3$

## REFERENCES


1. Environmental Radon. 1987. Edited by C.R. Cothern and J.E. Smith, Jr. Plenum Publishing Corp., 233 Spring St., New York, New York 10013.
2. L. Stein. 1983. The chemistry of radon. *Radiochim. Acta.* 32:163-171.
3. International Commission on Radiological Protection. 1979. Limits for Intakes of Radionuclides. ICRP Publication 30, Pergamon Press, New York, New York.
4. United Nations Scientific Committee on the Effects of Atomic Radiation (UNSCEAR). 1982. Ionizing Radiation: Sources and Biological Effects. United Nations Press, New York, New York.

*Environmental Radon: Occurrence, Control and Health Hazards.* Edited by S. K. Majumdar, R. F. Schmalz and E. Willard Miller 

# *Chapter Six*

# GEOLOGY AND GEOCHEMISTRY OF RADON OCCURRENCE

**ARTHUR W. ROSE, JOHN W. WASHINGTON and DANIEL J. GREEMAN**

Department of Geosciences
Pennsylvania State University
University Park, PA 16802

## INTRODUCTION

Prior to about 1984, geological aspects of radon abundance were relevant mainly for uranium mining and exploration, earthquake prediction, and atmospheric research. However, since the recognition in 1984 of extremely high radon concentrations in the Watras home near Boyertown, PA, the health hazard of radon to the general population has become the major focus of interest. The levels in the Watras home were far higher than allowable in uranium mines, and the possible hazard of radon-induced lung cancer in the general population was suggested. Geologists for the State of Pennsylvania immediately recognized that the Watras home was in a geological province having above average uranium concentration, namely the "Reading Prong." This recognition plus more widespread testing of houses and field studies of soil gases has provided extensive evidence for the view that much variability of radon in homes results from geological factors.

This paper is intended to summarize evidence for geological controls on radon in homes, and to give examples of geological controls. In addition, the paper discusses some basic principles related to radon abundance in soil gases, and some results of research on radon conducted by the authors. The emphasis is on Pennsylvania, but some examples are from nearby regions. Useful previous summaries of geologic controls for radon in homes include Tanner (1986), Nazaroff et al. (1988b), and Sextro et al. (1987).

## BASIC CONCEPTS

Radon (Rn-222) is produced in the radioactive decay series of uranium (U-238). Several intermediate decay products lead to radium (Ra-226), from which radon forms by emission of an alpha particle (Figure 1). Several of the radionuclides between U-238 and Rn-222 have half-lives that are relatively long relative to weathering and soil formation processes (U-234, half life 250,000 yrs; Th-230, 80,000 yrs; Ra-226, 1600 years). It is possible that these nuclides, especially Ra, can migrate during weathering and be separated from the location of the parent U-238. However, radon is most commonly generated close to the site of uranium, its ultimate long-lived parent. Radon is a noble gas and tends to migrate readily in air or water, but since it has a relatively short half-life, of 3.8 days, this migration is limited.

Several sources have been proposed for radon in houses:

1. Soils and rock, via migration of soil gas,
2. Degassing from water into household air,
3. Building materials,
4. Natural gas used in cooking or heating.

The first two sources are of significant interest and are discussed in more detail below. The third source is important for a few cases in which the waste products from uranium mining or unusually enriched rocks were used to manufacture concrete blocks, concrete, or wallboard, but the radon emission from most building materials is negligible. The fourth source is rarely important, because of decay of radon during the time between gas production in the well and consumption in the home.

Homes with private water wells, especially in areas of uranium-rich granite and similar rock, may receive significant amounts of radon by degassing from water supplies, but this source is generally subordinate, because most of these houses also receive major amounts of radon from soil. For average water usage and home construction, a ratio of Rn in water to Rn in indoor air of about 10,000 has been determined (Nero, 1988, p. 21-22; Nazaroff et al. 1988a). Since most water supplies contain a few hundred to a few thousand picoCuries/liter (pCi/l) of radon, the resulting contribution to indoor radon is less than 1 pCi/l in the air. However, in a few localities, 100,000 or more pCi/l of Rn occur in waters, leading to appreciable levels in houses. Rumbaugh (1983) found several such houses in a subdivision near Reading, PA. In municipal water supplies the storage time is usually sufficient that even if the source is ground water, most radon has decayed away prior to usage.

For most houses, soil gas is the major source of indoor radon. The concentration of Rn in soil gas ranges from about 50 to 5000 pCi/l, or higher, commonly with values in the range of 200 to 2000 pCi/l. Entry of only a small proportion of soil air into a house can create levels exceeding EPA's guideline of 4 pCi/l. Evidence for soil gas being the major source includes the rapid

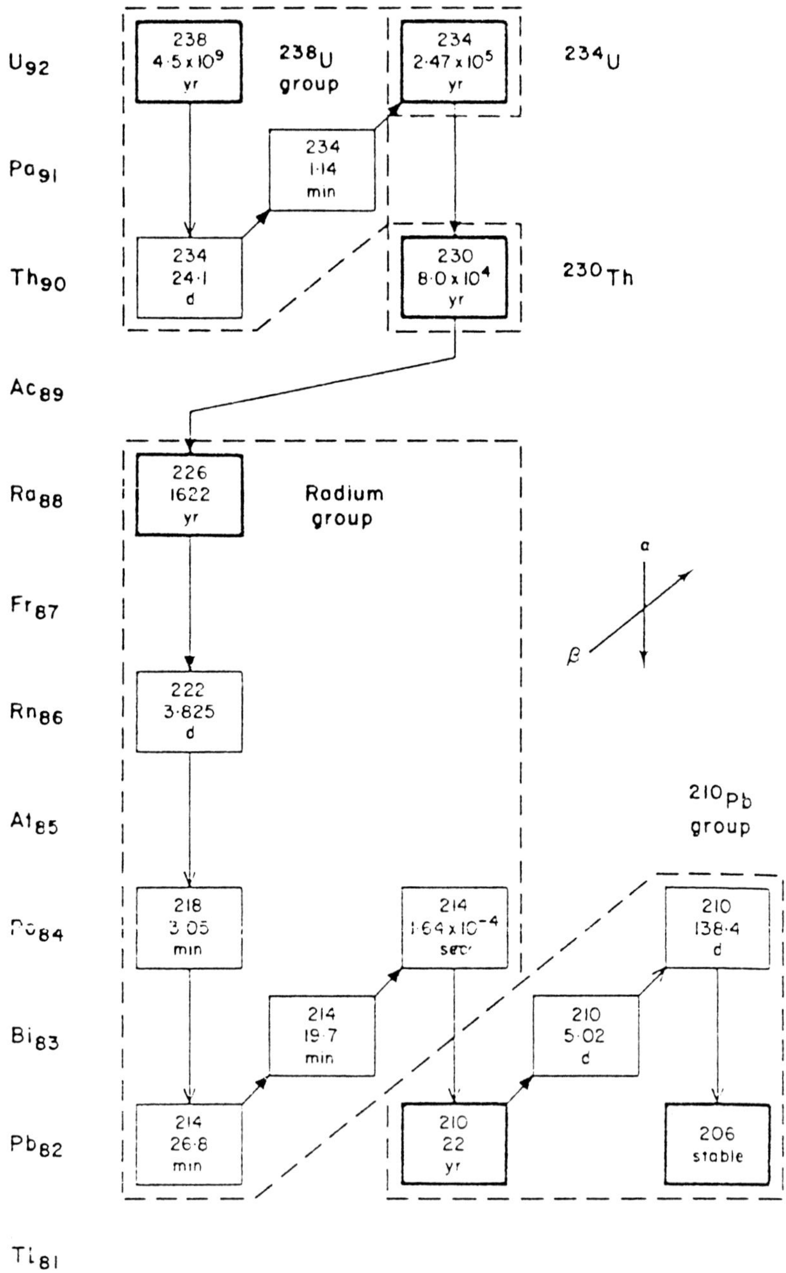


FIGURE 1. The U-238 decay series (after Khattab, 1970), showing half-lives and atomic weights of nuclides and decay type (vertical arrow, alpha decay; upward to right, beta decay).

response of indoor radon levels to pressure changes in houses (Nazaroff et al. 1985, 1987) and the major decreases commonly observed when effective methods of inhibiting soil gas entry are used for mitigation (EPA, 1986).

Figure 2 indicates the abundance of uranium in a variety of rocks. The average abundance of uranium in the earth's crust is about 2.5 ppm. Like other elements, traces of uranium occur in all natural materials, though some contain higher concentrations than others. In general, rocks with higher than average contents of uranium include some granites, many felsic alkalic igneous rocks, some shales (especially black or carbonaceous shales), bauxites and coals, and phosphate rock. Figure 2 illustrates the very large range of uranium in relatively common rocks (at least 3 orders of magnitude) and implies that, other things being equal, radon can be expected to show a similar range. Since some uranium ores contain only 500 to 1000 ppm U, there is essentially a continuum between normal rocks and ores.

Uranium and other ancestors of radon form relatively strong chemical bonds and occur at least partly dispersed as trace constituents within the crystal lattices of common minerals. When a Ra-226 atom decays to Rn-222 plus an energetic alpha particle, the Rn atom recoils away from the alpha particle. The recoil energy is enough to drive the Rn atom a distance of about 40 nm (several hundred atomic diameters) through crystalline materials (Tanner 1986). The radon can be emanated into pore space if its recoil path reaches the grain surface (Figure 3), and its energy is absorbed within the pore space. Since water absorbs

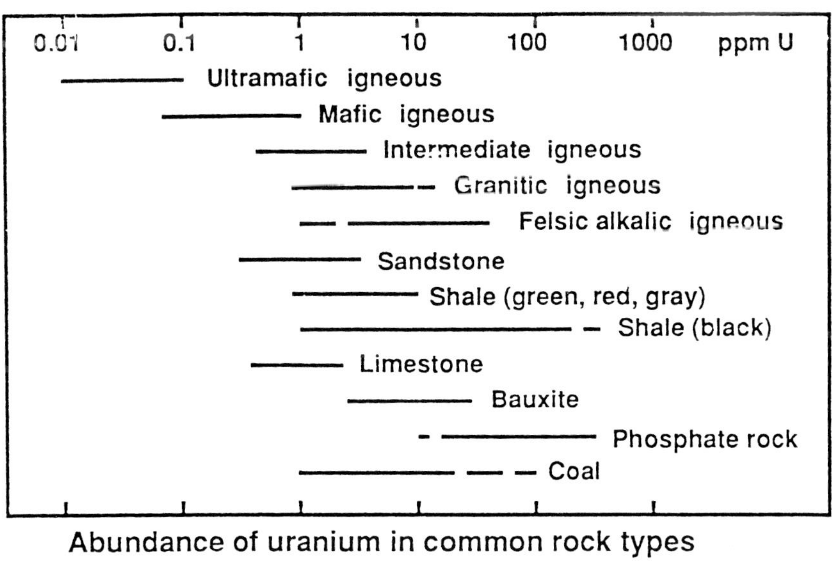


FIGURE 2. Abundance of uranium in common rock types. Data from Rogers and Adams (1970) and other sources.

recoil energy more efficiently than air, the presence of pore water tends to slow a recoiling radon atom within the pore rather than allowing it to embed in an adjacent grain, and thereby tends to promote radon emanation in soils. Water also tends to etch the recoil tracks, thereby tending to release radon embedded in adjacent grains. The radon atom is then free to migrate by diffusion in the water or air or to be carried by flow of the water or air until it decays to daughter nuclides.

The proportion of Rn atoms that escape their solid host and reach pore space is called the emanation coefficient. For well-crystalline materials with homogeneously distributed Ra, the emanation coefficient is typically 0.01 or less, but in the fine grained, poorly crystalline materials of soils, emanation coefficients of .15 to .35 are normal (Figure 4). Emanation coefficients this high indicate that much Ra in soils is either on the outer surface of grains, or in very small grains.

At depth in a dry soil where no net transport of Rn is occurring, the concentration of radon in the pore space may be calculated as

$$C_{Rn} = \frac{bC_U de}{p} \tag{1}$$

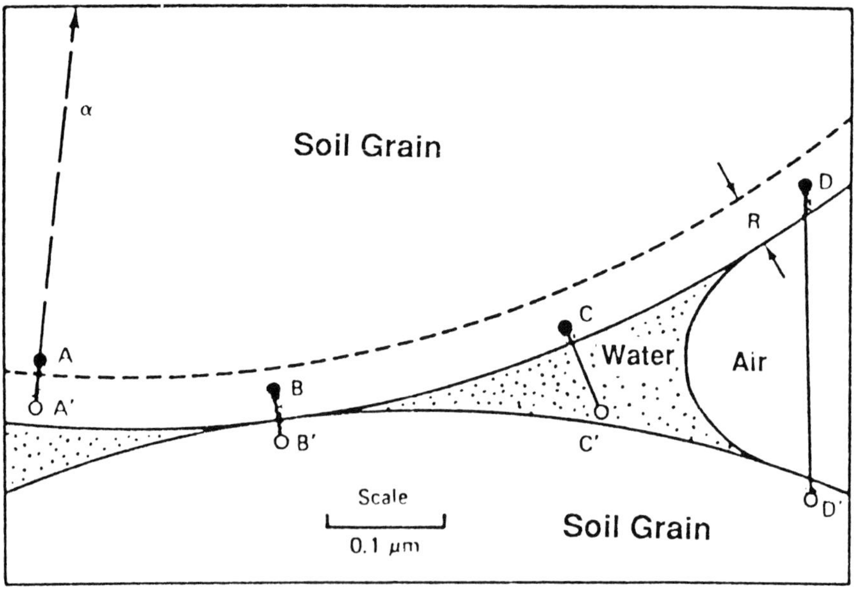


FIGURE 3. Emanation of radon from soil grains (modified after Tanner, 1980). Decay of a Ra-226 ( • ) creates a Rn-222 ( o ) and an alpha particle which recoil in opposite directions (A). The recoil range for Rn is about 40 nm, so at A, the Rn atom does not reach the pore space. At B, it embeds itself in the adjacent grain; at C the Rn loses its energy in pore water and is free to migrate in the pore space; at D the Rn loses little energy in the air-filled pore space and embeds itself in the lower grain, from which it may later escape into the pore space by etching and diffusion.

where $C_{Rn}$ is concentration of radon in the pore space, $C_U$ is concentration of uranium in the solid, d is density of the soil, e is the emanation coefficient, and p the volume proportion of air-filled pore space. The parameter b is the ratio of activity of radium to activity of uranium, and is equal to 1 if they have not been disturbed for many half-lives of the various U daughters. This condition is called secular equilibrium, and is normal in rocks and generally approximated in soils.

If the pores are partly filled with water, the radon will partition between air and water. The radon in the air phase will be higher than for dry soil, and the value for $C_{Rn}$ in equation 1 must be divided by $[(1\text{-}S) + (S/k)]$, where S is the proportion of water-filled pore space, and k is the concentration ratio of radon in air to radon in water for the soil temperature.

Given equation (1), the radon concentration of soil gas can be estimated. For example, for typical values in soils of U = 2.5 ppm (0.825 pCi/g, average crustal U), b = 1, d = 1.75 g/cm$^3$, e = 0.25, p = 0.3, and S=0, the calculated radon in soil air is 1.25 pCi/cm$^3$ or 1250 pCi/l. If S is 0.8 at 18°C (k=3.5), then $C_{Rn}$ in the soil air is 2920 pCi/l. These figures, confirmed by the level of observed soil gas Rn values, clearly show that Rn values several orders of magnitude higher than the EPA indoor air guideline of 4 pCi/l are to be expected in soil gases. One wonders at the health effects of these radon concentrations on animals living in the soil!

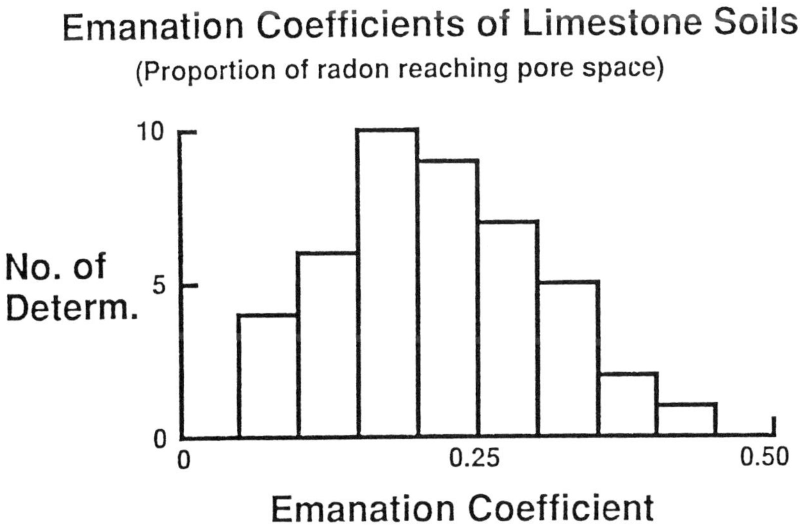


FIGURE 4. Histogram of emanation coefficients for 44 soils developed on carbonate rocks in Centre County, PA, in part from Hutter (1987). These coefficients were measured on soils that were disaggregated and wet or saturated with water.

In the outdoor air, Rn-222 concentration is about 0.2 pCi/l, so a large concentration gradient exists from soil air to open air. As a result, significant diffusion of Rn decreases soil air concentrations in the shallow soil. In typical soils, concentration gradually increases downward from 0.2 pCi/l at the surface to values of 100's and 1000's of pCi/l at a depth of a meter or so, as calculated in the preceding paragraph.

## DATA FOR SOME PENNSYLVANIA SOILS

The behavior of Rn with depth and time has been studied in detail at five sites in central Pennsylvania. Methods of study are described by Rose et al. (1988). Three sites are in soils developed on limestone or dolomite and two developed on shaly sandstone.

Several profiles of Rn vs. depth are shown in Figure 5. At this site in soil developed from limestone, profiles from June through early September show relatively simple patterns, generally increasing with depth in the top 100 cm as expected for diffusion, to values of several thousand picoCuries/liter. At greater depth, concentrations fluctuate somewhat, probably because of changes

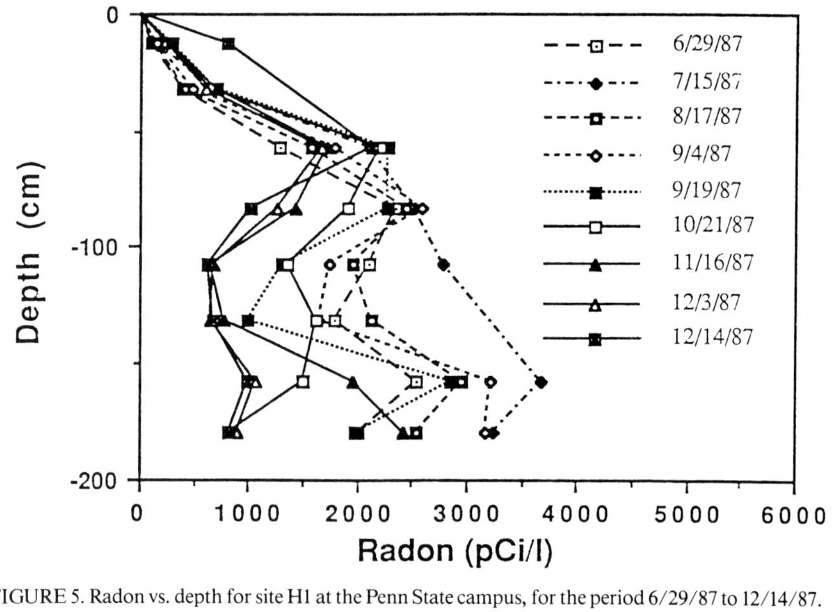


FIGURE 5. Radon vs. depth for site H1 at the Penn State campus, for the period 6/29/87 to 12/14/87. Note the complex behavior of the values at depths below 57 cm during fall and winter. Radon was extracted from tubing emplaced in the soil and measured with an EDA RD-200 radon detector (Rose et al., 1988).

in porosity, moisture, emanation coefficient and other soil properties. In contrast, profiles in winter are more complex, sometimes markedly reversing to decrease with increasing depth, and commonly restricted to lower concentrations.

The pattern with time at a site in soil developed from dolomite is illustrated on Figure 6. The deeper horizon (165 cm) shows a clear annual cycle, with high values in summer and low values in winter. The concentrations range by a factor of as much as 10 between summer and winter. Other researchers have found similar variations (Rose et al., in press; Schumann et al., in press) but more commonly these have highest values in winter. The large variability indicates that characterization of radon hazard at a site by measurement of radon in soil gas is subject to a large error, and needs further investigation.

The cause of the large variability remains uncertain. Possible causes include temperature-controlled adsorption of radon and air-water partition, moisture-controlled inhibition of diffusion into air-filled pores, and inhomogeneous radium distribution.

Limestone soils, or "terra rosa," are among the geological environments causing high radon values in houses. Data collected in this study show that the main cause is enrichment of uranium during weathering of rock to soil. Average uranium in central Pennsylvania limestone bedrock is 0.5 to 0.8 ppm (Table 1). During soil formation, the carbonate is dissolved away, but uranium is largely retained in the soil, along with Al, Fe and Si, to reach levels of 3 to 8 ppm U.

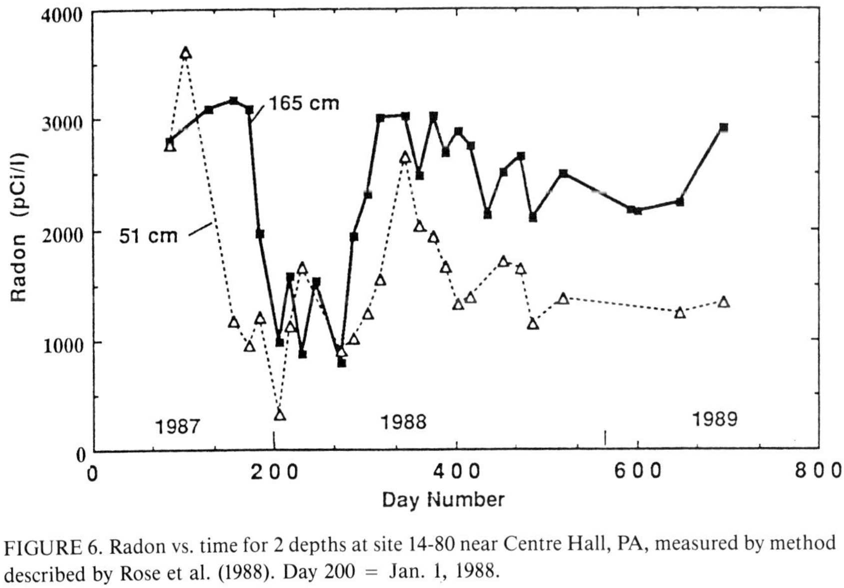


FIGURE 6. Radon vs. time for 2 depths at site 14-80 near Centre Hall, PA, measured by method described by Rose et al. (1988). Day 200 = Jan. 1, 1988.

## EXAMPLES OF GEOLOGIC CONTROL OF RADON

An interesting study of geologic controls for radon has been conducted by Gunderson et al. (1988) in the Boyertown area near the Watras house, and in Easton, PA. Concentrations of radon in both soils and houses near Boyertown are much higher along a zone of mylonite than in the adjacent hornblende gneiss and quartz-feldspar gneiss (Figure 7). In particular, the houses in the mylonite zone consistently have indoor radon levels exceeding 200 pCi/l, whereas houses in hornblende gneiss usually have less than 4 pCi/l. Houses on quartz-feldspar gneiss have intermediate radon levels near Boyertown, but relatively high radon near Easton, PA. Mylonite is characterized by intense granulation caused by shearing to form a fine-grained rock out of a medium or coarse-grained rock. In the process, uranium may be greatly enriched (Gunderson and Gates, 1988) and a directional permeability is generated. The result is a rock and overlying soil with high U content and considerable air permeability, allowing ready flow of soil gas into a house. Studies of the distribution of houses in the Reading Prong with elevated radon show that a majority are concentrated in certain structural-lithologic zones (R.C. Smith, personal communication, 1988).

A second example of geologic control is a study by Hand and Banikowski (1988) near Syracuse, NY. The area is underlain by gently dipping sedimentary rocks of Silurian and Devonian age. The Marcellus (black) Shale is known to have high uranium concentrations, the average for core samples from the area being 34 ppm. Radon concentrations in homes built on the Marcellus Shale average about 10 pCi/l, whereas homes overlying sandstone units average 2.5 pCi/l and those on a stratigraphically older group of shales average about 1.2 pCi/l (Figure 8). Homes built on limestone underlying Marcellus Shale also tend to have elevated radon, attributed to mobilization of uranium downward along fractures from the Marcellus Shale.

A third example is derived from data on radon in homes furnished to the Pennsylvania Department of Environmental Resources by major radon analytical labs (Figure 9). The analyses are classified by county and by radon concentration into 3 groups, < 4 pCi/l, 4-20 pCi/l and > 20 pCi/l. These data,

TABLE 1

*Uranium Enrichment in Soils Developed on Limestone/Dolomite (Centre Co., PA)*

ROCKS
U = 0.57 ppm (range 0.3 to 0.8 ppm)
(Average of 10 samples)

SOILS

| Location | U | n |
|---|---|---|
| Site 14-80 | 6.8 ppm | 9 |
| Site 14-81 | 3.9 | 10 |
| Site H1 | 3.9 | 12 |

especially the absolute values, cannot be regarded as unbiased, because most represent screening tests that overestimate annual radon levels, and they undoubtedly contain many clusters of analyses spurred by discovery of elevated values. Nevertheless, the consistent patterns seem worthy of discussion.

In general, counties in northwestern and northeastern Pennsylvania show low proportions of elevated radon values, though a few values exceeding 20 pCi/l occur even in these areas. The low radon in these counties seems related to the

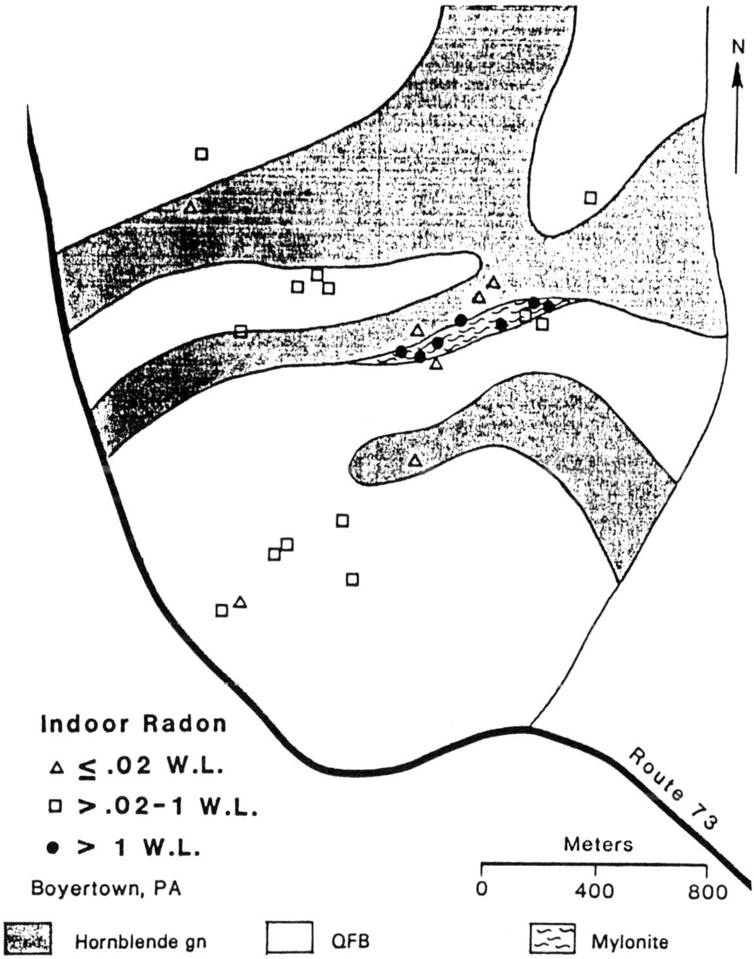


FIGURE 7. Radon concentrations in houses near Boyertown, PA, showing very high values in mylonite zone, low values in hornblende gneiss and moderate values in quartz-feldspar-biotile gneiss (QFB) (after Gunderson et al., 1988). One working level (W.L.) = 200 pCi/l; 0.02 W.L. = 4 pCi/l.

presence of glacial deposits and relatively immature soils in this region. A similar pattern is found in New Jersey (Muessig and Bell, 1988).

In contrast, a set of counties in southeastern and southcentral Pennsylvania shows consistently high proportions of elevated values. Several geologic causes can be suggested. Most of these counties have extensive areas of soils developed on limestone and dolomite, which have been shown in the previous discussion to have relatively high radon in soil gas. Other areas in these counties are underlain by black shales of the Martinsburg Formation. In addition, soils in these areas are well developed and relatively deep, so that much radium is probably on grain surfaces and emanation coefficients are probably relatively high. In Lehigh and Northampton Counties, houses associated with the U-rich granitic gneisses and mylonites of the Reading Prong probably combine to produce the observed pattern, though further more detailed studies are needed to confirm this set of mechanisms.

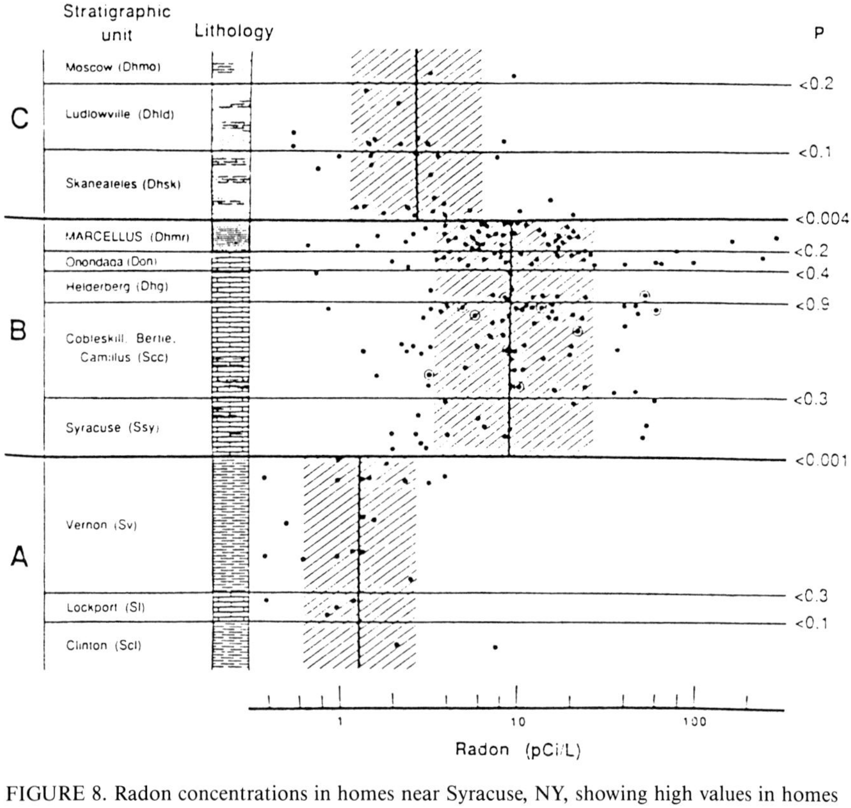


FIGURE 8. Radon concentrations in homes near Syracuse, NY, showing high values in homes built over Marcellus (black) shale and underlying limestones (Group B) compared with lower values associated with shales of Group A and sandstones of Group B (after Hand and Banikowski, 1988).

## CONCLUSIONS

Studies of indoor radon show many clear relations of high radon levels with geology. Probably the most important factor is the uranium (and radium) concentration of soil, which in most areas reflects the uranium concentration in bedrock. The main rock types tending to have high uranium contents, and thus high radon hazard, include granitic and alkalic igneous rocks, black shales (and to some extent other types of shales) and phosphate rock. Limestones and dolomite can weather to soils with relatively elevated uranium. Sheared zones (mylonites) in metamorphic rock can also furnish high uranium and radon concentrations.

Soil gas radon concentrations can be in the thousands of picoCuries/liter in these geologic environments, so that only a small proportion of soil air leaking into a home can create values exceeding the 4 pCi/l EPA guideline. However, it should be noted that the range of uranium, soil gas radon and indoor radon is very large for all the above rock types, so that not all houses in areas occupied by these rocks will contain elevated indoor radon. In addition, some mobility of uranium out of these source rocks is possible, so that houses adjacent to these rock bodies can also have high indoor radon.

Within soil, several factors modify the availability of radon for entry into a house. In highly weathered soils, radium usually tends to be concentrated on grain surfaces, leading to a higher proportion of radon atoms being emanated

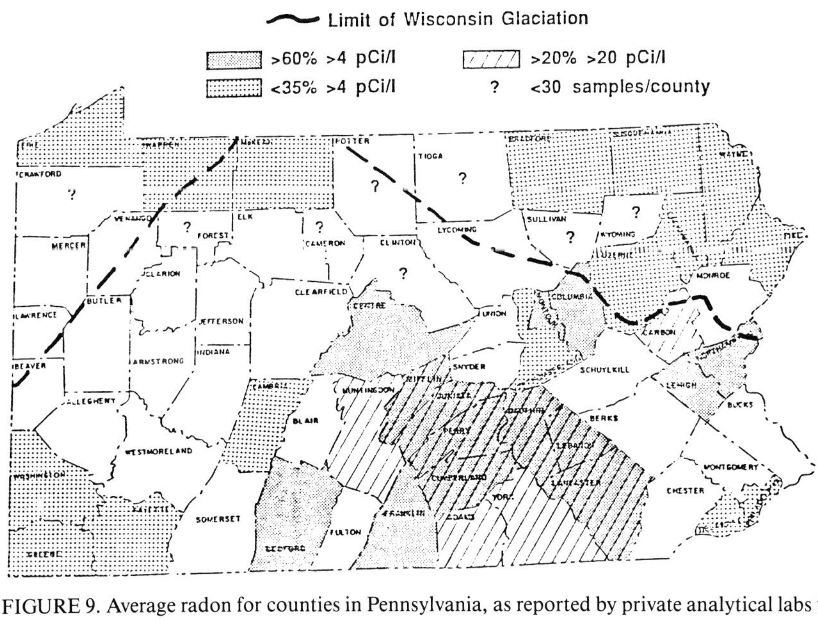


FIGURE 9. Average radon for counties in Pennsylvania, as reported by private analytical labs to the PA DER in January 1989. Only counties with more than 30 samples are classified.

into the pore space. Moisture also modifies the emanation, tending to increase the radon in pore spaces compared to totally dry conditions, and possibly inhibiting its mobility at high moisture contents. Large seasonal variations in soil gas radon have been observed in some soils in Pennsylvania, with the highest values occurring in summer and the lowest in winter.

County-wide averages of indoor radon in Pennsylvania show that northern counties have a low proportion of houses with elevated indoor radon, probably because of the glacial overburden and its relatively immature soil development. In contrast, mature soils in southcentral and southeastern Pennsylvania, especially in areas of granitic rocks, limestone and dark shale, tend to have relatively high proportions of houses with elevated indoor radon. However, some houses in all regions show values well above the EPA's 4 pCi/l guideline, at least for screening measurements. The indoor levels are obviously a complex function of areal geology, local soil and geologic features, house construction, and house ventilation.

## REFERENCES

EPA. 1986. Radon reduction methods. Report OPA-86-005, U.S. Environmental Protection Agency, 24 pp.

Gunderson, L.C.S., and A.E. Gates. 1988. Redistribution and enrichment of uranium in mylonite zones as a function of deformational processes. Geol. Soc. Amer. Abst. with Prog., v. 20, p. A179.

Gunderson, L.C.S., G.M. Reimer, and S.S. Agard. 1988. Correlation between geology, radon in soil gas, and indoor radon in the Reading Prong. *In:* Geologic causes of natural radionuclide anomalies, (eds.) M.A. Marikos and R.H. Hansman. Special Publ. 4, Missouri Dept. of Natural Resources, Div. of Geology and Land Survey, p. 91-102.

Hand, B.M. and J.E. Banikowski. 1988. Radon in Onandaga County, New York. Paleohydrology and redistribution of uranium in Paleozoic sedimentary rocks. Geology, v. 16, p. 775-778.

Hutter, A.R. 1987. Radon variability in soil gas over fracture traces. M.S. Thesis, Pennsylvania State University, 156 pp.

Khattab, K.M. 1970. Analysis of uranium and thorium and their daughter nuclides in uranium-thorium ores by high resolution gamma-ray spectroscopy. Ph.D. Thesis, Pennsylvania State University, 112 pp.

Muessig, K. and C. Bell. 1988. Use of airborne radiometric data to direct testing for elevated indoor radon. Northeastern Environmental Science, v. 7, p. 45-51.

Nazaroff, W.W., H. Feustel, A.V. Nero, K.L. Revzan, D.T. Grimsrud, M.A. Essling, and R.E. Toohey. 1985. Radon transport into a detached one-story house with a basement. Atmospheric Environment, v. 19, p. 31-46.

Nazaroff, W.W., S.R. Lewis, S.M. Doyle, B.A. Moed, and A.V. Nero. 1987. Experiments on pollutant transport from soil into residential basements by pressure-driven airflow. Environmental Science and Technology, v. 21, p. 459-466.

Nazaroff, W.W., S.M. Doyle, A.V. Nero, and R.G. Sextro. 1988a. Radon entry via potable water. *In*: Radon and its decay products in indoor air, W.W. Nazaroff and A.V. Nero, (eds.) Wiley, New York, p. 131-157.

Nazaroff, W.W., B.A. Moed, and R.G. Sextro. 1988b. Soil as a source of indoor radon: Generation, migration and entry. *In:* Radon and its decay products in indoor air, (eds.) W.W. Nazaroff and A.V. Nero, Wiley, New York, p. 57-112.

Nero, A.V. 1988. Radon and its decay products in indoor air: An overview. *In:* Radon and its decay products in indoor air, (eds.) W.W. Nazaroff and A.V. Nero, Wiley, New York, p. 1-53.

Rogers, J.J.W. and J.A.S. Adams. 1970. Uranium. *In:* K.H. Wedepohl, (ed.). Handbook of Geochemistry, Chapter 12, Berlin, Springer-Verlag.

Rose, A.W., A.R. Hutter and J.W. Washington. In press. Sampling variability of radon in soil gases. J. Geochemical Exploration.

Rose, A.W., J.W. Washington, and D.J. Greeman. 1988. Variability of radon with depth and season in a central Pennsylvania soil developed on limestone. Northeastern Environmental Science, v. 7, p. 35-39.

Rumbaugh, J.O., III. 1983. Effect of fracture permeability on radon-222 concentration in ground water of the Reading Prong, Pennsylvania. M.S. Thesis, Pennsylvania State University, 111 pp.

Schumann, R.E., D.E. Owen, and S. Asher-Bolinder. 1989. Weather factors affecting soil-gas radon concentrations at a single site in the semi-arid western U.S. Proceedings of the 1988 EPA Symposium on Radon and Radon Reduction Technology, v.2, EPA/600/9-89/006B, p 3.1-3.13.

Sextro, R.G., B.A. Moed, W.W. Nazaroff, K.L. Revzan, and A.V. Nero. 1987. Investigations of soil as a source of indoor radon. *In:* Radon and its decay products. P.K. Hopke, (ed.). American Chemical Society Symposium Series 331, p. 10-29.

Tanner, A.B. 1980. Radon migration in the ground: A supplementary review. *In:* Natural Radiation Environment III, (eds.) T.F. Gesell and W.M. Lowder. U.S. Dept. of Energy Rept. CONF-780422, v. 1, p. 5-56.

Tanner, A.B. 1986. Geological factors that influence radon availability. *In:* Indoor Radon. Lewis Publishers, Pittsburgh, PA. (Publ. SP-54, Air Pollution Control Assoc.), p. 1-12.

*Environmental Radon: Occurrence, Control and Health Hazards.* Edited by S. K. Majumdar, R. F. Schmalz and E. Willard Miller © 1990, The Pennsylvania Academy of Science.

# *Chapter Seven*

# GEOLOGY AND OCCURRENCE OF RADON PRECURSORS

**ROBERT F. SCHMALZ**

Professor of Geology
The Pennsylvania State University
University Park, PA 16802

## INTRODUCTION

The discovery that radioactive radon gas may occur as a significant indoor contaminant in houses and in the work-place has had far-reaching consequences in public health, real estate marketing, the construction industry, health and liability insurance underwriting, and in legislation at the federal and state levels. Many factors are known to affect radon level inside a building—its location, construction, ventilation, and substructure; the climate of the region in which it is located and the life styles of its occupants, for example (Nero et al. 1986). Despite the importance of assessing the hazard radon contamination may represent, the economic cost and the time required to screen hundreds of millions of individual buildings make such an effort impracticable. It is not surprising, therefore, that a nation-wide effort has been initiated to identify those regions where unacceptable indoor radon levels are most likely to occur, and where detailed testing will yield the most immediate and significant benefits (Muessig et al. 1988; Peake 1989; Tanner 1986). The effectiveness of large-scale regional screening to evaluate radon potential depends on an understanding of the chemical and physical properties of the gas, and of the geological and geochemical factors which control the distribution of its radioactive progenitors, radium, uranium and thorium. It is the purpose of this section to review and summarize our present knowledge of these large-scale controls on radon occurrence.

## RADON

*Properties*: Radon, element number 86, is the heaviest of the noble gases. Both odorless and colorless, it is produced by the radioactive decay of several isotopes of radium, which, in turn, are among the daughter nuclides found in uranium- and thorium-bearing rocks of the earth's crust. Chemically, radon shares the properties of other elements of the noble gas group (Wedepohl 1970). It forms few chemical compounds, the best known, perhaps, being radon fluoride (Fields et al. 1962). Radon has the highest boiling point (−62°C) and the highest freezing point (−71°C) of the noble gases. It is the most soluble of the noble gases in water and is very soluble in organic liquids including crude petroleum. In near-surface environments it occurs most commonly as a component of soil gases, in solution in ground- or surface-water or in solution in petroleum and associated oil field brines. Ordinarily the gas diffuses to the earth's surface where it is quickly dispersed in the atmosphere. Its concentration in dry air at the surface is on the order of 0.01 picoCurie/liter ($10^{-10}$ ppb by weight), and diminishes in abundance rapidly with altitude as a result of the combined effects of dilution and radioactive decay (Rankama 1954; Rankama & Sahama 1950; Vinogradov 1966).

There are three naturally occurring isotopes of radon; all are intermediate progeny in decay series starting with uranium or thorium and ending in stable isotopes of lead (Figures 1a, -b, -c).

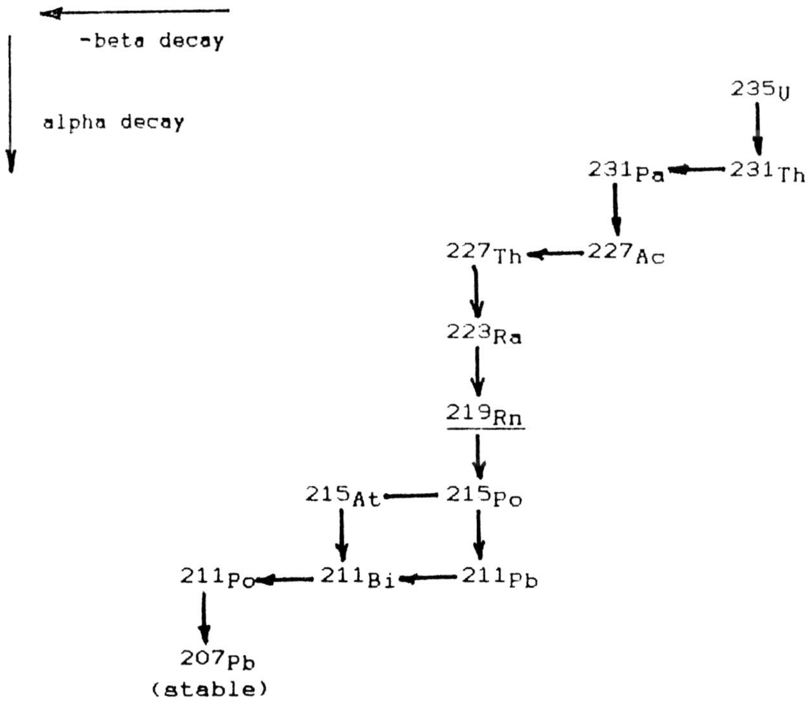


FIGURE 1a. The Uranium-235 "Actinium" Decay Series.

Radon-219 (historically "actinon"), is produced by alpha decay of radium-223, in the uranium-235 ("actinium" or [4n+3]) decay series (Figure 1a). Because of its very short half-life (3.96 seconds) and the relatively small natural abundance of uranium-235 (0.7% of naturally occurring uranium), radon-219 is of minor environmental significance and may be neglected.

Radon-220 (historically "thoron"), with a half-life of 56 seconds, is the daughter of radium-224 in the thorium-232 ([4n]) decay series (Figure 1b). Although the half-life of radon-220 is short, thorium is more abundant in nature than uranium-238, (see Table 1) and thorium-232 is the only naturally occurring isotope. Radon-220 may therefore make up a significant fraction of natural radon in some environments.

By far the most important isotope is radon-222. Its half-life of 3.82 days, although short by geological standards, is sufficiently long relative the rates of soil gas diffusion, atmospheric convection or ground-water flow to allow wide dispersal of the gas even in the absence of its radium-226 parent. It is an intermediate daughter in the [4n+2] uranium decay series headed by the most abundant naturally occurring uranium isotope, uranium-238 (Figure 1c). Together, its relatively long half-life and the abundance of its parent nuclide ensure that radon-222 and its progeny will be the dominant contributors to environmental radon contamination.

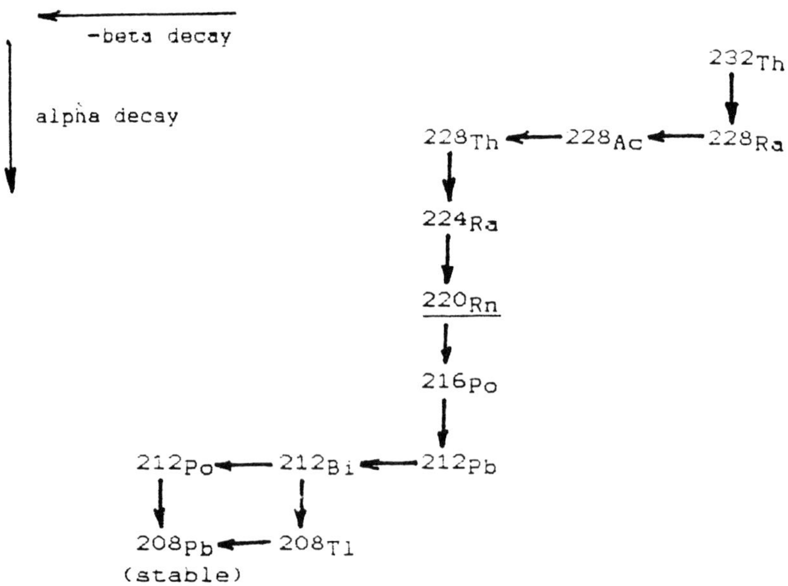


FIGURE 1b. The Thorium-232 Decay Series.

*Occurrence and Distribution:* With a maximum half-life of less than four days, appreciable concentrations of radon can exist only where continuously replenished by the decay of its radium progenitors. In the absence of radium-226, for example, less than 0.5% of the original concentration of radon-222 will remain after thirty-one days. The shorter-lived isotopes would vanish in less than an hour. Therefore, the natural occurrence of radon will be determined primarily by the distribution of radium in the rocks, soil and fluids of the earth's crust; the distribution of radium being controlled, in turn, by the natural occurrence of its progenitors, uranium and thorium. The mobility imparted to radon gas by its physical state, by its solubility, and by its chemical stability, however, implies that under certain circumstances the gas may migrate from the immediate area of its formation, and accumulate at a substantial distance from its radium source. This fact introduces a significant uncertainty into regional "radon potential" surveys based on the distribution of uranium, thorium or radium (see among others, Peake 1989; Muessig 1988).

Radon in soil gas is mainly derived *in situ* by the decay of radium contained in or on the solid particles of the soil. Where this occurs, the radon concentration in the soil gas will be directly related to the local abundance of radium, although it will be affected by soil porosity and permeability, by soil mineralogy,

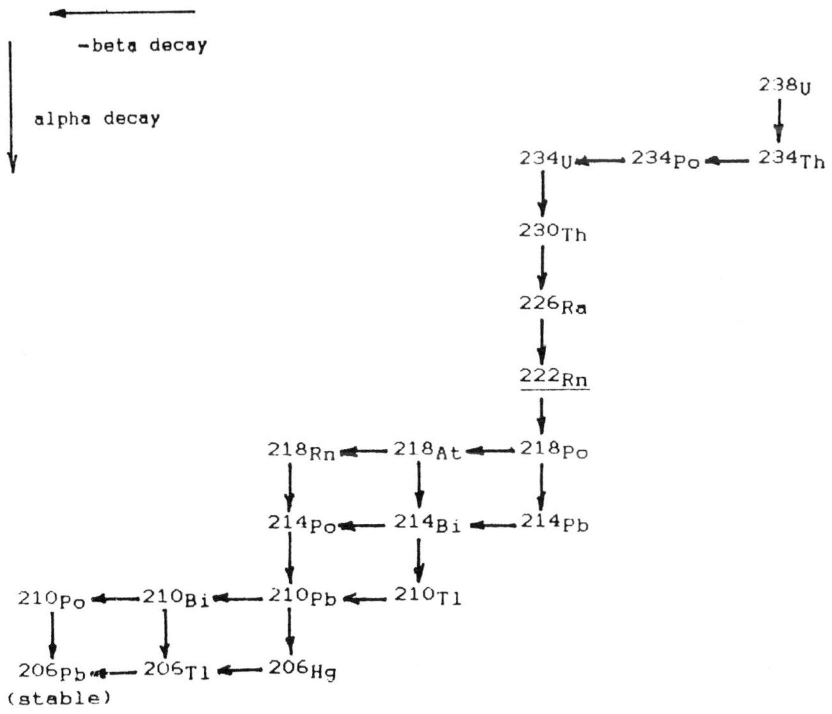


FIGURE 1c. The Uranium-238 Decay Series.

soil moisture, and by the emanation coefficient (Nazaroff et al. 1988). However, because the half-life of radon-222 may be long relative to the rate of movement of soil constituents under certain conditions, the gas may become separated from its parent nuclide by diffusion or convective flow in the gas phase, or by transport in aqueous solution in moving ground- or surface-water. If the soil gas concentration is not ephemeral, it must be sustained by a continuous transfer of radon from the source area to the site of accumulation. The transfer may take place along faults or fractures, or through caverns in bedrock (Hutter 1987), through high permeability (and high porosity) zones in soil or bedrock, or in flowing surface or sub-surface water (Banwell 1986; Graves 1987; Horner 1985). In many such cases, the path of radon migration can be identified and the isolated soil gas concentration linked to a radium source. In some cases, the fractures are thought to tap radon sources deep in the crust or possibly in the upper mantle.

Isolated high radon levels may also occur in wells and springs supplied by aquifers rich in radium (Banwell 1986). Because the dissolved gases will be released where subjected to pressure reduction, turbulence or heating, domestic use of such radon-charged water can contribute to indoor radon pollution, although most investigators regard this as a relatively minor source of contamination except under unusual circumstances (Nazaroff et al. 1988).

Transport of radon gas, whether in solution or in the gas phase, may be affected by barometric pressure or ground-water "pumping" — pulses of increased radon

TABLE 1

*Uranium, Thorium and Radium Concentrations in Common Rocks (parts per million or grams/metric ton)*

| | Uranium | Thorium | Radium |
|---|---|---|---|
| Iron meteorites | 0.001 - 0.01 | $< 0.001$ | $< 10^{-6}$ |
| Stony meteorites | 0.01 - 0.4 | 0.05 - 4.4 | $< 10^{-6}$ |
| Basic igneous rock | 0.6 - 1.0 | 1.0 - 4.0 | $0.6 \times 10^{-6}$ |
| Int. igneous rocks | 2.0 - 3.0 | 5.0 - 10.0 | $0.9 \times 10^{-6}$ |
| Granite/pegmatite | 4.0 - 5.0 | 10.0 - 50.0 | $1.4 \times 10^{-6}$ |
| Sandstone | 0.5 - 3.0 | 1.5 - 8.0 | $0.7 \times 10^{-6}$ |
| Shale | 3.0 - 4.0 | 10.0 - 14.0 | $1.08 \times 10^{-6}$ |
| Black shale | 8.0 - 80.0* | no data | no data |
| Limestone/dolomite | 1.0 - 3.0 | 1.0 - 2.5 | $0.42 \times 10^{-6}$ |
| Evaporite salts | $<0.1$ | $<0.01$ | trace |
| Marine phosphorite | 50.0 - 300.0 | 1.0 - 5.0 | trace |
| Crude oil | $1 - 50 \times 10^{-3}$ | no data | $0.02 \times 10^{-6}$ |
| Ash | 50.0 - 100.0 | no data | no data |
| Oilfield brine | 10.0 | no data | $8.85 \times 10^{-3}$ |
| Seawater | $3.0 \times 10^{-3}$ | $5.0 \times 10^{-5}$ | $1.0 \times 10^{-10}$ |
| Clarke | | | |
| Average crust: | 2.5 - 80.0 | 8.0 - 20.0 | $1 - 2 \times 10^{-6}$ |

* Uranium concentration in Swedish (black) alum shale: 160 - 200 ppm.

Sources: Boyle 1982; Horne 1969; Rankama et al. 1950; Vinogradov 1966; Wedepohl 1970.

flow will usually result from a rising ground-water table or by falling barometric pressure; the reverse changes will usually lead to diminished radon flow (Schuman et al. in press; Rose et al. 1988). Complex interactions among related changes in absorptivity, downward percolation of meteoritic water, emanation coefficient, dissolution, frost action and thermal degassing may reinforce or suppress these effects.

In the general case, however, the occurrence of radon is determined primarily by the distribution of its immediate progenitors, the isotopes of radium.

## RADIUM

*Properties:* Radium is a divalent alkaline-earth metal (Group IIA of the Periodic Table). It shares the chemical properties of berylium, magnesium, calcium and strontium, and exhibits very strong geochemical coherence with barium. Radium is produced by the radioactive decay of uranium and thorium, and it occurs primarily in rocks bearing those elements, most notably the sialic (alkaline) igneous rocks, pegmatites, and certain hydrothermal veins. Like the other elements in the group, radium forms a wide range of minerals, most of which have more familiar calcium, magnesium or barium analogues. The radium-bearing minerals are usually characterized by high melting points and very low solubility in water. Radium commonly replaces the other alkaline-earth metals diadochically, substituting for calcium, magnesium and barium in common carbonate, sulfate and phosphate minerals (Wedepohl 1970; Vinogradov 1966; Palace, et al. 1942).

There are four naturally-occurring radium isotopes, three of which have relatively short half-lives (radium-223, 11.4 days; radium-224, 3.6 days; radium-228, 5.75 years). The fourth and most abundant isotope, radium-226, has a half life of 1620 years, long enough to allow significant concentrations of radium to persist in geologically Recent deposits without a supporting concentration of uranium or thorium. Thus, the radium concentration in certain post-Pleistocene marine deposits and in some modern marine organisms is greater than the amount in equilibrium with the coexisting uranium.

*Occurrence and Distribution:* Except for some Recent deposits, the distribution of radium is determined by the occurrence of uranium or thorium, modified somewhat by the chemical properties of radium itself. In Pleistocene and older rocks, the element is invariably found in association (and usually in radioactive equilibrium) with its parent nuclides.

The largest concentrations of radium are found in association with uranium and thorium ores, and in some of the more mafic minerals in alkaline igneous rocks, pegmatites and hydrothermal veins. Uranium or thorium ores like those at Joachimstal, Czechoslovakia, or Great Bear Lake in Canada, for example, are rich in radium and the mines from which the ores are extracted are inevitably severely contaminated by radon gas.

Shales typically show the highest concentrations of radium among common sedimentary rocks, with values on the order of $10^{-6}$ parts per million.[1] Coarser clastic rocks (sandstone, greywacke, conglomerate) are somewhat less radium rich, and pure limestones and dolomites usually contain the least radium (Table 1). Crude oil is radium-poor, but the associated oil field brines are often very rich in radium. Radium in oil field brines is probably the source of elevated radon levels observed in many petroleum crudes.

On weathering, radium follows the alkali-earth metals into the marine realm. Several factors remove radium from seawater selectively, however, producing radium-enriched deposits and causing the depletion of radium (relative to uranium and thorium) in modern seawater. First, thorium-220 (the immediate progenitor of radium-226 in the uranium-238 series) is "scrubbed" from seawater by adsorption on precipitating oxides and hydroxides of iron and manganese. Recent marine manganese nodules may contain up to 135 x $10^{-6}$ ppm. radium (Rankama & Sahama 1950). Secondly, radium is selectively absorbed on clay minerals, particularly potassium-bearing clays and degraded micas. Recent marine muds and clay-rich carbonate oozes are therefore enriched in radium. Thirdly, radium carbonate and radium sulfate are very much less soluble than calcium carbonate and barium sulfate, respectively, and this difference may lead to differential precipitation the radium phases where carbonate or sulfate minerals are deposited, as in evaporite basins or on sub-tropical carbonate platforms. Selective resolution of the calcium and barium phases below the carbonate compensation depth or where submarine volcanoes release acid gases into solution may augment this effect. Finally, many marine organisms, particularly the phytoplankton, concentrate radium biologically (up to one thousand fold) relative to seawater. Any of these processes will yield sediments far richer in radium than the concentration in equilibrium with the coexisting uranium and thorium. The radium in such enriched deposits will decay to secular equilibrium within a few thousand years, but in geologically young environments, they may give rise to radon levels far higher than would be predicted on the basis of uranium or thorium concentrations alone (Horne 1969).

Radium may also be enriched in Recent soils and regolith, usually as the result of the decay of uranium or thorium minerals concentrated in the soil by the weathering process. Additional radium may accumulate in soil when radium in ground- or surface-water is fixed by adsorption on clay particles or co-precipitated with calcium in carbonates.

[1] Geochemical concentrations are expressed in parts per million (ppm.) throughout the text. The units are numerically equivalent to grams per metric ton.

## URANIUM

*Properties:* Uranium, element number 92, is the most massive of the naturally-occurring elements (sp. gr. 18.8 gm/cm$^3$). It is the fourth element in the actinide series which is analogous to the lanthanide series of rare earth elements structurally and geochemically. Uranium appears to be concentrated in the upper lithosphere, particularly in sialic igneous rocks and in late magmatic fluids. Uranium does not occur as a native element, nor does it form sulfides, tellurides, or arsenides. It behaves geochemically as a strongly oxyphile element.

The three naturally occurring isotopes of uranium have atomic weights of 234, 235 and 238. Uranium-234 is one of the intermediate progeny in the uranium-238 [4n+2] decay series (Figure 1c). It has a half-life of 2.5 x $10^5$ years, and breaks down by alpha-decay to yield thorium-230. Uranium-234 makes up a very small fraction (approximately 0.0055%) of natural uranium, and is important only as a critical intermediate nuclide in the generation of radon-222 by the breakdown of uranium-238.

Uranium-235 makes up approximately 0.7% of natural uranium. This isotope has a half-life of 7.1 x $10^8$ years, and is one of the "primordial" nuclides in the earth's crust. Uranium-235 will undergo fission with slow neutrons and because it can sustain a chain fission reaction with the release of large amounts of energy, it is important as a nuclear fuel. Uranium-235 yields radon-219 as an intermediate daughter in the [4n+3] decay series (Figure 1a).

Uranium-238 is by far the most abundant uranium isotope, and makes up more than 99.2% of natural uranium. Uranium-238 has a very long half-life (4.5 x $10^9$ years), and like uranium-235 is among the "primordial" nuclides of the crust. The [4n+2] radioactive decay series headed by uranium-238 yields the most important and longest-lives radon isotope, radon-222, as shown in Figure 1c.

Uranium most commonly combines to form chemical compounds in either a tetravalent (reduced) or a hexavalent (oxidized) state. Hexavalent uranium usually occurs as the uranyl ion $(UO_2)^{+2}$ which is water-soluble, and exhibits the properties of a large (0.9 Å) divalent metallic ion. Uranyl ion may precipitate directly to form any of several complex hydrated minerals including particularly the potassium uranyl vanadate, carnotite. In the presence of organic matter, the uranyl ion is reduced to tetravalent uranium which immediately precipitates in the form of highly insoluble $UO_2$, the mineral urananite.

*Occurrence and Distribution:* Uranium occurs naturally in meteorites, and in igneous, metamorphic and sedimentary rocks. It is associated primarily with the late magmatic phase, and is enriched in granites, late stage pegmatites and hydrothermal veins, and in metamorphic rocks of similar composition (Table 1). Because of geochemical similarities between uranium and thorium, zirconium, calcium, barium and yttrium, diadochic replacement by uranium is common in mineral which normally contain these elements. For example, uranium may replace zirconium in zircon, barium in barite, or calcium in apatite and fluorite. Diadochic substitutions of this kind are responsible for the concentration of uranium in some evaporite and marine phosphate deposits.

Because of its low solubility in the tetravalent state, uranium oxide (urananite) is often associated with organic matter, especially in sedimentary rocks and secondary ores. Many of the uranium ore bodies of the Colorado Plateau were formed where ground-water solutions rich in uranyl ion came into contact with wood fragments or other organic matter in Tertiary channel sandstones and tuffs (Garrels et al. 1959). In black (organic rich) shales like the Martinsburg Shale of Pennsylvania or the Chattanooga Shale of Tennessee, uranium enrichment is accompanied by increased potassium content, as would be expected if carnotite (a potassium uranyl vanadate) were an important uranium-bearing phase. Uranium may also be associated with other mica-like "uranmicas" or with degraded potassium mica or clay minerals in shales. Humus and ordinary coal ash may contain as much as 1000 ppm uranium. The high ash Swedish "kolm" coal, which is interbedded with uranium-rich (200 ppm) alum shale, may carry as much as 1% uranium (10,000 ppm).

The largest uranium deposits are the hydrothermal pitchblende (urananite) ores of Katanga, Great Bear Lake (Canada) and Joachimstahl (Czechoslovakia). Uranium deposits of all types are discussed in detail by Boyle (1982), Tatsch (1976) and De Vivo, et al. (1984).

## THORIUM

*Properties:* Thorium-232 is the only naturally occurring isotope of thorium. It has a half life of $1.405 \times 10^{10}$ years and, like the naturally occurring isotopes of uranium, it is one of the primordial nuclides. There is a strong geochemical coherance between thorium and uranium, and diadochic substitution of one element for the other is common. Like uranium, thorium is concentrated in the late magmatic igneous phases, sialic igneous rocks, pegmatites and hydrothermal veins, as shown in Table 1 (Boyle 1982; Wedepohl 1970). In such rocks, the concentration of thorium may be three to four times that of uranium; in soils and many sedimentary rocks the thorium:uranium ratio may exceed 10:1 (Rankama & Sahama 1950; Wedepohl 1970).

Thorium is the second element of the actinide series, and exhibits geochemical similarities to both uranium (and other actinides) and the lanthanides. It also resembles elements in Group IVB, particularly hafnium and zirconium. Thorium in the crust occurs primarily in diadochic substitution in monazite and zircon, both of which are common minerals in granite and pegmatite, and in thorianite (uranium-thorium oxide) and thorite (a thorium silicate isostructural with zircon).

*Distribution and Occurrence:* The primary concentrations of thorium, like uranium, occur in sialic igneous rocks, pegmatites and hydrothermal veins. On weathering, thorium-bearing minerals are often preserved in soils or sedimentary deposits. Selective thorium enrichment in limestone soils has been observed and may contribute to elevated radon levels observed in soil gases in some limestone valleys of the central folded Appalachian Mountains of Pennsylvania (A.W. Rose, personal communication). The high thorium:uranium ratio observed in igneous rocks is reversed in seawater where the Th:U ratio is typically on the order of 1:3 (Horne 1969; Vinogradov 1966). Thorium in seawater appears to be selectively "scrubbed" by iron and manganese precipitates. This may help to explain why uranium is more abundant than thorium in marine phosphate deposits and in some marine evaporite and carbonate deposits, as well.

Thorium is a common diadochic constituent in minerals such as zircon and monazite, both of which are heavy resistate minerals often found in lag deposits and sedimentary placers (Boyle 1982). A lag deposit of thorium-bearing monazite is a major ore, for example, in India. In other sedimentary rocks, thorium accompanies uranium, and in shales, the uranium content, the carbon content and the thorium-uranium ratio are almost invariably closely related.

## SUMMARY

Under exceptional circumstances, elevated radon levels may be observed in the absence of radium, uranium or thorium. If persistent, such radon concentrations must be maintained by the radioactive decay of radium at some remote location. The radon daughter must then be carried to the site of accumulation either by convection or diffusion in the gas phase, or in solution in ground- or surface-water. The short half-life of the principal radon isotope requires that the transit time be short, generally no more than a few days. Isolated radon concentrations are therefore associated with fractures, fault zones, caverns or highly permeable strata which link the source with the accumulation site. Springs and hydrocarbon seeps are also suspect.

Non-equilibrium radon concentrations lasting for hundreds or thousands of years may occur where supported by radium-enriched fluids, soils or rocks. Recent marine precipitates (carbonates, sulfates, phosphates or salts), marine ferro-manganese nodules or continental bog iron, marine organic oozes and oil field brines all may be enriched in radium. Radon produced by decay of the radium may accumulate in adjacent hydrocarbons, ground water or soil gases. Radium may also be concentrated in Recent marine and continental colloidal sediments, shales and mudstones.

In Pleistocene and older terranes, radioactive equilibrium will ordinarily have been attained, and elevated radon levels will be observed only in association with high concentrations of uranium and thorium. Igneous rocks rich in silica (granites, rhyolites), late magmatic phases (pegmatites) and hydrothermal deposits may carry substantial concentrations of uranium and thorium, particularly in the more mafic mineral phases (micas, zircon, sphene, monazite, magnetite, rutile). Lag deposits, residual soils and sedimentary placer deposits in which these minerals may be concentrated are likely to be enriched in uranium, thorium and radium as a result. Metamorphic rocks, particularly the more sialic gneisses and slates, will usually retain elevated uranium and/or thorium levels which may have characterized the parent rock. They may also be enriched in uranium or thorium during metamorphosis. In sedimentary rocks, uranium and thorium are commonly concentrated in shales, particularly black shales rich in organic matter. Coal and even dark-colored carbonate rocks are often found to be rich in uranium and thorium, as are the residual soils formed upon them. Marine evaporites (both salt and sulfate deposits), and phosphorites, including nodular phosphate deposits may be enriched in radon progenitors. Finally, uranium and thorium may be concentrated in coarse-grained clastic sediments of marine or continental origin where uranium-bearing ground water has reacted with local concentrations of organic matter causing the precipitation of urananite. All of these uranium- or thorium-rich rocks are potential sources of environmental radon.

Elevated levels of radon in ground water or soil gases may result from any of a range of diverse conditions. High levels of uranium and thorium in soil or bedrock will almost certainly be accompanied by elevated radon levels. Suspect areas can be identified in a general way by geologic mapping, and described with greater accuracy by regional gamma-ray surveys. The absence of identified (or suspected) concentrations of uranium or thorium, however, does not ensure that a region will be radon-free. Elevated radon levels may be supported in Pleistocene (and younger) terranes by non-equilibrium radium concentrations. Local radon anomalies may also occur where favorable structural or lithologic features give radon-rich gases or fluids access to the surface of the earth.

Broad areal surveys of geology, structure, lithology or gamma-ray emission can be used separately or together to estimate the minimum regional radon potential. Detailed observations in individual structures extending over more than a single season are essential for an accurate assessment of the actual radon hazard.

## REFERENCES

Banwell, G.M. 1986. Helium-4 and Radon-222 Concentrations in Ground Water and Soil Gas as Indicators of the Extent and Depth of Fracture Concentration in Rock. Unpub. M.S. Thesis, The Pennsylvania State University, University Park, PA. 118 pp + 4 appendices.

Boyle, R.W. 1982. *Geochemical Prospecting for Thorium and Uranium Deposits.*

Developments in Economic Geology, 16. Elsevier Scientific Publishing Co., New York, NY. 392 pp + bibliography, index.

De Vivo, B. et al. 1984. *Uranium Geochemistry, Mineralogy, Geology, Exploration and Resources.* Institution of Mining and Metallurgy, Brookfield, VT. 188 pp. + index.

Fields, P.R. et al. 1962. Radon Fluoride. J. Am. Chem. Soc., Vol. 84, pp. 4164-4165.

Garrels, R.M. & E.S. Larsen. 1959. *Geochemistry and Mineralogy of the Colorado Plateau Uranium Ores.* Professional Paper 320. U.S. Geological Survey, Washington, D.C. 236 pp.

Graves, B. ed. 1987. *Radon, Radium and Other Radioactivity in Ground Water.* Proceedings of the NWWA Conference, April 7-9, 1987. Lewis Publishers, Somerset, NJ. 546 pp.

Horne, R.A. 1969. *Marine Chemistry.* Wiley-Interscience, New York, NY. 472 pp. + appendix, index.

Horner, J.K. 1985. *Natural Radioactivity in Water Supplies.* Studies in Water Policy, No. 6. Westview Press, Boulder, CO. 346 pp. + 9 appendices + index.

Hutter, A.R. *Radon Variability in Soil Gas over Fracture Traces.* Unpub. M.S. Thesis, The Pennsylvania State University, University Park, PA 156 pp.

Muessig, K. and C. Bell. 1988. Use of airborne radiometric data to direct testing for elevated indoor radon. Northeastern Environmental Survey, Vol. 7. pp. 45-51.

Nazaroff, W.W. and A.V. Nero. 1988. Eds. *Radon and Its Decay Products in Indoor Air.* John Wiley, New York, NY. 518 pp. + index.

Nero, A.V., M.B. Schwehr, W.W. Nazaroff and K.L. Revzan. 1986. Distribution of airborne radon-222 concentrations in U.S. homes. Science, Vol. 234, pp. 992-ff.

Palache, C.H. Berman and C. Frondel. 1942. *The System of Mineralogy,* 7th Edition, Volume 1. John Wiley & Sons, Inc., New York, NY. 808 pp. + index.

Peake, R.T. 1989. Assessing Radon Potential of the United States. Professional Geologist, Vol. 26, No. 5, p. 8, (Abstract).

Rankama, K. 1954. *Isotope Geology.* Pergamon Press, New York, NY. 426 pp. + 2 appendices, bibliography, index.

Rankama, K. and T.G. Sahama. 1950. *Geochemistry.* The University of Chicago Press, Chicago, IL. 788 pp. + 7 appendices, bibliography, index.

Rose, A.W., J.W. Washington and D.J. Greenman. 1988. Variability of radon with depth and season in a central Pennsylvania soil developed on limestone. Northeastern Environmental Science, Vol. 7 pp. 35-39.

Schumann, R.E., D.E. Owen and S. Asher-Bolinder. In press. Weather factors affecting soil-gas radon concentrations at a single site in the semi-arid western U.S. Proceedings of the 1988 EPA Symposium on Radon and Radon Reduction Technology.

Tanner, A.B. 1986. Geological Factors that Influence Radon Availability.

Chapter 1 in *Indoor Radon.* Lewis Publishers, Pittsburgh, PA. pp. 1-12.
Tatsch, J.H. 1976. *Uranium Deposits: Origin, Evolution and Present Characteristics.* Tatsch Associates, Sudbury, MA. 242 pp. + bibliography, index.
Vinogradov, A.P. 1966. *Chemistry of the Earth's Crust.* Translated by N. Kaner. Jerusalem. Israel Program for Scientific Translation. (2 volumes).
Wedepohl, K.H. 1970. Ed. *Handbook of Geochemistry.* Springer-Verlag, New York, NY. (5 volumes, ring-bound).

*Environmental Radon: Occurrence, Control and Health Hazards.* Edited by S. K. Majumdar, R. F. Schmalz and E. Willard Miller © 1990, The Pennsylvania Academy of Science.

# *Chapter Eight*

# RADIOACTIVE ELEMENTS IN PENNSYLVANIA WATERS

**ARTHUR W. ROSE,[1] WILLIAM A. JESTER[2] and BONNIE C. FORD[3]**

[1]Department of Geochemistry
[2]Department of Nuclear Engineering
[3]Low Level Radiation Monitoring Laboratory

The Pennsylvania State University
University Park, PA 16802

## INTRODUCTION

The first recognition of radioactive elements in natural waters dates back many years, but interest has accelerated in recent years with the advent of concern about the health effects of radioactivity. At the present time, extensive monitoring of public water supplies for radioactive substances is mandated by federal and state law, and monitoring near nuclear facilities is required by federal regulations, so that a great deal of information is accumulating on the amount and distribution of radioactivity in natural waters. These results reveal that small amounts of radioactive elements are universally present in natural waters, and that the concentrations vary over an appreciable range as a result of natural processes and human activities.

The purpose of this chapter is to summarize the origin, behavior, abundance and hazard of the main radioactive species in Pennsylvania surface and ground waters. This treatment is intended to provide background to the interested reader in comprehending questions such as the hazard of radon in homes with private wells and pollution related to the nuclear power cycle.

A number of terms and concepts are necessary to understand radioactive phenomena. Atoms are classified chemically into 92 naturally occurring elements and an additional dozen or so man-made elements, based on the number of protons in their nucleus (the atomic number). Atoms of the same element may occur as *isotopes*, which differ in the number of neutrons accompanying the protons in the nucleus. The sum of the number of protons and neutrons in the nucleus is termed the *mass number* and can be indicated by a superscript preceding the chemical symbol of the element. For example, three isotopes of hydrogen (1 proton) are known, denoted as $^{1}H$ (no neutrons), $^{2}H$ (1 neutron, also known as deuterium), and $^{3}H$ (2 neutrons, also known as tritium). In speaking of isotopes of several elements, the different nuclear species are referred to as *nuclides*, such as $^{12}C$, $^{18}0$ and $^{32}S$. Some nuclides are radioactive, such as $^{238}U$ and $^{137}Cs$, and spontaneously decay to a different nuclide with the emission of characteristic energetic particles and/or electromagnetic energy. These radioactive nuclides, or radionuclides, are the main focus of this chapter.

The time at which an individual radionuclide atom decays is indeterminate, but for a large number of such atoms, a characteristic *half-life* can be measured during which half the atoms of the radionuclide spontaneously decay to a daughter nuclide. The decay is normally accompanied by the emission of energetic alpha and/or beta particles, gamma rays, and recoil of the daughter nucleus. *Alpha particles* are identical to nuclei of the helium atom ($^{4}He$, composed of 2 protons and 2 neutrons), *beta particles* are identical to electrons (or positrons), and *gamma rays* are a form of electromagnetic radiation originating in the nucleus. An additional form of nuclear breakdown is called *fission*, in which the nucleus breaks into two nuclides approximately half the size of the parent.

Four general groups of radionuclides can be distinguished. One group has such a long half-life that they were formed along with all other elements during primordial nucleosynthesis prior to the origin of the earth about 4.5 billion years ago. The main representatives of this group are $^{238}U$, $^{235}U$, $^{232}Th$, and $^{40}K$, plus trace amounts of $^{87}Rb$, $^{143}Sm$ and others. These radionuclides necessarily have half lives of about $10^9$ years or more. A second group consists of the many shorter lived daughters of U and Th, such as Ra and Rn, which form as a result of the decay of their long-lived parents. The third group originates by continuing natural nuclear transformations driven by cosmic rays or natural sources of neutrons or energetic particles. The best examples of this group are $^{14}C$ and $^{3}H$, which are formed in the upper atmosphere by cosmic rays. Many other nuclides are formed in trace quantities by such processes. The fourth group is formed by nuclear weapons, nuclear reactors and similar man-made devices. Important members of this group are $^{90}Sr$, $^{137}Cs$, $^{14}C$ and $^{3}H$. Note that many members of the third group are also formed in this fourth fashion, such as $^{14}C$ and $^{3}H$.

Radioactivity is commonly measured in units of *curies* (Ci) or *bequerels* (Bq). A curie is $3.7 \times 10^{10}$ radioactive disintegrations per second, and a bequerel (the newer unit of the international system) is one disintegration per second. Because the curie is so large, concentrations in natural waters are commonly expressed in picocuries per liter (pCi/l) equal to $10^{-12}$ curies per liter. The newer unit is the bequerel per cubic meter, where 1 pCi/l equals 37 $Bq/m^3$.

The amounts of the various radioactive nuclides in natural waters depend on overall geochemical abundance of the species, its chemical properties, and its half life. Nuclides with low abundances are inherently limited to low concentrations in natural waters. In contrast, near a concentrated source such as a uranium deposit or a source of radionuclide pollution, the concentration in water may be high if other factors are favorable. If the element is insoluble or strongly adsorbed on natural materials, it will generally attain only low concentrations in solution (though it may occur as suspended particles). Conversely, elements that are soluble in solution have the potential for high concentrations if a source is available. Finally, if the nuclide has a very short half life (an hour or less), it will generally occur in water only if it originates from a longer lived radioactive parent that is also present in the water.

## REGULATIONS GOVERNING CONTAMINANTS IN PENNSYLVANIA WATER SUPPLIES

The Pennsylvania Safe Drinking Water Act (P.L. 206, No. 43) (PA SDWA) was signed into law on May 1, 1984. The PA SDWA implemented regulations, the Pennsylvania Safe Drinking Water Regulations, Chapter 109, which became effective on December 8, 1984.[1]

The current limits for radioactive materials are intended to limit the human intake of radioactive substances so that the total radiation exposure of any individual will not exceed that defined in the radiation protection guides recommended by the Federal Radiation Council.

A Maximum Contaminant Level (MCL) is the maximum permissible level of a contaminant which may be delivered in drinking water to a user of a public water system. All MCL's for radiological contaminants are primary, indicating that they govern contaminants which may adversely affect public health, and that specific monitoring requirements must be followed. Radiological contaminants are divided into two groups, natural and man-made. The primary MCL's are as follows:

| | |
|---|---|
| *Natural* | |
| Gross alpha | 15 pCi/l |
| Combined $^{226}Ra$ and $^{228}Ra$ | 5 pCi/l |
| *Man-Made* | |
| Gross beta | 50 pCi/l |
| Tritium | 20,000 pCi/l |
| $^{90}Sr$ | 8 pCi/l |

A community water system is defined as a public water system which serves at least 15 service connections used by year-round residents, or regularly serves at least 25 year-round residents. This includes systems such as municipalities, water authorities, mobile home parks, institutions and associations, as well as bottled water and bulk water hauling systems. Monitoring requirements for the various categories of community water systems are listed below:

For bottled water systems and bulk water hauling systems, both surface and ground water sources must be monitored annually for gross alpha. Surface water systems must be monitored annually for gross beta, tritium and $^{90}Sr$. Follow-up requirements are the same as discussed below.

For all other community water systems, surface and/or ground water sources must be monitored every 4 years for gross alpha. When the gross alpha exceeds 5 pCi/l, the same or an equivalent sample must be analyzed for both $^{226}Ra$ and $^{228}Ra$. Surface water sources serving more than 100,000 people must be monitored every 4 years for gross beta, tritium and $^{90}Sr$. When the gross beta activity exceeds 50 pCi/l, an analysis of the sample must be performed to identify the major radioactive constituents present and the appropriate organ and total body doses should be calculated.

Cothern et al[2] have estimated that drinking water accounts on average for between 0.1 and 2.5% of the radiation exposure of persons on community water systems. Radon and radium are the major contributors to this exposure, with the major radon exposure being acquired by inhaling radon degassed from the water. The exposure from radon entering houses via soil gas is much larger, accounting for about 50% of the average radiation exposure.

## LONG-LIVED NATURAL RADIONUCLIDES AND THEIR DAUGHTERS

As noted above, the major parent nuclides in this group are $^{238}U$, $^{232}Th$, and $^{40}K$. The nuclide $^{235}U$ is the parent of another decay series, but the abundance of $^{235}U$ is so low that these daughters are rarely significant in waters. The U and Th nuclides are each the parent of a series of shorter lived daughter nuclides (Fig. 1). In typical common rocks, U, Th and K (including their daughters) account for nearly all the natural radioactivity, and are each responsible for about a third of the natural radioactivity. Among the daughters in the U and Th series are Ra (radium) and Rn (radon), which are perhaps the most important nuclides of this series in terms of health risk.

Thorium is chemically very insoluble and strongly adsorbed in essentially all natural environments. It is rarely found in significant amounts in natural waters. This generalization applies both to long-lived $^{232}Th$ and to shorter-lived radionuclides such as $^{234}Th$ and $^{230}Th$ that form in the $^{238}U$ decay chain.

In contrast, uranium is relatively soluble in most oxidizing surface and ground waters, and tends to be present in dissolved form, though it can be strongly

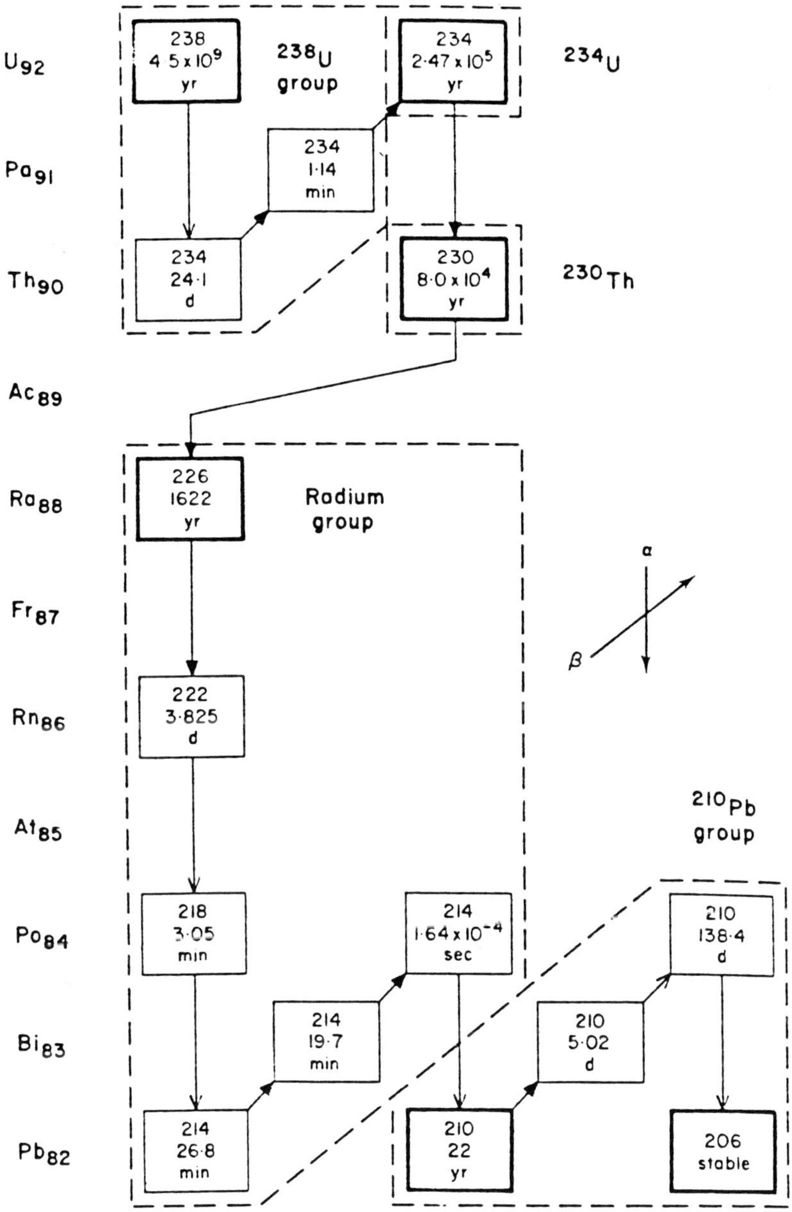


FIGURE 1. The [238]U radioactive decay series, showing half-lives and atomic mass numbers (in boxes) and decay type (alpha decays downward, beta decays upward to right). Dashed boxes enclose nuclides that tend to occur as a group headed by a long-lived nuclide.

adsorbed from near-neutral solutions onto organic matter and iron oxides. As a result, most waters contain detectable amounts of uranium. Typical levels in Pennsylvania waters are in the range of 0.05 to 0.5 parts per billion (ppb) (refs. 3 through 12).

*Gross alpha and gross beta determinations*

Measurements of "gross alpha" and "gross beta" activity are used as screening tests for water supplies. In these methods, up to 1 liter of water is evaporated to dryness and the counting rates of alpha and beta particles are measured. Volatile radionuclides such as $^{222}$Rn and $^{3}$H are lost from the sample. In order to allow for absorption of some alpha and beta particles by the solid residue and for less than 100% detection efficiency, the observed count rate is calibrated by comparison with the count rate for samples of natural uranium (in equilibrium with its daughters) or $^{241}$Am for gross alpha, and with $^{137}$Cs or $^{90}$Sr-$^{90}$Y for gross beta. Gross alpha gives an estimate of the $^{238}$U and $^{232}$Th and their alpha-emitting daughters, including $^{234}$U, $^{226}$Ra, $^{228}$Ra, and others. Gross beta provides an estimate of beta-emitting U and Th daughters plus species such as $^{137}$Cs, $^{90}$Sr, $^{40}$K and others formed as fission and cosmic ray products. Because of differing detection efficiencies for different radionuclides, the determinations are only semi-quantitative expressions of alpha and beta activity.

For this paper, most data on gross alpha and gross beta have come from determinations in the Low Level Radiation Monitoring Laboratory at Penn State.[13] These have been supplemented by determinations of Horton[4] done as part of an E.P.A. survey of water supplies, and by the older values of Scott and Barker[3] of the U.S. Geological Survey. In addition, the U.S. Geological Survey has furnished a compilation of data from their WATSTORE database,[14] mainly for surface waters of the state, and the Pennsylvania DER furnished a printout of its files for radionuclides in water supplies.[15]

Figure 2 shows the gross alpha values for a series of samples of major rivers (mainly the Allegheny and Susquehanna Rivers) plotted against specific conductivity. The gross alpha values are seen to correlate well with specific conductivity, which is a measure of the total content of dissolved ions in the water. The specific conductivity is dominated by Ca, $HCO_3$ and $SO_4$ in these samples. Thus, as has been reported by other workers,[16] uranium and its daughters correlate with major solutes in the waters. Since ground water in a given area commonly contains higher solids than most surface water in the area, because of its longer contact with rock and soil,[17] it is expected that ground-water supplies will tend to contain higher contents of alpha-emitting nuclides than surface waters in the same vicinity. Data to demonstrate this generalization does not seem to be available.

The frequency distribution of gross alpha values for 95 public water supplies scattered across the state is plotted in figure 3. The majority of values are less than 4 pCi/l, the most common value is in the range 0.0 to 0.5 pCi/l, and the mean value is 1.8 pCi/l. The frequency distribution is highly skewed, with a few samples of 10 to 21 pCi/l. For a much larger set of 2163 samples analyzed at the Penn State laboratory, the mean is slightly lower at 0.97 pCi/l, and the maximum value is 72.7 pCi/l.

Classification of the gross alpha values by county (Fig. 4) and by geologic unit (Table 1) indicates that the distribution of high values is not random. The most distinct area of high values is in Montgomery and Bucks Counties in southeastern Pennsylvania, where average values are 3.3 and 1.6 pCi/l, respectively. Examination of the data shows that the high values are localized in areas underlain by rocks of Triassic age, and that waters in areas of Triassic rocks are high in gross alpha not only in Montgomery and Bucks Counties but also southwestward along their outcrop belt in Berks, Lebanon, Dauphin, York, and Adams Counties. The average gross alpha for this group of samples is 3.72 pCi/l, with about 20% of the values exceeding 5 pCi/l. The major rock types in this belt are red sandstones and shales, though some gray shale and diabase are present. Numerous small uranium prospects are known in these rocks in Montgomery and Bucks Counties.

Another set of relatively elevated gross alpha values is found in the Devonian Catskill Fm. in northeastern Pennsylvania. Although the mean value for samples from the area underlain by the Catskill is only slightly greater than the average for the entire data set, 32% of the values in this group exceed 1 pCi/l, a distinctly higher proportion than several groups with similar means (e.g., Great

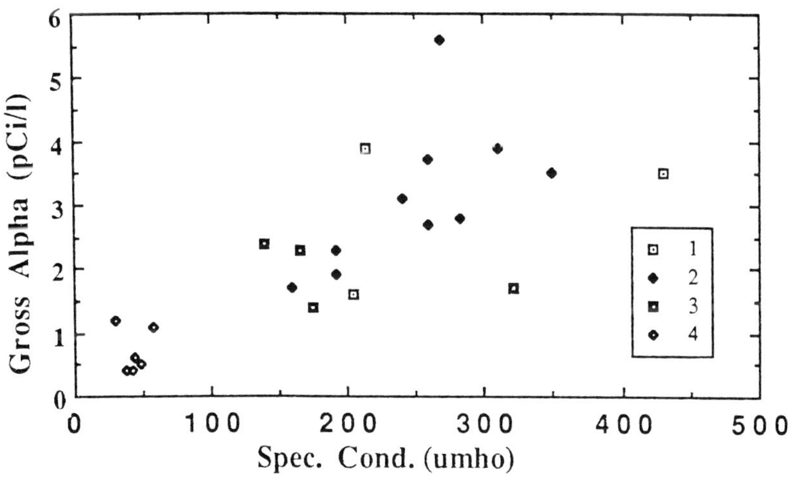


FIGURE 2. Gross alpha vs. specific conductance for river water samples.[14] The specific conductance is approximately proportional to total dissolved solids. 1 = Allegheny R., Oakmont; 2 = Allegheny R., New Kensington; 3 = Susquehanna R., Harrisburg; 4 = Young Womans Cr. near Renovo.

Valley carbonates, Lancaster Valley carbonate-sandstone). The major rock type of the Catskill Fm. is red sandstone and shale, as for the Triassic, suggesting that uranium and its daughters tend to be more available and more mobile in red sediments. A number of small uranium prospects are also known in the Catskill Fm.

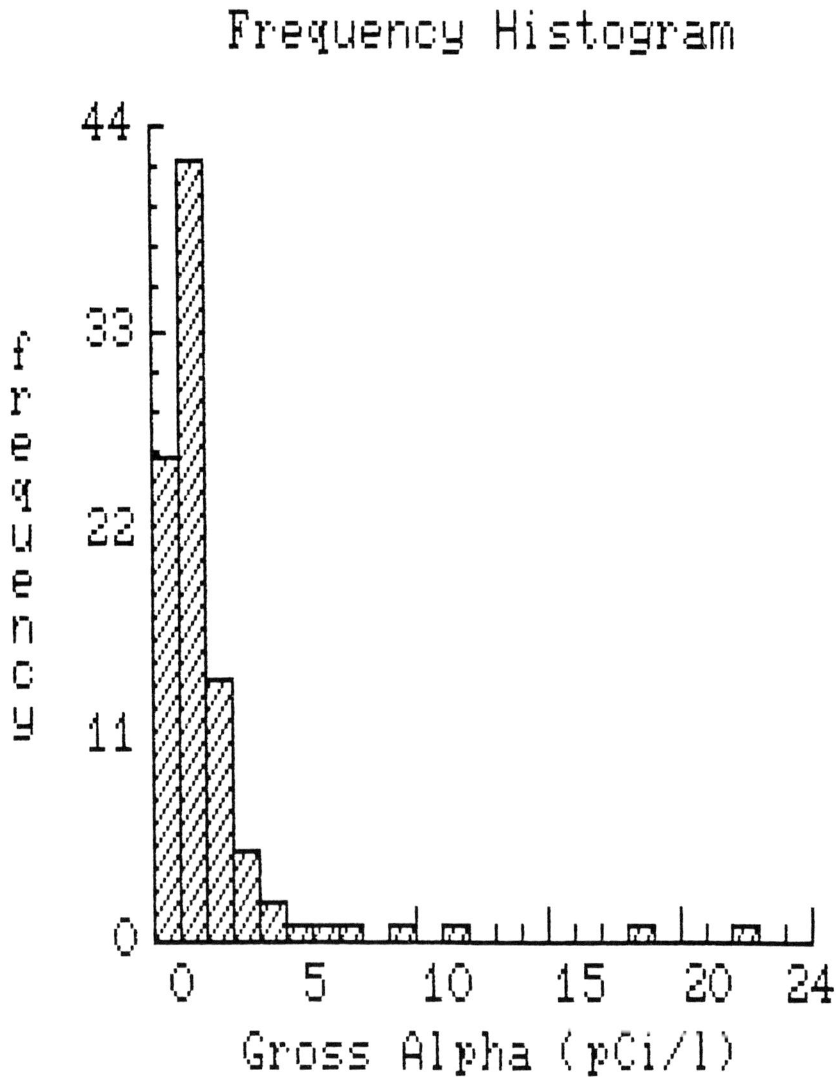


FIGURE 3. Histogram of gross alpha values for public water supplies for Pennsylvania.[4] Some negative determinations are recorded because of counting statistics.

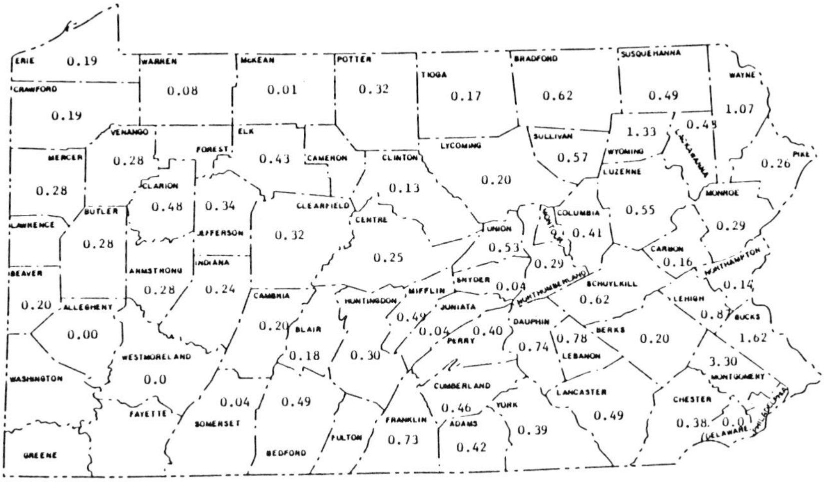


FIGURE 4. Median gross alpha values (pCi/L) for counties, based on 2163 samples analyzed at the Low Level Radiation Monitoring Lab, Penn State University.[13] Determinations were averaged by town before determining the median value for each county.

TABLE 1

*Mean Values of Gross Alpha for Geologic Units.*

| Geologic Unit | N | Mean | >1 pCI/l | Area |
|---|---|---|---|---|
| Triassic Basin | 54 | 3.72 pCi/l | 69% | SE PA |
| Precam. Granitic Gneiss | 11 | 6.37 | 36 | Del., Ches., Lanc. Cos. |
| Wissahickon Schist | 19 | 0.38 | 5 | SE PA |
| Reading Prong | 5 | 0.17 | 0 | E PA |
| Lancaster Valley Carb. | 22 | 0.80 | 22 | Lanc., York Cos. |
| Great Valley Carb. | 34 | 1.04 | 26 | SE PA |
| Martinsburg Shale | 14 | 0.56 | 14 | SE PA |
| Catskill Group | 68 | 0.98 | 32 | NE PA |
| Silurian Sequence | 19 | 0.36 | 5 | NE PA |
| Miss.-Penn. Sequence | 31 | 0.48 | 13 | Anthracite Basin |
| Ridge & Valley Prov. | | 0.30 | | 10 Cos. in province |
| Appalachian Plateau | | 0.20 | | 20 Co. in W PA |
| All samples | 2163 | 0.97 | | Entire State |

Source of data: Reference 13. Data for each water supply were averaged, and plotted on a geologic map; then supplies in the geologic units were averaged to obtain the above results. For the Ridge & Valley and Appalachian Provinces the means for selected counties falling largely or wholly in the province were averaged.

One might expect that the Precambrian area of the Reading Prong would contain water supplies with elevated gross alpha, in view of the known enrichment of uranium in this province,[18] but this is not the case. However, water supplies in areas of Precambrian granitic rocks and gneisses in Chester, Delaware and Lancaster Counties do tend to have high gross alpha. In contrast, supplies in the surrounding Wissahickon schist have very low gross alpha.

The Cambro-Ordovician carbonate and sandstone belts of the Lancaster-York and Great Valley areas (Easton-Allentown-Reading-Lebanon-Cumberland-Chambersburg) show gross alpha values slightly above average. In the Lancaster-York zone, the high values result from elevated radium in ground waters from the Chickies quartzite.[19] These waters are strongly acid and slightly reduced because of pyrite oxidation, and apparently the acidity accentuates leaching of radium. In particular, these waters have high values of $^{228}$Ra, a daughter of $^{232}$Th. In the Great Valley, no such correlation can be noted.

In contrast to these relatively high gross alpha values near 1 pCi/l and above in eastern Pennsylvania, values in the Appalachian Plateau and Ridge and Valley Province of western and central Pennsylvania are distinctly lower (Table 1, Fig. 4). The Mississippian and Pennsylvanian rocks of the Anthracite Basins also have a markedly lower average gross alpha value.

Figure 5 illustrates the distribution of gross beta values for the 95 samples from public water supplies of the EPA study. The most frequent class is 0.5 to 1.0 pCi/l, the mean is 2.0 pCi/l, and the data are distinctly skewed as for the gross alpha values. The maximum value of the EPA study is 9.5 pCi/l. In the larger set of determinations from Penn State,[13] the mean is 2.33 pCi/l and the maximum is 35.6, based on 223 samples (including the above-mentioned EPA values). However, this data set may be biased because the majority of the gross beta values are limited to surface water supplies. The gross beta values in rivers from the U.S.G.S. set correlate reasonably well with specific conductivity (Fig. 6), as for the gross alpha values.

A principal reason for gross alpha determinations is to detect elevated radium values. As noted above, radium concentrations above MCL values have been found in ground waters from the Chickies quartzite of southeastern Pennsylvania.[19] Elevated radium values, up to 5000 pCi/l, have been measured in brines produced by gas wells in western Pennsylvania.[20] These waters are too saline to constitute drinking water, but may be mixed with potable waters locally.

### *Radon*

Of all the radionuclides found in Pennsylvania water, $^{222}$Rn occurs at the highest concentrations, specifically in certain well waters. Radon-222 is a radioactive daughter of $^{226}$Ra in the $^{238}$U decay series (Figure 1). It is a noble gas with minimal tendency for chemical reaction, so that when it is generated by decay of $^{226}$Ra in soil or rock, it can migrate either in air-filled spaces or dissolved in water and move much more freely than other radionuclides. However, its half-life is only 3.8 days, limiting the distance it can travel from its parent $^{226}$Ra before it decays. It decays to a sequence of four short-lived daughters, namely $^{218}$Po and $^{214}$Po, both alpha emitters, and $^{214}$Pb and $^{214}$Bi, which emit both beta and gamma radiation.

Most ground waters contain much higher $^{222}$Rn than its parent $^{226}$Ra, because the $^{226}$Ra is fixed in solid phases, but when it decays to $^{222}$Rn its chemical bonds

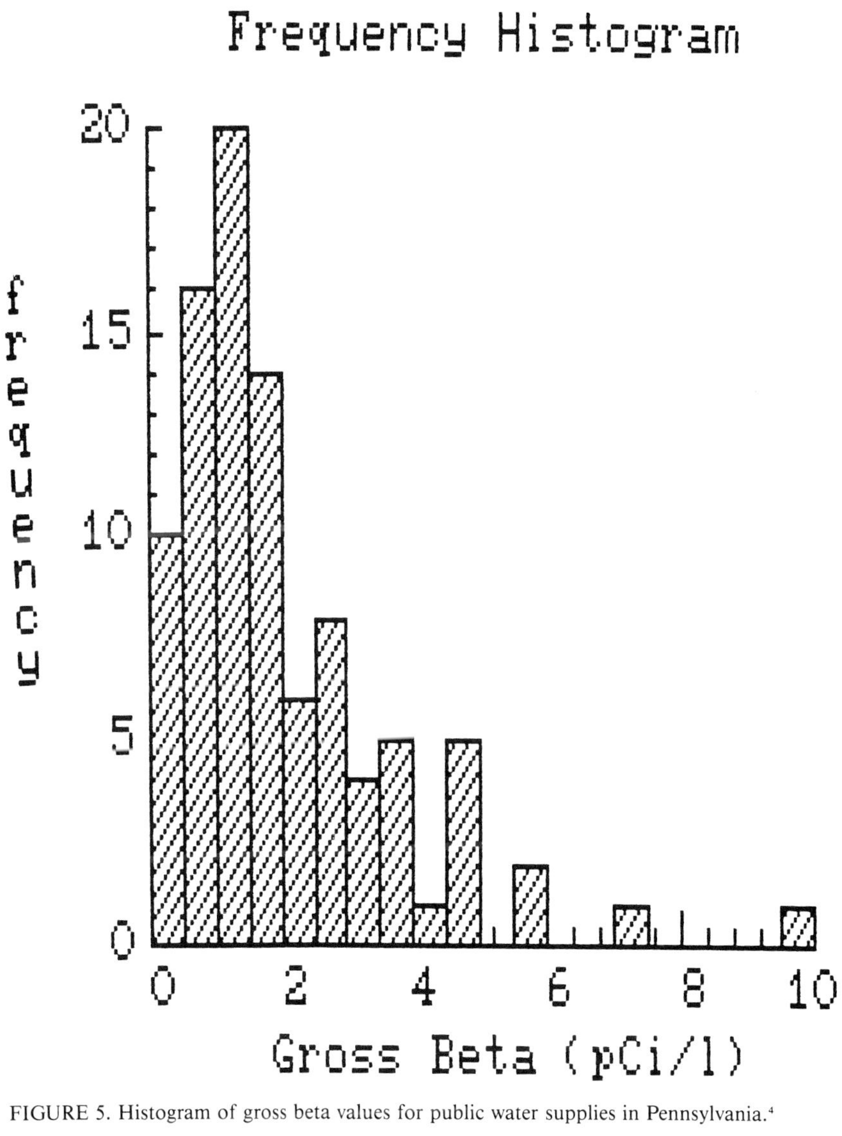


FIGURE 5. Histogram of gross beta values for public water supplies in Pennsylvania.[4]

disappear and it readily enters the aqueous or air phase. In surface waters, Rn is rapidly degassed into the atmosphere, so few surface waters contain appreciable Rn concentrations. In contrast, ground waters commonly contain hundreds to thousands of pCi/l Rn. Also, because of its volatility, Rn and its daughters are not usually measured in gross alpha determinations. Since the radon mostly escapes from water into the air, $^{222}Rn$ is considered to be much more of an airborne lung exposure hazard than a drinking water hazard. In a recent study by EPA,[21] it was estimated that in a typical home it takes about 25,000 pCi/l of dissolved radon to raise the radon level of indoor air by 1 pCi/l. Thus, in most cases the radon entering a home in soil air is a more serious hazard than the radon dissolved in water.

The abundance of Rn in 95 relatively typical Pennsylvania public water supplies is illustrated in Figure 7. The mean value is 757 pCi/l, the most common class is 100 to 200, and the maximum is 4180 pCi/l. Note that these public water supplies may have been aerated or stored in contact with air, so that they may have lost a good deal of their original Rn, or they may have originated as surface waters that contained low Rn initially.

Several studies of Rn in household well waters have been made, but none appear to approximate representative samples for the state. Data on Rn in 1006 well waters from northeastern Pennsylvania[22] are summarized in Figure 8. These

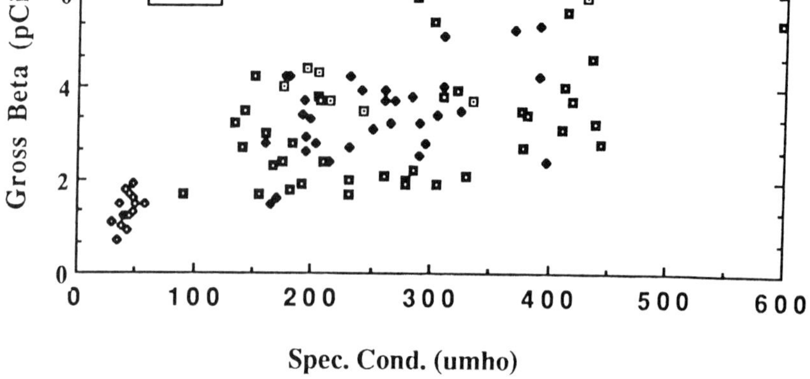


FIGURE 6. Gross beta vs. specific conductance for river water samples.[14] The specific conductance is approximately proportional to total dissolved solids. For explanation of symbols see Figure 2.

well water values are clearly much higher in radon than the public water supplies, and include two values exceeding 100,000 pCi/l, both in the Reading Prong area of Bucks County.

Two types of geological areas are known to have much higher Rn values than normal ground water. Rumbaugh[12] found Rn values averaging 14,000 pCi/l (std.

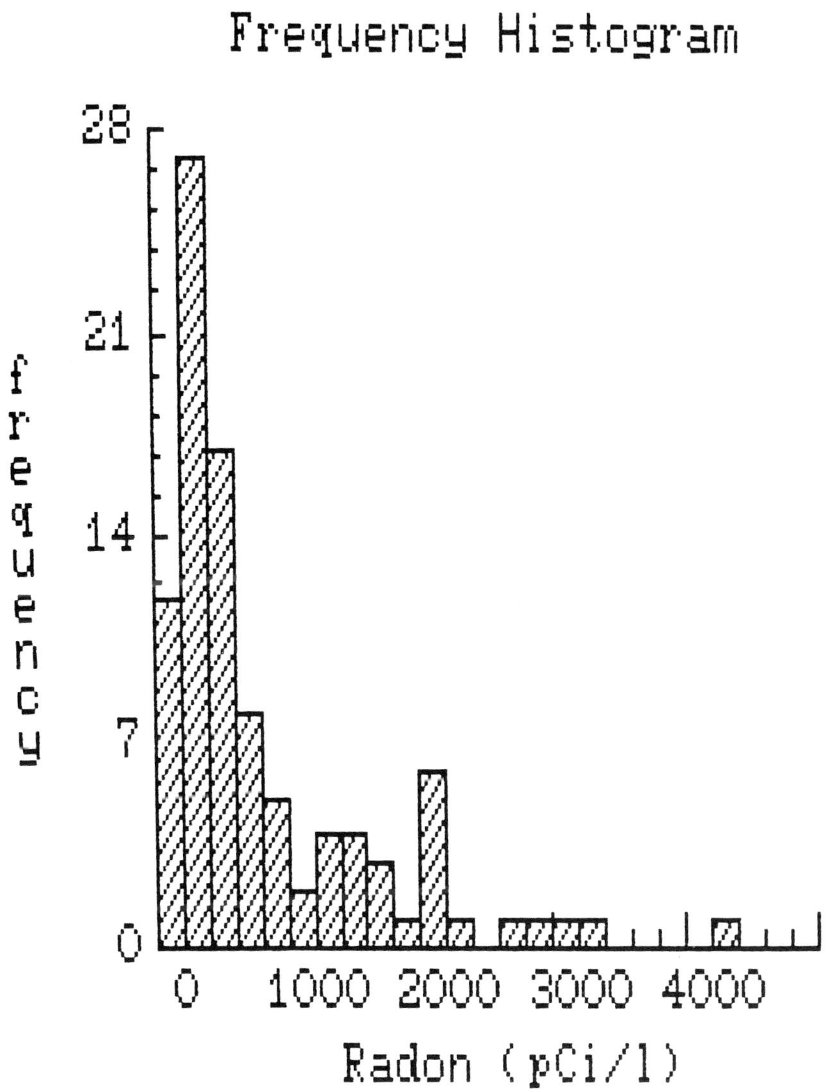


FIGURE 7. Histogram of radon values for public water supplies in Pennsylvania.[4] Some negative values are recorded because of counting statistics.

dev. 33,600) for 131 samples from a small area of Precambrian grantic gneiss near Alsace Manor in the Reading Prong, Bucks County. Radon values as high as 241,000 pCi/l occur in this area, and are accompanied by elevated U (one sample has 450 ppb U).

A second area with high Rn is the Devonian Catskill Fm. with its small uranium occurrences. Korner and Rose[10] studied Rn in ground waters of several areas of Catskill redbeds and uranium occurrences, and a number of non-mineralized areas (Fig. 9). In non-mineralized areas, a majority of well waters contain less than 1000 pCi/l, with a median of about 500 pCi/l, except in areas of non-mineralized redbeds, where values extend up to 2500 pCi/l. Areas with known U prospects have Rn values from about 500 to 7000 pCi/l, with a median of about 1500 pCi/l. In a set of 37 stream waters, Rn ranged up to about 500 pCi/l for non-mineralized areas, with a median of less than 100. Influx of ground water into the stream near the sample site may be responsible for the higher values. Areas with uranium mineralization did not show higher values in stream water.

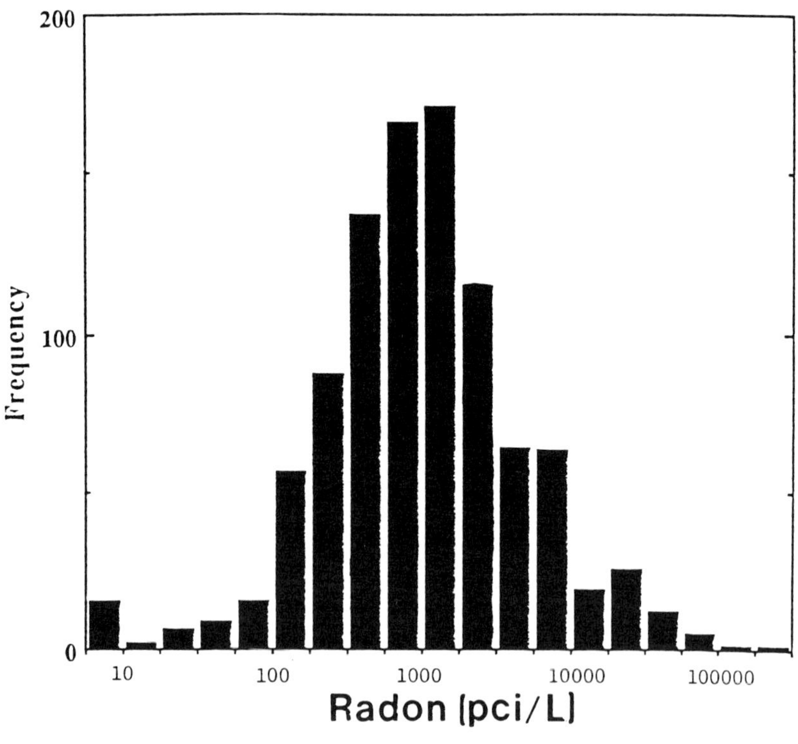


FIGURE 8. Histogram of radon values for 1006 well water samples from southeastern Pennsylvania, including the Reading Prong,[21] showing much higher values than for public water supplies (Fig. 7).

Banwell and Parizek[23] have measured radon in ground waters in Lehigh County and found higher values along regional fracture zones. Kratchman and Richards[24] report radon contents of 420 to 5300 pCi/l for a few household wells in Bucks County, PA.

## *Tritium ($^3H$)*

Tritium (T or $^3H$) is the radionuclide present in the highest concentration in most surface waters and in many ground water samples. Tritium is the isotope of hydrogen which has two neutrons in its nucleus. It is primarily found in the environment as tritiated water (HTO). It has a half life of 12.3 years and decays by emitting a very low energy beta particle. Because of its low energy, it is much less of a hazard than most other radionuclides found in the environment.

Tritium is generated mainly from three sources, namely by cosmic ray interaction with atmospheric air, atmospheric nuclear weapons testing, and nuclear power plant emissions.[25]

Prior to the initiation of large-scale atmospheric testing, tritium came almost entirely from the interaction of cosmic ray produced neutrons with nitrogen and oxygen of the atmosphere. The resulting tritium levels in natural waters were between 6 and 24 pCi/l. Atmospheric testing of nuclear weapons between 1952 and 1973 caused an increase in tritium levels in rainfall and surface waters

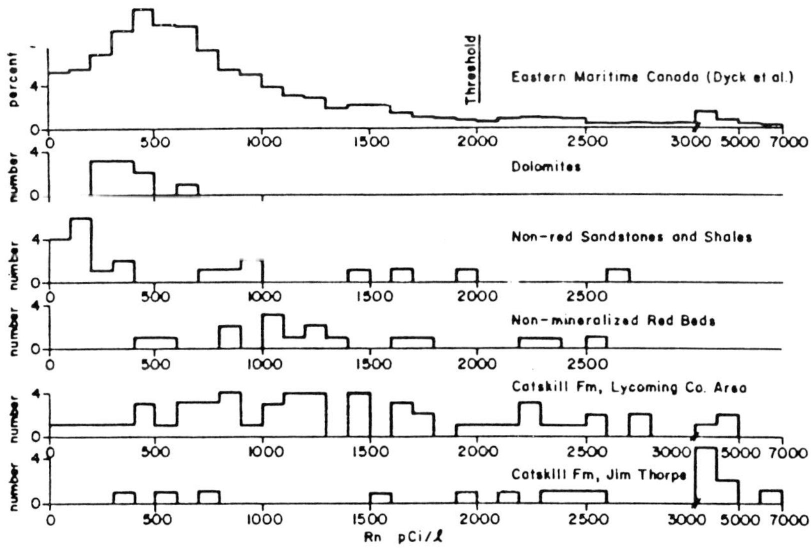


FIGURE 9. Histograms of radon contents in well waters for several areas in Pennsylvania[10] and for a survey in eastern Maritime Canada. Values tend to be slightly higher in redbeds, and still higher near uranium prospects.

to between 300 and 5000 pCi/l. Since that time, through processes of dilution and decay, tritium levels in public water supplies in Pennsylvania, as determined by both DER[15] and Penn State's LLRML,[13] are now typically between 30 and 300 pCi/l, depending on the effective time at which the water entered the hydrologic cycle from the atmosphere.

Nuclear power plants also produce tritium, and levels as high as 2500 pCi/l have been detected near a Pennsylvania nuclear power plant. Such tritium levels are well below the Maximum Contaminant Levels considered as safe for public drinking water by PA DER, EPA and NRC.

### *Carbon-14*

Carbon-14, like tritium, is produced by cosmic radiation, nuclear explosives, and nuclear reactors. Carbon-14 has a half life of 5730 years, and decays by emitting a very low-energy beta particle. While its rate of production in the upper atmosphere is about a factor of 10 greater than the tritium production rate, the global increase in $^{14}C$ produced by atmospheric testing has increased the worldwide concentration by only about 10%, and levels in the northern hemisphere (where most of the tests have occurred) by about 65%.

Since carbon in many of its various forms such as methane is relatively insoluble in water, and since only a small fraction of the carbon is $^{14}C$, the maximum concentration of $^{14}C$ in Pennsylvania waters, even near nuclear power stations, should be less than 1 pCi/l.

## OTHER RADIONUCLIDES PRODUCED BY COSMIC RAYS

There are more than 20 radionuclides produced by interaction of cosmic rays with the atmosphere, but they have sufficiently lower production rates than $^{14}C$ and tritium that their presence is not readily seen in environmental waters.

## OTHER RADIONUCLIDES PRODUCED BY FISSION

Over 350 different fission products are formed by thermonuclear explosions and nuclear power reactors. The two most significant of these are $^{137}Cs$ and $^{90}Sr$.

Cesium-137 has a half-life of 30.1 years and is a strong gamma and beta emitter. Strontium-90 and its daughter yttrium are both beta emitters. The major part of these radionuclides found in the environment comes from past atmospheric nuclear weapons testing, and very little from operating reactors or from reactor accidents (e.g., TMI and Chernobyl). Also, since both of these radionuclides are almost completely removed from water as it passes through most soils, typical levels in waters are usually less than 1 pCi/l. Values as high as 5 pCi/l have occasionally been found in waters close to nuclear power plants.

Periodically, low concentrations of $^{131}I$ are found in surface waters. These occurrences can usually be traced to a sewage treatment plant and results from the medical use of this radionuclide.

## NEUTRON ACTIVATION PRODUCTS FROM NUCLEAR POWER PLANTS

Occasionally, one can detect low levels of neutron activated metallic elements near operating nuclear power plants. These include radionuclides in the range of a few picoCuries per liter of $^{51}Cr$, $^{54}Mn$, $^{59}Fe$, $^{60}Co$, and $^{65}Zn$. These radionuclides are produced by the irradiation of dissolved metal atoms in the reactor coolant. The levels of these radionuclides in such waters are too low to constitute a health hazard.

## CONCLUSIONS

A wide range of radionuclides is found in natural waters of Pennsylvania and other regions. The radionuclides present at the highest levels of radioactivity are typically $^{222}Rn$ and $^{3}H$. Radon-222, generated from natural decay in the uranium series, is abundant in some ground waters. Tritium, generated from nuclear weapons tests during the 1950's through 1970's, reaches its highest levels in ground waters that entered the earth during this period. Extensive testing of public water supplies has led to reduction of radioactivity for supplies with values exceeding MCL's and thereby limited the hazard from public supplies, but levels in some homes supplied by wells may still be cause for concern, especially in certain areas where radon and radium are more abundant than average, such as in areas of red sandstones and shales, granitic rocks, or the Cambrian Chickies quartzite.[13]

## ACKNOWLEDGMENTS

We are indebted to Barry Greenawald and Carl Granlund of PA DER and to David Click of the U.S. Geological Survey for furnishing files of data.

## REFERENCES

1. PA Dept. of Environmental Resources, Div. of Water Supplies. 1986. Public Water Supply Manual, Part 1, Guide to the Safe Drinking Water Program, p. 11-123.

2. Cothern, C.R., W.L. Lappenbusch, and J. Michel. 1986. Drinking water contribution to natural background radiation. *Health Physics.* 50:33-47.
3. Scott, R.C. and F.B. Barker. 1962. Data on uranium and radium in ground water in the United States from 1954 to 1957. U.S. Geological Survey Professional Paper 426, 115 p.
4. Horton, T.R. 1985. Nationwide occurrence of radioactivity in public water supplies. U.S. Environmental Protection Agency Rept. EPA 520/5-85-008, 208 p.
5. Ferguson, R.B. and P.L. Jones. 1979. Scranton 1° x 2° NTMS area, New Jersey, New York and Pennsylvania, Preliminary basic data report, NURE hydrogeochemical and stream sediment reconnaissance. U.S. Dept. of Energy Rept. GJBX-2(79), 226 p.
6. Heffner, J.D. 1979. Harrisburg 1° x 2° NTMS area, Pennsylvania, Data release, Hydrogeochemical and stream sediment reconnaissance. U.S. Dept. of Energy Rept. GJBX-31(79), 227 p.
7. Heffner, J.D. 1979. Williamsport 1° x 2° NTMS area, Pennsylvania, Data report, NURE Hydrogeochemical and stream sediment reconnaissance. U.S. Dept. of Energy Rept. GJBX-152(79), 219 p.
8. Heffner, J.D. 1980. Newark 1° x 2° NTMS area, New Jersey, New York and Pennsylvania, Data report, Hydrogeochemical and stream sediment reconnaissance. U.S. Dept. of Energy Rept. GJBX-128(80), 214 p.
9. Price, V., R.B. Ferguson, and E.I. Baucom. 1976. Orientation study in the Williamsport, Pennsylvania area. U.S. Dept. of Energy Rept. GJBX-53(76), 12 p.
10. Korner, L.A. and A.W. Rose. 1977. Radon in streams and ground waters of Pennsylvania as a guide to uranium deposits. U.S. Dept. of Energy Rept. GJBX-60(77), 151 p.
11. Applin, K.R. and D. Langmuir. 1978. Ground water geochemistry as a prospecting tool for uranium deposits in Pennsylvania. U.S. Dept. of Energy Rept. GJBX-132(78), 145 p.
12. Rumbaugh, J.O. 1983. Effect of fracture permeability on radon-222 concentration in ground water of the Reading Prong, Pennsylvania. M.S. Thesis, Penn State Univ., University Park, PA, 111 p.
13. Ford, B.C. 1988. Compilation of radionuclide determinations on public water supplies, as analyzed at the Low Level Radiation Monitoring Lab, Penn State Univ., University Park, PA.
14. U.S. Geological Survey. 1988. Printout of data on radionuclides in Pennsylvania waters from WATSTORE database, furnished by David Click. See also Water Resources Data for Pennsylvania, U.S.G.S., various years from 1968 to 1980, for chemical data on these same waters.
15. PA Dept. of Environmental Resources. 1988. Printout of radionuclide data for community water supplies in Pennsylvania, obtained from Barry Greenawald.

16. Rose, A.W. and R.J. Wright. 1980. Geochemical exploration models for sedimentary uranium deposits. *J. Geochemical Exploration*, 13:153-179.
17. Hem, J.D. 1985. Study and interpretation of the chemical characteristics of natural water. U.S. Geological Survey Water Supply Paper 2254, p. 39.
18. Smith, R.C. and J.H. Barnes. In press. Geology and mineralogy of uranium and thorium in the Reading Prong of Berks, Lehigh and Northampton Counties, Pennsylvania. PA Geological Survey, Mineral Resource Report.
19. Cecil, L.D., R.C. Smith, M.A. Reilly, and A.W. Rose. 1987. Radium-228 and radium-226 in ground water of the Chickies Formation, Southeastern Pennsylvania. *In:* Barbara Graves (Ed.), Radon, radium and other radioactivity in ground water (Proc. of National Water Well Assoc. Conf.). Lewis Publ., Pittsburgh, PA. p. 437-447.
20. Dresel, P.E. and A.W. Rose. In press. Chemistry and origin of oil and gas well brines in western Pennsylvania. Pennsylvania Geological Survey Bulletin.
21. Partridge, T.R., T.R. Horton, and E.L. Sensintuffer. 1979. A study of Radon-222 released from water during typical household activities. U.S.E.P.A., Eastern Environmental Radiation Lab., Rept. ORP/EERF-79-1, 24 p.
22. PA Dept. of Environmental Resources. 1988. Listing of data on radon in waters and houses, supplied by Carl L. Granlund, 8/24/88.
23. Banwell, G.M. and R.R. Parizek. 1988. Helium-4 and radon-222 concentrations in ground water and soil gas as indicators of zones of fracture concentration in unexposed rock. *J. Geophysical Research.* 93:355-366.
24. Kratchman, J., and A.J. Richards. 1956. Radon in ground water in Hunterdon County, New Jersey and Bucks County, Pennsylvania. U.S. Atomic Energy Comm. Rept. RME-4076, 33 p.
25. Fritz, P. and J.C. Fontes. 1980. Handbook of environmental isotope geochemistry, volumes 1 and 2. Elsevier, Amsterdam, 545 p.

*Environmental Radon: Occurrence, Control and Health Hazards.* Edited by S. K. Majumdar, R. F. Schmalz and E. Willard Miller © 1990, The Pennsylvania Academy of Science.

# *Chapter Nine*

# DETECTION AND MEASUREMENT OF RADON AND RADON DECAY PRODUCTS

**AHMAD E. NEVISSI**

Laboratory of Radiation Ecology
University of Washington
Seattle, WA 98195

## INTRODUCTION

In detection and measurement of radon and radon decay products a number of general issues should be explained that are important for the understanding of the basic detection methodologies and the choice of devices to be used for the measurements.

The three naturally occurring isotopes of radon are decay products of uranium and thorium that occur in virtually all types of rock and soil. Radon-219 "actinon" and its decay products are of minor importance because of low abundance and very short half-life of this isotope. Radon-222 and radon-220 have almost comparable activities in rocks and soils, but the short half-life of radon-220 results in low concentration of this isotope in air. For these reasons, the measurements methodology of radon-222, or simply "radon" and its decay products are discussed here and only in specific cases reference is made to radon-220, or simply "thoron" and its decay products. The specific decay sequences of radon are shown in Table 1.

Radon arises from trace concentrations of radium in soils and rocks. Once formed by the decay of radium, radon atoms may diffuse into pore spaces of soil and become a minor constituent of the soil gas. The radon emanates into the atmosphere mainly from the top layers of soil. The radon emanation rate varies from place to place due to differences in radium concentrations, soil permeability, water content, temperature, and a number of other parameters. Thus there are variations in radon atmospheric concentration with time as well as with location. In the air, turbulent diffusion and mixing processes disperse the radon. Diurnal and seasonal variations of radon takes place as a result of changes in temperature and meteorological conditions.

In a closed isolated space, radon and its short-lived decay products reach a state of radioactive equilibrium, all having the same activity, within approximately three hours (Fig. 1). If the time period is shorter than three hours, the ratio of activity of a daughter product to that of the radon would be less than unity. An overall measure of daughters to radon ratio is termed the equilibrium or factor F. Generally, in indoor air, radioactive equilibrium (F=1) is rarely, if ever, reached because the concentration of decay products are further reduced relative to that of radon due to removal processes in addition to radioactive decay.

Radon is a noble gas, chemically not reactive, and as such is present in the air as individual atoms. When an atom of radon decays in the air, the first decay product, Po-218, is also formed initially as an individual atom, but, is in the form of a positively charged ion. Very rapidly, most of the Po-218 atoms are attached to atmospheric gases, water vapor, and aerosols to form clusters. The attachment process continues as Po-218 decays further to other radon decay products. However, a small portion of these decay products stay unattached or become detached from the aerosol and remain as such in the air. These two components of radon daughter products in the air are termed attached fraction and unattached fraction.

In undisturbed air with no circulation, the radon decay products will build up and the state of equilibrium between radon and its decay products will depend primarily on the ingrowth time for progeny and secondarily on the physical processes that remove the radon daughters from the air such as plating out on the surfaces.

Radon being an inert gas is likely to be exhaled during respiration before decaying and thus is not considered to be hazardous. The radon daughters are solids and chemically reactive. Therefore, if inhaled, they tend to adhere to lung tissue where they may decay, exposing the tissue to damaging radiation. Thus,

TABLE 1

*Decay Chain from Radium-226 to Lead-206*

| Name & Symbol | Half-Life* | Radiation & Energy (MeV) |
|---|---|---|
| Radium, Ra-226 | 1,600 y | a 4.785 (94%) 4.602 (6%) |
| Radon, Rn-222 | 3.82 d | α 5.490 (100%) |
| Polonium, Po-218 | 3.11 min | α 6.003 (100%) |
| Lead, Pb-214 | 26.8 min | β 0.65 (50%)<br>γ 0.295 (19%) 0.352 (37%) |
| Bismuth, Bi-214 | 19.8 min | β up to 3.26<br>γ 0.609 (46%) 1.120 (15%) |
| Polonium, Po-214 | 1.6x10-4 s | α 7.687 (100%) |
| Lead, Pb-210 | 22.3 y | β 0.015 (81%) |
| Bismuth, Bi-210 | 5.01 d | β 1.161 (100%) |
| Polonium, Po-210 | 138 d | α 5.297 (100%) |
| Lead, Pb-206 | stable | none |

* y=years; d=days; min=minutes; s=seconds

it may be expected that measurement of radon daughter concentrations in the air would uniformly characterize the hazards of radon progenies. However, it is more difficult to measure daughter concentrations than radon concentrations, because of the short half-lives of the daughters. Thus, it is common to measure the radon concentration and use an assumed value of the equilibrium factor, F, to convert to concentration of decay products. However, a distinction should be made between the determination of radon daughter concentrations that require measurement of the sample with very little delay and determination of radon concentration through measuring of radon daughter decays that grow in a radon sample and reach radioactive equilibrium within 3-4 hours (Fig. 1).

Generally speaking, radon concentration in the air varies with time, as well as with location. Hence, based on the type of information desired and the available time, a choice must be made with respect to the length of the sampling period. In addition, the environmental concentration of radon and its decay products for most outdoor and many indoor measurements are low, typically

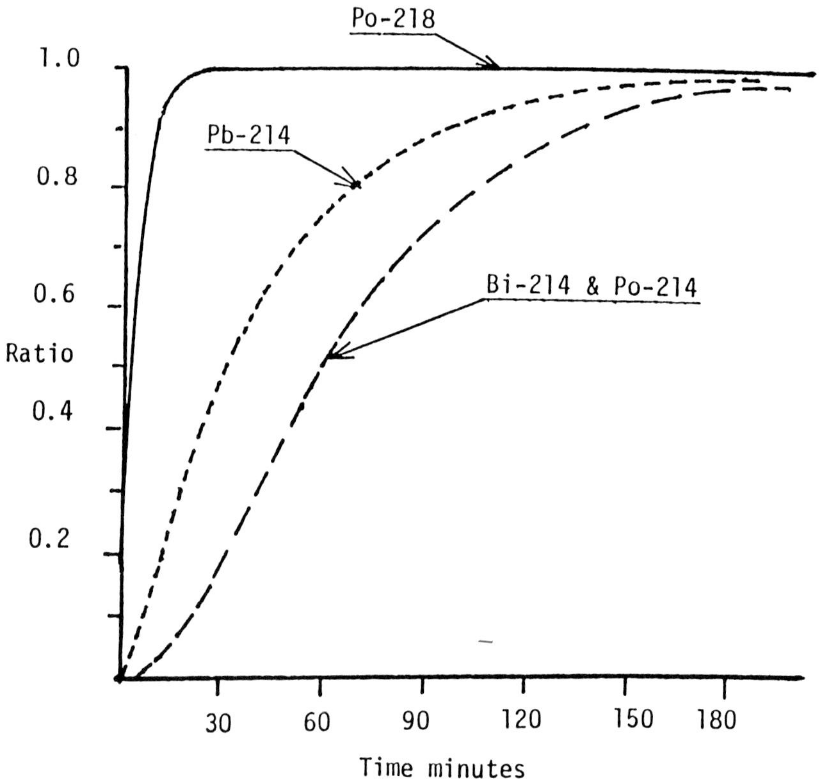


FIGURE 1. Ingrowth of short-lived radon daughters in initially pure radon.

less than one pCi/l. Hence, the problems of low counting rates are important issues. To minimize these problems, higher counting rates are achieved by sampling a large volume of air and later concentrating the radon or by sampling for extended time periods.

Basically, three different sampling methods are used for measurement of radon in indoor air. These are instantaneous, continuous, and integrating sampling periods. Instantaneous or grab sample provides information on the radon concentration during the instant of sample collection. The method is suitable for screening large number of homes in a short time. Continuous sampling is used for evaluation of the fluctuations over time such as diurnal and seasonal variations of radon concentration. Integrating or passive sampling gives the average concentration of radon during the sampling period that may range from few days to a year.

## MEASUREMENT OF RADON CONCENTRATION IN AIR: INSTANTANEOUS METHODS

In the instantaneous or grab sampling technique, filtered air can be sampled directly into the evacuated detection devices if the radon concentration is high enough so that no pre-concentration is necessary. Otherwise, usually between 5 to 20 liters of air is sampled in plastic bags, metal cans, or glass jars and brought back to the laboratory for pre-concentration and analysis. Pre-concentration is achieved by passing the air through an activated charcoal trap at liquid nitrogen temperature, where radon is retained but air passes through. The trap is then warmed and the radon is transferred to the detection devices using a small amount of helium.

The instruments currently used for instantaneous radon measurements are: scintillation cells, ionization chambers, and liquid scintillation counters. In all three instruments, radon measurement is commenced after it reaches equilibrium with its decay products inside the instrument. Thus, each radon-222 decay leads to the emission and detection of three alpha particles.

### *Scintillation Cells*

Alpha-particle scintillation counters, commonly known as Lucas cells, are the instrument of choice for many applications. The device developed by Lucas[1] is a cylindrical counting cell of about 100 $cm^3$ with a transparent window at one end and a sampling port at the other end for filling and evacuation. The inside wall of the cell is covered with silver-activated zinc sulfide, ZnS(Ag), except the end window which is left transparent for optical coupling to a planar photomultiplier tube. The cell and the photomultiplier tube are put in a light-tight housing. When an alpha particle strikes the wall of the cell, a flash of light is emitted from the phosphor coating, sensed by the photomultiplier tube, and translated into an electrical signal by the associated electronics.

For analysis, a sample of air is admitted to the evacuated cell and counted after a delay of at least three hours to allow for equilibrium between radon and its decay products. The efficiency of typical Lucas Cells is 70% to 80% and the background rate is about 0.1 counts per minute. The lower limit of detection[2] for Lucas cells is reported to be 0.025 pCi/l.[3]

The Lucas cells are used both as collector and detector of radon. Much lower detection limits may be achieved by extracting radon from a large volume of air and transferring it to the cell. Alternatively, higher sensitivity can be achieved using scintillation cells with larger volume. Cells as large as 3 liters have been developed for direct measurement of radon as low as 0.1 pCi/l with a statistical accuracy of about 11% for a two-hour counting period.[4]

### *Ionization Chambers*

Internal ionization chambers are instruments in which an electric field is applied across a volume of gas to measure the ions produced in the gas by the alpha particles emitted from radon and its progeny.[5] The ionization current or electrical pulses from the decay events are registered by the associated electronics of the instruments. Ionization chambers are highly accurate measurement devices used basically in laboratory calibration and intercomparisons. The International Bureau of Weights and Measures (BIPM) maintains ionization chambers for national and international calibrations. Ionization chambers with suitable current meters are used to make relative measurements of radioactivity with 0.02% precision. However, for routine radon measurements they do not appear to have major advantage over Lucas cells.

### *Liquid Scintillation Counting*

Alpha and beta particles produced by decay of radon and its daughter products interact with scintillating fluids to emit light. The emitted light is detected by photomultiplier tubes to provide information on the quantity of radon present in the sample.

In the preparation of the radon sample for liquid scintillation counting, the air is passed through an organic scintillation solution at low temperature. Radon, being highly soluble at low temperature, is retained in the solution whereas air passes through.[6] At equilibrium, radon and its decay products produce a total of five alpha and beta particles that are collectively detected with high efficiency.

This technique is suitable in the assay of a large number of samples and permits the use of large air samples, together with counting in a vial of small volume. Several hundred vials can be measured automatically in a conventional liquid scintillation counter. However, for accurate measurements, in addition to instrument calibration, the efficiency of the radon extraction under field sampling conditions should be measured experimentally.

## MEASUREMENT OF RADON CONCENTRATION: CONTINUOUS METHODS

In continuous or flow-through radon measurement, a constant flow of air is passed through the instrument and the recorded signal is integrated periodically. The measurement integration interval or the time resolution of these instruments, depending on the measurement of relatively short-lived or long-lived radon decay products, range from 30 minutes to a few hours. The major classes of instruments in this category are flow-through scintillation cells, flow-through ionization chambers, and diffusion cells.

### *Flow-Through Scintillation Cells*

For continuous monitoring of radon in mines, mine tailings, and radon chambers, the flow-through scintillation cells are instruments of choice. Basically, a steady stream of filtered air is drawn through a two-port scintillation cell. Alpha particles from radon and its decay products, generated within the cell, produce light pulses that are proportional to the radon concentration, its residence time in the cell, and the production of radon daughters in the cell.[6] The time resolution, signal integration, of these devices can be reduced to 30 minutes or less by applying a set of mathematical equations to correct contributions from the previous signal integration intervals.[6] In addition, deposition of radon daughters on the cell wall increases the cell background, requiring frequent replacing of the cells. The used cells are evacuated and stored for a few days until short lived radon daughters have decayed. Pb-210 and Po-210, continue to build up in the cell.

The signal integration time of a flow-through scintillation cell can be further reduced if the radon daughter products within the cell are removed before they hit the phosphor coated wall of the cell. This can be achieved by electrostatic removal of radon decay products within the cell.[7]

### *Flow-Through Ionization Chambers*

Flow-through internal ionization chambers are two-port chambers of varying sensitive volumes. Filtered air is drawn at a constant rate through the chamber where radon reaches partial equilibrium with its daughters. The degree of the equilibrium depends on the residence time of radon within the chamber. To increase the residence time and hence the degree of the equilibrium, chambers with sensitive volumes up to several hundred liters have been used.[8] However, the measurement integration intervals, as in flow-through scintillation cells, remains relatively long. The chambers are mainly used in research.

### *Diffusion Cells*

The method is based on the diffusion of radon through porous barriers and measurement of the progeny formed inside the cell. The detector is a centrally-located phosphor-covered light pipe coupled to a photomultiplier tube. The radon daughters, newly formed inside the cell, are charged ions that are directed to the detector by means of an electric field. The high-voltage electric field is established between the wire mesh in the inner wall of the cell and the mylar cover of the phosphor. Positively charged polonium-218 ions are electrostatically collected on the mylar where they decay and their alpha particles interact with phosphor to generate light pulses. The photomultiplier tube and associated electronics register these pulses.[9] Due to the passive diffusion, there is always a time lag between the instrument reading and the changes in radon concentration in the ambient air. This time lag can be measured experimentally. The detector is sensitive to humidity and has a relatively long signal integration time. These limitations are overcome by the use of desiccants and replacing the detection apparatus with a solid-state detector.[10]

## MEASUREMENT OF RADON CONCENTRATION IN AIR: INTEGRATIVE METHODS

The integrative or passive monitors yield a single value that constitutes an integration of radon concentration during the entire exposure period. In most cases, the technique is based on diffusion of radon to the detector and the measurement of alpha particles emitted from radon or its progeny. Since the diffusion time of radon through the detector wall is much longer than the half-life of thoron, this isotope has decayed before it enters the detector. Passive detectors are commonly used for long periods of monitoring, usually several months to a year. One exception is the charcoal detector which is used for integration periods of several days. Passive monitoring devices most commonly used are: etched track detectors, charcoal detectors, thermoluminescent detectors, and electret detectors.

### *Etched Track Detectors*

The method is based on the microscopic track of damage produced by alpha particles in certain materials. Currently the most widely used detector material is CR-39, an allyl diglycol carbonate.[11] When the alpha damages are enlarged by etching, the tracks become visible under a microscope. The number of tracks in a given area of the detector is proportional to the number of alpha particles and therefore to the radon concentration in the surrounding air. Most commonly, radon daughters are excluded by placing the detector in a small plastic cup and covering it with filtering membrane which admits radon only. The detectors are well suited for obtaining the annual-average indoor concentration of radon.

A different version of the method, employing cellulose nitrate solid state nuclear track detector, is used for personnel monitors and surveys. The alpha sensitive cellulose nitrate film is placed in a light-tight container with a membrane-covered small hole. Radon diffuses through the membrane into the container, where it decays and the resulting alpha particles produce tracks on the film.[12] The tracks are processed and counted as in the etched track detectors.

### *Charcoal Adsorption Detectors*

The property of activated-charcoal to adsorb radon is utilized to develop a detection system which is useful for integrating radon concentrations over a period of about one week. The activated-charcoal is placed in a metal can with a small membrane covered hole to allow radon diffusion. The detector is sealed with tape and sent to the measurement site. Before the exposure, the tape is removed and the detector is left in ambient air for about one week. After the exposure, the tape is replaced, and the detector is returned to the laboratory for measurement.[13]

The counting is usually performed by measuring the gamma rays of Pb-214 and Bi-214 using a NaI detection system. The detectors are reusable by baking them in an oven to re-activate the charcoal. To some extent, the measurement is sensitive to humidity and temperature. The technique is widely used for indoor radon surveys.

### *Thermoluminescent Detectors*

Ionizing radiation can induce long lasting metastable, "excited levels", in some materials such as glass and crystals. If an extra amount of energy is supplied to these materials, usually by heating, a de-excitation process is triggered. Light is then emitted from these materials that can be converted into an electrical signal by a photomultiplier tube. Thermoluminescent detectors (TLD) for measuring ionizing radiation are based on this property.

Im most devices, the TLD chip is exposed to alpha particles, beta particles, and gamma-rays emitted from radon and progeny, as well as to the general background radiation. For this reason, usually an alpha sensitive lithium fluoride (LiF) TLD chip is used to measure Po-218 and Po-214 alpha particles, Pb-214 and Bi-214 beta particles, plus the gamma-rays, whereas, a second TLD chip protected from radon and its daughters, encased in a metal container, measures the gamma-ray exposure alone.

In one system, based on this method, the passive environmental radon monitor (PERM), radon diffuses into a diffusion cell where decay products are formed. An electrostatic field within the cell carries the charged polonium ions to a dose integrating TLD chip.[14] The TLD reading, after correction for gamma-ray background, is directly proportional to the time integrated radon exposure rate.

Devices using an electret instead of an electrical power source and TLD for measurement of environmental gamma-ray and radon have been developed recently. One such device is the environmental gamma-ray and radon detector (EGARD) that measures the combined gamma-ray and cosmic-ray exposure, as well as, the radon concentration.[15]

### *Electret Detectors*

An electret is a piece of dielectric material, such as Teflon, capable of maintaining constant electrostatic charges over a long period of time.[16] If a known electrostatic charge is put on an electret, the reduction of this charge will be due to the collection of ions of opposite sign from the surrounding gas. Passive electret detectors for radon detection are based on this property.

The electret passive detector consist of a thin insulated Teflon disc mounted inside the lid of a small can and positively charged to several hundred volts. Radon diffuses into the can through a membrane covered hole. Inside the can, alpha particles from decay of radon and its progeny ionize the air by production of ion pairs. The negative ions are collected on the Teflon disc causing a reduction in the electric charge. The surface potential of the disc is measured before and after the exposure by means of a voltmeter. The charge reduction, after correction for background gamma-rays, is directly proportional to the time integrated radon exposure rate.[17] The detector is reusable until the surface potential drops to a low value.

Before the detector is sent to the measurement site the opening to the diffusion membrane is closed by means of a screw cap. At the site the cap is unscrewed and the detector is exposed to ambient air. The exposure period can range from few days to a year.

## MEASUREMENT OF RADON PROGENY IN AIR

Radon daughters that are attached to dust particles or other surfaces in the air become an integral part of the atmospheric aerosol. These daughter products are removed from the air by radioactive decay and atmospheric aerosol removal processes.[18] The attached radon daughters are sampled in the same way as the atmospheric aerosol.

Unattached radon daughters are ions or neutral species with a high diffusion coefficients. Sampling methods of the unattached fraction are based either on their electrostatic properties (electrostatic collectors) or on their high diffusion coefficients (diffusion tubes, screen samplers, or impactor samplers). The determination of the unattached radon daughters will not be reviewed here. A discussion of measurement of unattached radon daughters is presented in NCRP report No. 97.[3] This report also presents a comprehensive review of techniques for detection of radon and its daughters.

Like radon, sampling for the progeny can be separated into instantaneous measurement, continuous measurement, and integrative measurement periods. A review of these sampling techniques is given in this section.

## MEASUREMENT OF RADON PROGENY IN AIR: INSTANTANEOUS METHODS

Instantaneous or grab sampling of radon daughters is based, in most cases, on drawing a known volume of air through a filter and counting the deposited activity by alpha particle scintillation or alpha spectroscopy. A single counting of the sample yields only a rough estimate of the total concentration of the radon daughters in the air. However, in general the individual daughter concentrations in the air is needed and this requires more comprehensive counting.

### *Alpha-Particle Scintillation Counting*

The filter containing radon daughters is covered with a thin sheet of zinc sulfide (ZnS) phosphor and placed on the photomultiplier tube for counting. Because of the differences in the half-lives of radon daughters, their concentrations can be individually calculated if counting of the sample is performed in three or more successive time periods.[19] This method, originally developed by Kusnetz, is often used in mines to monitor occupational exposure levels. A summary of variations of this method is given in NEA 1985.[20]

### *Alpha-Particle Spectroscopy*

In alpha-particle spectroscopy a solid-state detector is used to distinguish between Po-218 and Po-214 alpha energies, 6.00 MeV and 7.69 MeV, respectively. As in alpha-particle scintillation counting, the filter containing radon daughters can be counted once, to obtain an estimate of Po-218 and Po-214 concentrations,[21] or at two or more sequential time periods to obtain the concentrations of all the daughter products.[22]

Alpha particle spectroscopy can be used for measurement of mixed radon and thoron decay product concentrations in the air.[23] The technique is based on the simultaneous measurement of the 8.78 MeV alpha particles of thoron progeny, Po-212, and alpha particles of radon progenies, Po-218 and Po-214.

## MEASUREMENT OF RADON PROGENY IN AIR: CONTINUOUS METHODS

Continuous measurement of radon daughters is achieved by collection of successive discrete grab samples on automatically changeable filters, or by continuous collection on a stationary filter. Basically, a constant flow of air is drawn through the filter and radon daughters deposited on the filter are measured either continuously or discretely by alpha scintillation or alpha spectroscopy.

A variety of instruments are available in this category. In some devices, electrostatic collectors have replaced the air filter collection. Some instruments are designed for sequential sampling and counting intervals whereas others are for overlapped sampling and counting intervals.[5]

## MEASUREMENT OF RADON PROGENY IN AIR: INTEGRATIVE METHODS

For integrative measurement of radon progeny, air is drawn through a filter that catches the radon daughters. The dose integrating element is a TLD chip located near the surface of the filter.[4] It is also possible to obtain integrative values from continuous measurement of radon progeny.

## RADON MEASUREMENT IN WATER

Radon concentrations in water can be measured using liquid scintillation counting. In this method, the water sample and a mineral oil based scintillation cocktail are put in a glass vial where they form two phases. Radon diffuses to the organic phase and interacts with scintillating fluids. The resulting light pulses are detected by photomultiplier tubes. This method lends itself to large scale monitoring with automation.

## REFERENCES

1. Lucas, H.F. 1957. Improved Low Level Alpha Scintillation for Radon. Rev. Sci. Instrum. 28, 68.
2. Pasternack, B.S. and N.H. Harley. 1971. Detection Limits for Radionuclides in the Analysis of Multi-Component Gamma-Ray Data. Nucl. Instrum.

and Methods, 91, 533.

3. Measurement of Radon and Radon Daughters in Air. NCRP Report No. 97, Table 7.1 (National Council on Radiation Protection and Measurements, Bethesda, MD. 1988).
4. Cohn, B.L., M. Granayni and E.S. Cohen. 1983. Large Scintillation Cells for High Sensitivity Radon Measurement. Nucl. Instrum. and Methods, 212, 403.
5. A Hand Book of Radioactivity Measurements Procedures. NCRP Report No. 58, (National Council on Radiation Protection and Measurements, Bethesda, MD, 1978).
6. Thomas, J.W. and R.J. Countess. 1979. Continuous Radon Monitor. Health Phys. 36, 734.
7. Chittaporn, P., M. Eisenbud and C.R. Harley. 1981. A Continuous Monitor for the Measurement of Environmental Radon. Health Phys. 41, 405.
8. Hogberg, S.A.C. and J.E. Gustavsson. 1973. An Ionization Chamber for Radon-222 Measurements. Nucl. Instrum. and Methods, 113, 583.
9. Wrenn, M.E., H. Spitz and N. Cohn. 1975. Design of a Continuous Digital-Output Environmental Radon Monitor. IEEE Trans. Nucl. Sci., NS-25, 645.
10. Negro, V.C. and S. Watnick. 1978. A Radon Measuring Instrument with Fast Response. IEEE Trans. Nucl. Sci., NS-25, 757.
11. Casso, R.M. and E.V. Benton. 1978. Properties and Applications of CR-39 Polymeric Nuclear Track Detectors. Nucl. Track Detection, 2, 173.
12. Urban, M. and E. Pietsch. 1981. Low Level Environmental Radon Dosimetry with a Passive Track Etch Detector Device. Radiat. Prot. Dosimetry, 1, 97.
13. Cohn, B.L. and R. Nason. 1986. A Diffusion Barrier Charcoal Adsorption Collector for Measuring Radon Concentrations in Indoor Air. Health Phys. 50, 457.
14. Nyberg, P.C. and D.E. Bernhard. 1983. Measurement of Time Integrated Radon Concentration in Residences. Health Phys. 45, 539.
15. Maiello, M.L. and N.H. Harley. 1987. EGARD: An Environmental Gamma-Ray and Radon Detector. Health Phys. 53, 301.
16. Sessler, G.M. and G.E. West. 1980. Electrets, Topics in Applied Phys. 33, Springer-Verlag, New York, NY.
17. Kotrappa, P., J.C. Dempsy, J.R. Hickey and L. R. Stieff. 1988. An Electret Passive Environmental Rn-222 Monitor Based on Ionization Measurement. Health Phys. 47, 56.
18. Nevissi, A.E., J.N. Beck and P.K. Kuroda. 1974. Long-Lived Radon Daughters as Atmospheric Radioactive Tracers. Health Phys. 27, 181.
19. Kusnetz, H.L. 1956. Radon Daughters in Mine Atmospheres: A Field Method for Determining Concentrations. Am. Ind. Hyg. Assoc. J. Qurt. 17, 85.
20. NEA (Nuclear Energy Agency, OECD). 1985. Metrology and Monitoring

of Radon, Thoron, and Their Daughter Products. Report by a Group of Experts (Paris OECD).

21. Revzan, K.L. and W.W. Nazaroff. A Rapid Spectroscopic Technique for Determining Potential Alpha-energy Concentration of Radon decay Products. Health Phys. 45, 509.
22. Nazaroff, W.W. 1983. Radon Daughter Carousel: An Automated Instrument for Measuring Indoor Concentrations of Po-218, Pb-214 and Bi-214. Rev. Sci. Instrum. 54, 1227.
23. Zhang, C. and D. Lou. 1983. Measurement of Radon and Radon Daughter Concentrations in Air. Nucl. Instrum. and Methods, 215, 481.

*Environmental Radon: Occurrence, Control and Health Hazards.* Edited by S. K. Majumdar, R. F. Schmalz and E. Willard Miller © 1990, The Pennsylvania Academy of Science.

# *Chapter Ten*

# RADON DETECTORS AND MEASUREMENT TECHNIQUES

**WILLIAM A. JESTER[1] and JAMES LIVINGSTON[2]**

[1]Professor of Nuclear Engineering and Technical Supervisor
of the Low Level Radiation Monitoring Laboratory
The Pennsylvania State University
University Park, PA 16802
and
[2]Graduate Assistant
Nuclear Engineering
The Pennsylvania State University
University Park, PA 16802

## INTRODUCTION

This chapter contains information about many of the various types of commercially available radon monitoring equipment and briefly describes how they are used. It also lists some of the advantages and disadvantages associated with each equipment type and gives examples of those systems with which the authors are familiar.

## BACKGROUND

Radon is an inert radioactive gas which is naturally occurring in the soil, ground water, and some building materials. Radon will migrate into living areas where it may pose a health hazard. The magnitude of this health risk is dependent on the radon and radon progeny concentrations. The determination of the radon and/or radon progeny concentrations is necessary to evaluate the need for, or importance of, mitigative actions in reducing the health risk.

Radon concentrations in a home will fluctuate dramatically with changes in ventilation. Concentrations change with the operation of home heating and air conditioning systems, which can cause air dilution and in-house pressure differences. The radon influx into the home will also be a function of diurnal and seasonal fluctuations in barometric pressure, temperature, windspeed, and fluctuations in outside soil conditions.

The concentration of airborne radon-222 progeny, which include polonium-218, lead-214, bismuth-214, and polonium-214, varies with the amount and size distribution of airborne particulates. These are affected by factors such as tobacco smoking, cooking, cleaning, and heating activities within the home. A majority of the radon detectors employed in the various radon monitoring systems detect primarily alpha radiation. Thus they detect alphas from polonium-218, polonium-214, and possibly from radon-222 itself. Other detectors monitor the gamma and/or beta radiation coming from lead-214 and bismuth-214.

Some detectors are used primarily for making initial screening measurements under closed house (winter) conditions over a one to seven day period. Other detectors are more suited for making long-term measurements, from three months to a year, and are used primarily for follow-up measurements which are made under normal living conditions.

## EQUIPMENT CLASSIFICATIONS

The equipment classifications used in this chapter follow closely those used by the U.S. Environmental Protection Agency in its materials used to describe participation in the National Radon Measurement Proficiency (RMP) Program.[1,2,3] The classification used are as follows:

Activated Charcoal Adsorption (AC) and Charcoal Loaded Liquid Scintillators (CLS)
Electret Ion Chambers (EIC)
Alpha-Track Detectors (ATD)
Continuous Radon Monitors (CRM)
Grab-Sample Radon Monitors (GRM)
Continuous Working Level Monitors (CWLM)
Grab-Sample Working Level Monitors (GWLM)

AC, CLS, EIC, and ATD are passive detector systems using no power source; the radon enters and leaves the detectors by diffusion. Such monitors require some type of additional reader system to convert the information collected by the passive detector into radon concentration or working level units. The remaining detector types listed are called active detectors, since they use some type of pumping system to pump radon-containing air into or through the detector. The radon detector itself may be powered so that output readings can be obtained during data collection or immediately at the end of a monitoring period.

The first series of detectors or detection systems described in this chapter measure the radon-222 concentration and thus give results in terms of radon-222 concentration in picocuries per liter of air (pCi/l) or Becquerels per cubic meter of air ($Bq/m^3$). The last series of detector systems covered in this chapter are used to monitor the concentration of radon-222 progeny in the air, and thus give their results in units of working levels (WL), a term originally developed to compute radon progeny exposure to miners. It is frequently assumed, under conditions normally found in a home, that 1 working level of exposure is associated with a radon concentration of 200 pCi/l. This number assumes that the concentration of radon-222 progeny in the air is 50% of its equilibrium value.

## ACTIVATED CHARCOAL ADSORBERS (AC) AND CHARCOAL LIQUID SCINTILLATION (CLS)

### (A) *Charcoal Canisters*

Currently the most commonly employed passive monitors used for two to seven day screening measurements are tin plated steel canisters containing activated charcoal which is held in place by a wire mesh screen. These passive devices are inexpensive ($2 to $5 each - not including analysis charges) and easy to use. They lend themselves to being sold in stores, used by the homeowner, and then mailed to a laboratory for analysis. Unfortunately, these canisters have a potential for considerable measurement error.

As received from the monitoring company, the canisters contain no radon adsorbed on the charcoal. When the canisters are opened, radon gas in the room air enters the can and diffuses into the charcoal, where it becomes adsorbed at activated sites. If the radon concentration in the air remains constant during the exposure period (from a few hours to many days, depending on the canister design), the radon concentration in the canister will reach an equilibrium value which is a linear function of the airborne radon concentration. Since airborne radon levels seldom remain constant, additional radon atoms will be adsorbed when the room radon concentration rises, and conversely some of the radon atoms will be desorbed from the charcoal when the room radon concentration falls. Thus a changing radon concentration is a source of error in the measurement. The rate at which the radon adsorbs and desorbs from canisters is a function of canister design, room temperature, and humidity. These effects can also bring errors into the measurements.

Some canister designs employ a diffusion barrier to make the monitors better integrators of the radon concentration[5] and to reduce the effects of air movement. The diffusion barrier (e.g., filter paper, small holes or membranes) restrict the amount of radon that can reach and be adsorbed on the charcoal. Its purpose is to reduce the amount of occupied activated sites such that the charcoal appears as an infinite sink (i.e., no desorption). This obviously means that the radon concentration in the canister will not be in equilibrium with the surrounding air. The radon adsorbed will then be more proportional to both the radon concentration and the duration of the exposure. Since diffusion barriers decrease the amount of radon which can be adsorbed on the charcoal, they reduce the radon detection sensitivity of the canisters.

Water molecules also adsorb onto activated charcoal and interfere with the charcoal's ability to adsorb radon. Some canisters employ a desiccant to remove moisture from the radon-containing air prior to its entering the charcoal region. Procedures have also been developed whereby canister weight gain caused by water adsorption can be used to correct for the resulting loss in radon adsorption efficiency.[6,7]

At the completion of the exposure time, the canisters are closed with a radon-proof cover and returned to the supplier for analysis. At the lab a technician uses a gamma-ray spectroscopy system, usually employing a sodium iodide (NaI(Tl)) detector to measure three gamma rays of lead-214 and the single gamma-ray of bismuth-214 decay. These results are then used to estimate the average airborne radon concentration to which the canister has been exposed.

(B) *Charcoal Loaded Bags*

A variation of the charcoal canister approach is the charcoal-loaded paper bags which look like oversized tea bags. These bags have no diffusion barriers and thus very quickly become in equilibrium with the radon in the air. Thus they are useful only for short measurements of no more than one day duration. These bags are also analyzed using NaI(Tl) gamma-ray spectroscopy.

(C) *Charcoal Loaded Liquid Scintillation Devices (CLS)*

Another method which utilizes activated charcoal adsorption of radon is called the charcoal loaded liquid scintillation device.[5] While a typical canister may hold between 25 to 100 grams of activated charcoal, a CLS device contains only one to three grams of charcoal in a capped liquid scintillation vial. These devices use a diffusion barrier to regulate the rate of radon adsorption and desorption from the charcoal. Some versions include a few grams of desiccant to reduce the adsorption of moisture by the charcoal. Before and after the exposure period, the devices are sealed in radon-proof enclosures.

When returned to the analysis lab a liquid scintillation solution is added to the sample vial. Within a few hours the radon desorbs from the charcoal and dissolves in the solution. Once the radon is in equilibrium with its daughters, the alpha radiation in the sample is counted using a liquid scintillation counter.

While the combined use of a smaller amount of charcoal and a diffusion barrier results in a considerably lower amount of radon being adsorbed in the

CLS as compared with most AC canisters, the efficiency for liquid scintillation detection is much higher than for gamma-ray spectroscopy. Thus both of those techniques have similar sensitivities.

(D) *Integral Response*

Because of the decay of radon-222 and possible desorption associated with each of these methods, there will be proportionally fewer radon atoms remaining on the charcoal at the time of analysis from the beginning of the exposure period than from the end of the period. Thus this type of monitor is not considered to be a true integrator of the radon concentration during the exposure period. If the radon concentration is significantly lower at the end of the exposure period than at the beginning, the final results can be unacceptably low.

(E) *Use of Charcoal Devices for Follow-Up Measurements*

EPA recommends the use of activated charcoal devices for follow-up measurements under the following conditions. If the screening results are less than 20 pCi/l but higher than 4 pCi/l, then the homeowner can estimate the annual average concentration by using one AC or CLS measurements for each quarter for one full year. These four measurements would all be made for 2 to 7 days under normal living conditions and are averaged together to give an average annual radon concentration. If the screening measurements indicate that the radon concentration exceeds 20 pCi/l, then EPA will accept as a follow-up measurement a second 2 to 7 day measurement made under closed house conditions. If the screening measurement is less than 4 pCi/l, no follow-up measurement is necessary. If the follow-up measurement indicates that the annual average radon concentration exceeds 4 pCi/l, EPA recommends that the homeowner begin remediation actions to lower it below this limit.

(F) *Advantages of Charcoal Devices*

1. The cost of using one of these devices is usually quite low if the homeowner deploys the device then mails it to the analysis lab ($15-$30).
2. No special skills are needed to deploy the devices, so that trained technicians are not needed to collect data in the home.
3. The collecting devices are completely passive and need no external power source.
4. The detectors are small and thus can be placed in an appropriate location where they will not be disturbed by household activities.
5. If the detectors are properly located and the home is kept in closed house conditions including the 12 hours before the start of measurement, and

if the humidity is low and no major significant changes in atmospheric conditions take place during measurement period, then good sensitivity with acceptable precision can be achieved.

6. These devices are insensitive to the presence of radon-220 (thoron) and its daughters since because of its short half-life (55 seconds), there is little time for the radon-220 to reach and become adsorbed in the charcoal.
7. Charcoal canisters can be reused if before each use they are dried to a constant weight and checked to make sure there is no residual radon from their previous use.

(G) *Disadvantages of Charcoal Devices*

1. Since they are not true radon activity intergrators, they may give erroneous results if the average daily radon concentration changes significantly during the measurement period.
2. Unless a desiccant is used or proper weight gain corrections have been made, the canisters will give false low readings when they adsorb water vapor.
3. Changes in room temperature will change the radon adsorption rate for short term measurements (3 days or less).
4. Poor air circulation near the canister can result in an inaccurate (usually low) measurement of average room radon concentration.
5. The laboratory analysis of the devices require expensive gamma ray spectroscopy or liquid scintillation equipment ($5,000 - $15,000) and a trained technician for analysis.
6. The devices cannot be read in the field, so it takes several days to obtain the results.

(H) *Examples of Equipment Available For Activated Charcoal Measurement*

There are currently a large number of companies supplying charcoal canisters. The Low Level Radiation Monitoring Laboratory (LLRML) employs 4″ diameter open face canisters of EPA design obtained from F&J Specialty Products from Miami Springs, Florida. These canisters have performed satisfactorily for 5 to 7 day measurements with humidity corrections.[7] This company also sells canisters of other sizes, with and without moisture barriers and diffusion barriers. Some other suppliers of various types of canisters include Air Sciences in Crystal City, Missouri; Sun Corporation, Pittsburgh, PA; TMA Eberline, Albuquerque, New Mexico; Teledyne, Inc., Westwood, New Jersey; and Pylon Electronic Development Company, Ottawa, Ontario.

The most inexpensive approach for multichannel gamma-ray spectroscopy of activated charcoal devices is to use personal computer-based multichannel pulse height analyzers along with NaI(Tl) detectors. The system used at the

LLRML is an IBM-XT containing a Nucleus pulse height analysis card and a Nucleus Digital Multiplexer Router which allows the one computer-based analyzer to simultaneously service up to ten NaI(Tl) detectors. The Nucleus, Inc. is located in Oak Ridge, Tennessee. Similar systems are available from EG&G ORTEC of Oak Ridge, Tennessee and Canberra Industries of Meriden, Connecticut. Inexpensive systems are also available which use a single channel analyzer rather than a multichannel analyzer. Such systems are more difficult to set up and use than the multichannel systems in which the user has a gamma spectrum to aid in calibration and use.

## ELECTRET ION CHAMBERS (EIC)

### (A) *Background*

The Electret Ion Chamber is a relatively new concept in radon monitoring and its use as a screening monitor is growing rapidly.[8] It is available from only one manufacturer, Rad Elec, Inc., Frederick, Maryland, under the trade name E-PERM.

The E-PERM is a passive device which is easy to use and can be employed by a homeowner by following a few simple instructions. Each device includes a reusable plastic container costing $35 and an Electret costing $15.

### (B) *Theory of Operation*

The EIC Chamber is made of a conducting plastic and contains a spring operated screw-on cap. When the device is turned on by raising the cap, radon-containing air can enter and leave the inner chamber by means of a series of holes containing dust filters located beneath the cap. The electret is screwed onto the bottom of the chamber. The electret contains a teflon disk in which the molecules have been polarized in order to maintain a quasi-permanent electric static charge. When the electret is screwed onto the chamber, the combination becomes an ion chamber in which the electret has a positive charge with a corresponding voltage of up to a maximum of about 700 volts and the interior chamber walls have a corresponding negative charge.

Since the pathway into the chamber contains a filtered diffusion barrier, only radon-222 enters. Within a few hours the radon in the chamber reaches equilibrium with its short-lived decay products. With each decay event, especially with the alpha decay of radon-222, polonium-218 and polonium-214, the air within the chamber is ionized, i.e., air molecules are converted to negative and positive ions. The positive ions are drawn to the chamber walls where they pick up electrons and are neutralized. The negative ions are attracted to the electret where they are neutralized by giving up an electron. In this process the electric charge on the electret is reduced in proportion to the number of ions collected and thus in proportion to the radiation exposure in the chamber. Since the radon concentration within the chamber follows the radon concentration of the surrounding air, this type of device is very close to being a true integrator of radon exposure. The response time of the device to changes in room radon concentration is on the order of one to two hours; thus the minimum measurement time should be one day or longer to assure that the time it takes for radon to reach equilibrium with its daughters is only a small fraction of the total monitoring time.

(C) *Electret Reader*

The electret read-out system is a portable, battery-operated device called a surface potential electret reader (SPER). This unit can read the voltage of an electret to within one volt in less than a minute, and this measurement, along with the radon concentration calculations, can be accomplished in the field. The SPER costs about $1,500 and requires a minimal amount of training to use. A caution which must be taken in reading the surface voltages on the electrets is that one must be careful not to touch the exposed teflon surface, since this will cause a large reduction of the surface voltage and thus invalidate the measurement and possibly render the electret useless for subsequent measurements.

There are short term and long term electrets currently available. The short term electret has a teflon thickness of about 1.3 mm and has a calibrated voltage loss of about 1.6 volts per picocurie of radon, per liter of air, per day of exposure. It is this electret which is recommended for use under closed house conditions for the two to seven day screening measurements. Electrets have a useful operation voltage between 700 and 200 volts. Thus, on the average, each electret can be used for many measurements before it must be replaced.

The long term of electret has a teflon wall thickness of 0.13 mm and has a voltage loss of about 0.18 volts per picocurie of radon per liter per day. The manufacturer recommends their use for follow-up measurements, but EPA has not yet listed E-PERMS for this use.

The EIC has a small sensitivity to environmental background gamma radiation typically equivalent to a radon concentration of about 0.8 pCi/l. There is also some slight variability in the internal volume of the ion chamber and the thickness of the teflon on the electrets. This variability has been found to contribute a $\pm 7\%$ uncertainty to any measurement.

(D) *Advantages of EIC's*

1. The cost to the customer of using one of these devices for screening measurements is low ($10 to $30) assuming that the home owner employs

the device and mails it for analysis.

2. No special skills are needed to employ this device.
3. The device is completely passive and needs no external power source.
4. The detectors are small and unobtrusive. Thus they can be placed in an appropriate location where they will not be disturbed by the household.
5. The systems are very close to being true radon integrators and are insensitive to temperature and humidity changes, air drafts, and changes in atmospheric conditions. Thus under most conditions they should give accurate results.
6. The delay time it takes for radon gas to diffuse into the ion chamber means that these devices are insensitive to the presence of radon-220.
7. The chamber section of each monitor should last for hundreds of measurements, while the short term electrets should last on the average of tens of measurements until they discharge to near 200 volts.
8. The monitor can be quickly read in the field and the average radon concentration calculated by a technician shortly after the completion of a measurement.

(E) *Disadvantages of EIC's*

1. The device is relatively new, thus there are not many companies currently experienced in using these devices, and even EPA has not completed their evaluation of them.
2. If an EIC is damaged in shipment or in use, its replacement cost will be \$15 to \$50.
3. Electrets must be read and calculations made by trained staff.
4. A gamma background correction factor is needed and could be a source of error if it is significantly different from the assumed value.
5. Once the electrets have been discharged to below 200 volts, they can no longer be used. Currently there is no cost-effective way to recharge the devices.
6. There is a measurement error ($\sim \pm 7\%$) associated with variations in the teflon thickness, chamber volume, etc.

(F) *Available Equipment and Comments on Their Use*

As stated previously, this equipment is available only from Rad Elec, Inc. LRML personnel have been testing the equipment available from Rad Elec, Inc., and have found it easy to use. The short term electrets give good agreement with other environmental monitors tested in the lab, and have been especially useful as a research tool.

## ALPHA-TRACK DETECTORS (ATD)

### (A) *Background*

ATD's comprise the remaining major class of passive radon monitoring devices to be described in this chapter. This radon detection concept has been used for about 20 years[9] and finds its major use as the main device for measurements of 3 months to a year in duration. As with the other passive devices, the ADT's are easy to use and relatively inexpensive to manufacture, costing only a few dollars each, not including their processing. It is the processing step which makes up the majority of the cost for this device, and there are only a few laboratories in the country that are set up to perform this type of analysis.

### (B) *Theory of Operation*

These detectors consist of a small piece of alpha particle sensitive plastic film usually placed inside a plastic cup which is closed by a particulate filter diffusion barrier. Radon-containing air diffuses into the container through the filtered opening. Thus radon-222 gas is separated from its airborne daughters. The film does not respond to gamma or beta radiation, but only to alpha particles having an energy below about 4 MeV.[9,10] Thus the film will be responsive to the partially attenuated alpha particles of radon-222 and its polonium-218 and -214 daughters. Alphas in this energy range damage the film such that in a subsequent chemical etching step the alpha tracks are visible under a microscope. The number of alpha tracks per unit area is counted manually or by an optical machine. The resulting alpha track density is a true integrator of the radon concentration during the film's exposure time. Cosmic rays can also make tracks on the film. This will occur when the film is in storage prior to its use. Thus unless the plastic is preannealed prior to use, (something not done by most suppliers of ADT) or freshly manufactured, there will be a need to make background corrections, especially for conditions resulting in low alpha track densities.

### (C) *Use as a Screening Method and Follow-Up Method*

Some laboratories are starting to advertise the use of short term ATDs for screening measurement under closed house condition. For short exposure times, (up to 7 days) the track density in most cases will be quite low and thus there will be a large uncertainty in the measurement. EPA recommends that ATD screening measurements be taken over a three month period under closed house conditions unless sufficient area of the detector is analyzed for tracks to achieve an adequate lower limit of detection.

EPA recommends that if a screening measurement results in a radon level of between 4 and 20 pCi/l, the ATD detectors can be used for a 12 month follow-up measurement made under normal conditions. If the screening result is between 20 and 200 pCi/l, the ATD can be used over a three month period as a follow-up measurement if during these three months the house is kept under closed house conditions (e.g., three winter months). If the results of a screening measurement exceeds 200 pCi/l, then a short term ATD follow-up measurement of days or weeks may be appropriate.

(D) *Advantages of ATD's*

1. This is a relatively low cost service in the range of $20 to $60 as long as the customer places and retrieves the detectors and mails them to the processing laboratory.
2. No special skills are required to use these detectors in the home.
3. They are frequently distributed by mail.
4. ATD's are completely passive and need no external power.
5. The cups containing these detectors are small and thus they can be put in an appropriate place where they are not disturbed by household activities.
6. If placed and processed properly, these detectors are true integrators of the average radon exposure and are unaffected by changes in temperature, humidity, and atmospheric conditions.
7. They can continuously measure long term average radon concentration over a 12 month period, a time period considered optimal by EPA.

(E) *Disadvantages of ATD's*

1. They are not very satisfactory when used for short duration screening measurements, since the shorter the exposure time and/or the lower the radon concentration, the greater the inherent variability of the result.
2. A skilled operator and expensive equipment are required to etch and read these detectors.
3. Accuracy is dependent on the etching and track counting procedures employed by the analysis lab and the skilled operator. Thus sizable errors are sometimes observed using ATD's.

(F) *Examples of Companies Providing Services*

Alpha track detectors are available from Tech/Ops Landover, Inc. and Terradex Corporation, both of Glenwood, Illinois.

Penn State's Health Physics Office has made extensive use of Terradex ATD's and have found them to perform satisfactorily for 3 month measurements.

## CONTINUOUS RADON MONITORS (CRM)

### (A) *Background*

There is a wide variety of continuous radon monitors on the market today sold for use in home radon monitoring. These are usually very sophisticated systems which often combine both the radon detector and the analysis unit in the same package. These instruments tend to be expensive (from $2,500 to $10,000) and are sufficiently complex that they are usually operated by a trained technician for home radon measurements. These instruments monitor the radon in filtered air and thus give results in picocuries per liter. EPA lists two types of detectors under this classification. One type employs a scintillation cell sometimes called a Lucas cell; the other type employs some type of ion chamber.

### (B) *Theory of Lucas Cell CRM's*[11]

The walls of the Lucas cell, excluding its transparent base or window, are coated with a zinc sulfide scintillator. The sample gas is collected inside the cell and when the alpha particles emitted by radon-222 and its daughters polonium-218 and polonium-214 contained in the air sample strike the scintillator, they produce flashes of light called scintillations. The light passes out of the cell, by means of the transparent window, into a photomultiplier tube. The photomultiplier tube in turn generates electrical pulses at a rate proportional to the sampled radon concentration. The number of pulses per unit operation time (usually 1 hour) is stored in a memory unit for subsequent recall and analysis. Thus these systems provide not only a measure of the average radon concentration over the sampling period, but also the time variation in the radon concentration.

The most common type of Lucas cell has two gas ports. Filtered air is slowly pumped into the cell at its inlet port and leaves the cell at its outlet port. If the air is pumped too fast, the radon-222 will not reach equilibrium with its daughters within the cell and false low results will be achieved. Thus the flow rate through the cell must be consistent with the method used in system calibration.

A variation on the above procedure is called a quasi-continuous mode. In this mode, the cell is periodically evacuated and then an air sample is drawn into the cell. Time is allowed for the radon in the air sample to reach equilibrium with its daughters and then the cell is counted. At the start of the measurement, the unit is programmed to repeat this operation at preset time intervals throughout the sampling period.

In a third variation, ambient air with its radon diffuses through a filter and into the scintillation chamber rather than being pumped through the chamber. It takes longer for this cell to reach equilibrium conditions and thus this system will respond more slowly to changes in radon concentrations than those using the previous mentioned cell types.

With all of these calls, the radon concentration as detected lags the ambient radon concentration because of the inherent delay caused by the time required for the new radon level to reach equilibrium with its daughters. Thus these detectors are not quite true integrators.

### (C) *Theory of Ion Chamber CRM's*

In this type of CRM, airborne radon is pumped or diffuses through a particulate filter and into an ion chamber. The decay radiation, especially alpha particles, interacts with the air inside the chamber, producing a burst of ions for each particle-air interaction. The positive and negative ions are collected at negative and positive electrodes respectively, resulting in the production of an electrical pulse for each decay event. The number of pulses per unit time is then proportional to the radon concentration in the chamber at that time. The output of some ion chambers is a current rather than pulses and the higher the radon concentration the higher the output current.

A variation of this type of system uses a silicon surface barrier detector as a solid state ion chamber. This detector produces pulses for each alpha particle entering the detector. The pulses are then recorded as a function of time and, if needed, of alpha energy.

### (D) *Screening and Follow-Up Measurements*

EPA recommends that these systems be used for screening measurements of 24 hour duration under closed house conditions. They can also be used for follow-up measurements under these conditions: If the screening result is less than 20 pCi/l, then the follow-up conditions would be four separate 24 hour measurements under normal living conditions once every three months for a year. If the screening result is greater than 20 pCi/l, then only a second 24 hour measurement, made under closed house conditions, is required as a follow-up measurement.

### (E) *Advantages of CRM's*

1. They require only a short duration (24 hours) measurement under closed house conditions for a screening measurement.
2. They usually obtain hourly results to allow the operator to track radon variations over a 24 hour period.
3. These devices can be used to find locations where radon is entering a home.
4. If properly calibrated, most models will give very precise results.
5. Most modern systems contain a small microprocessor so that results can be made available at the sampling site shortly after the completion of a measurement.

(F) *Disadvantages of CRM's*

1. They are much more expensive than some of the other methods used for screening measurements (on the order of $100 to $300 per measurement depending on the type of service provided).
2. A trained operator is required.
3. The devices require extensive calibration in a radon calibration chamber.
4. The operator must make frequent background measurements.
5. Many of these systems are heavy and difficult to move about.

(G) *Examples of Equipment Available as Continuous Radon Monitors*

The authors of this chapter have had considerable experience using the Pylon model AB-5 portable radon monitor with a model LCA-2 Lucas cell adapter and model 300 two port Lucas cell. The AB-5 system has its own air pump and can operate both in a continuous mode or quasi-continuous mode. Another attachment to the AB-5 is a Model PRD-1 passive radon gas monitor which allows the unit to operate as an integrating device for continuous radon monitoring.

The AB-5 radon monitor has a built-in battery pack which allows the system to operate from between 8 hours with the pump running at maximum capacity to 75 hours without the pump running before the battery has to be recharged. While this system can store the count data of up to 680 intervals, it does not determine the radon concentration directly. To do this, the system can be connected to a personal computer through an RS-232 connector. The data is transmitted to the computer for processing by a program supplied by the Pylon Electronic Development Company, Ltd. of Ottawa, Canada. A similar unit is the RDA-22 sold by EDA Instruments, Inc. of Toronto, Canada.

The authors have also had an opportunity to test a relatively new instrument called the ILMAradon Detector which uses a specially constructed ion chamber to measure the ion concentration in room air.[12] The detector is portable, weighing about 20 pounds, and can operate for about five hours of continuous use before the internal batteries need to be recharged. It also gave the most rapid response to changes in radon levels of any unit tested at the LLRML. The radon concentration is immediately displayed on an analog dial. The analog signal is also put to a BNC connection in the rear of the detector. This instrument should be helpful in detecting locations in the home where radon gas is leaking in. The high sampling rate can quickly disturb localized pockets of radon gas, so that while one can obtain an initial high reading at a location, the signal can quickly drop off. The ILMAradon detector is distributed by FMT Equipment Company of Hamden, Connecticut.

Probably the least expensive continuous radon monitor is the Honeywell Professional or Digital Radon Monitor Systems. These systems employ silicon detectors. One version, called a Portable Digital Radon Monitor, costs about $400 and will display, at a push of a button, both the current radon level (past 12 hours) and a long term average reading (average since last Reset). The second system, called a Portable Radon Detector Data Logger, stores up to 96 data points and can print the data out in both tabular and graphic form. This system with printer costs about $1,100. The systems are simple enough that the average homeowner, if they can afford them, will be able to operate them. These two instruments are manufactured by Honeywell, Inc. of Golden Valley, Minnesota and are distributed by several other companies.

## GRAB-SAMPLE RADON MONITORS (GRM)

### (A) *Background*

Because of their short sampling times, radon grab sampling techniques provide a near instantaneous measurement of the radon concentration and not an average value. Thus because of the inherent variability of radon levels in homes EPA does not recommend their use for either screening or follow-up measurements. The value of the procedures is that they can be used to make many measurements per day under conditions known to the sampler. These systems could be used for prescreening techniques to help decide the optimum location or locations for placing screening instruments.

### (B) *Radon Grab-Sampling Using a Lucas Cell*

Grab-sample radon monitors are usually not used for screening or follow-up measurements, but are primarily used to quickly determine the radon concentration at various locations of interest through a building or home. In one procedure a number of Lucas cells are evacuated and then over a five to ten minute period filtered air samples with their radon are allowed to slowly drain into each cell. The cells are either counted at the sampling location or returned to the lab for analysis. The detection equipment uses a photomultiplier tube based system. Usually a four hour wait is employed between collection and counting to allow the radon-222 to reach equilibrium with its daughters.

### (C) *Radon Grab-Sampling Using Activated Charcoal*

In a second type of grab sampling, the filtered sampled air is pumped through an activated charcoal cartridge for about one hour. The cartridge is weighed prior to and after sampling to determine the correction that should be applied to account for the reduced radon adsorption due to water absorption. The cartridges are then returned to the lab and analyzed using a NaI(Tl) gamma ray spectroscopy system.

(D) *Advantages of GRM*

1. Many results can be quickly obtained from a home.
2. The sampling equipment is usually light, portable, and operates, if needed, using a battery pack.
3. Since a trained operator is taking the measurements, he or she will be aware of the conditions which exist in the home prior to and during the measurements.
4. These procedures will give more accurate results than those obtained from grab-sample working level measurements.

(E) *Disadvantages of GRM*

1. These measurements are expensive, \$80-\$300 per measurement depending on the type and number of sources.
2. Sampling time is short (5 minutes to 1 hour) and thus susceptible to variations in radon concentration which can occur at different times throughout the day.
3. These measurements are not considered suitable by EPA for screening or follow-up measurements.
4. These instruments require a skilled operator to insure that the samples are properly collected and counted.

(F) *Examples of Equipment Available as Grab-Sample Radon Monitors*

The equipment sold for Lucas cell continuous radon monitors can usually be used as grab-sample radon monitors. These include the Pylon AB-5 and the EDA instruments RDA-200 previously mentioned. Since the Lucas cell counter need not be transported to the field, these types of measurements can be made by laboratory-based Lucus cell counters such as the Ludlund Model 2200 scaler with Lucas cell adapter which can be purchased from Ludlum Measurements, Inc. of Sweetwater, Texas. The gamma spectroscopy equipment sold for use with activated charcoal adsorbers would be used for analysis of radon grab samples using charcoal adsorption.

## CONTINUOUS WORKING LEVEL MONITORS (CWLM)

(A) *Background*

Radon progeny monitors of the continuous time-integrating type (CWLM) have become very popular in recent years.[14] In these systems, air is drawn at a slow rate (0.1 to 1 liter per minute) through a filter cartridge or filter paper. The filter is continuously monitored by an alpha detector. The detector might be an alpha scintillator such as ZnS(Ag) which is viewed by a photomultiplier tube, or it might be a silicon semiconductor detector such as a diffused-junction or surface barrier alpha detector. Since these detectors are responsive only to the alpha radiation of polonium-218 and -214 in the dust, the results will be computed in terms of working level. These systems often contain a microprocessor to keep track of the average count rate and record the count rate per unit time (usually hourly). Some of the systems used for continuous radon monitors can be adapted to simultaneously or separately determine the concentration of radon decay products (working levels).

EPA recommends that for screening measurements these detectors be used for a minimum of six hours under closed conditions. For follow-up measurements the following procedures are recommended: if the screening measurements gives results between 4 and 20 pCi/l, then one 24 hour measurements is to be made under normal living conditions every three months for a one-year period. If the screening level exceeds 20 pCi/l, then a 24 hour measurement made under closed house conditions can be used.

(B) *Advantages of Continuous Working Level Monitors*

1. These systems have one of the shortest EPA recommended sampling time periods for screening and follow-up measurements.
2. The determination of hourly results can track variations of the concentration of the radon decay products in a home.
3. Most systems have very small precision errors.
4. For most systems, the results are available on site either during the measurement period or shortly thereafter.
5. Some systems are capable of measuring individual radon and thoron decay products. Others can measure the percentage of radon and thoron decay products.
6. Some systems can act as both CRM and CWLM, either simultaneously or as separate measurement steps.

(C) *Disadvantages of Continuous Working Level Monitors*

1. These measurements tend to be more expensive than many of the other measurements used for screening and follow-up measurement (in the range of $100 to $300 per sample depending on the type of service offered).
2. The equipment is costly, since each monitoring unit may cost between $2,500 to $10,000 depending on the model and its capabilities.

3. Some models are heavy and awkward to move.
4. Most systems require a trained operator to use them.
5. To obtain accurate results, these systems must be extensively calibrated in a radon calibration chamber.
6. Because of the highly reactive nature of radon decay products, these systems are much more susceptible to sampling errors than continuous radon monitors.
7. The systems do not respond very quickly to changes in radon level and do not provide true readings until radioactive steady state conditions in the sampling filters have been reached.

### (D) *Examples of Equipment Available as Continuous Working Level Monitors*

Both the Pylon AB-5 and the EDA RDA-200 can be made to operate as a CWLM. EDA also sells two separate units, its WLM-30 and WLM-300 working level monitors. The WLM-300 employs a ruggedized silicon detector and has a microprocessor and sophisticated software program for the rapid calculation of results. The WLM-30 uses a solid state detector and built-n multichannel analyzer to produce an alpha spectra which allows for the detection and quantification of not only polonium-218 and polonium-214, but also the thoron daughters of polonium-216 and polonium-212.

Bicron Corporation is selling a small, lightweight CWLM produced by Thomson & Nielsen Electronics, Ltd. of Kahata, Canada which employs a silicon solid state detector. This system gives a visual readout in alpha counts and can be preset to sample from ½ to 24 hours. A manual mode can also be selected for continuous operation. This system can be attached to an IBM XT/AT computer for working level analyses.

A new and very sensitive continuous working level monitor and grab sampler has been reported by Tom Scurry Associates of Boulder, Colorado. Their instrument, called a CWL 430, employs a sensitive detector to monitor the air filter. This system can operate continuously up to one week. Each half hour count is stored in a protected memory. The computer in the system is menu driven for ease of operation.

## GRAB-SAMPLING WORKING LEVEL MONITORS (GWLM)

### (A) *Background*

Grab sampling working level techniques have been used for many years to determine the concentration of airborne radon daughters, especially polonium-218 and polonium-214. These procedures usually involve a five minute collection of dust on an air filter followed by a series of measurements using a detector which is sensitive to alpha particles. These measurements are usually made under closed house conditions. Because of the short sampling times, these techniques are useful for making measurements at a number of locations in a home over a few hours. But since measurements are usually not made around the clock, these techniques are not recommended by EPA for screening or follow-up measurements. Two such techniques are commonly used: the Kusnetz procedure[15] and the Modified Tsivoglov procedure.[16,17] The systems used for continuous working level monitors can usually be used as GWLM's, since both types of measurements need the same equipment, e.g. pump, air filter, and alpha particle counters. The detectors may be ZnS scintillators or silicon semiconductor detectors. The price of this equipment is in the range of $1,500 to $10,000.

(B) *Kusnetz Procedure*

In the Kusnetz procedure, 100 liters of air are passed through a filter over a 5 minute period and the alpha activity is counted at any time between 40 and 90 minutes after the end of the sampling period. The activity measured can then be converted into units of working level using the "Kusnetz factor" based on the time between the end of the measuring period and the middle of the counting period.

(C) *Modified Tsivoglov Procedure*

The sampling conditions and radiation detection procedures are the same for the Tsivoglov procedure as modified by Thomas,[17] as for the Kusnetz procedure, but the filter is counted between 2 and 5 minutes, 6 and 20 minutes, and 21 and 30 minutes after sampling. This procedure will not only give the working level value, but will also give the concentrations of polonium-218, lead-214, and polonium-214 in the air.

(D) *Advantages of GWLM*

1. The measurement time is short and the results are quickly obtained.
2. The equipment is usually light and portable, and usually operates using a battery pack.
3. These measurements can be made as part of a more complex system doing radon grab sampling measurements or continuous radon concentration measurements.
4. A large number of samples can be collected and analyzed each day.
5. Since a trained operator is taking the measurements, he or she will be aware of the conditions which exist in the home prior to and during the measurements.

(E) *Disadvantages of GWLM*

1. These measurements are expensive, \$80-\$300 per measurement depending on the type of services.
2. Sampling time is short and may not be representative of the long term average values.
3. These measurements are not considered suitable by EPA for screening or follow-up measurements.
4. These instruments require a skilled operator to insure that the samples are collected properly and counted correctly.
5. As with the other procedures involving the collection of radon decay products, these procedures are much more susceptible to sampling errors than corresponding radon as measurements.

(F) *Examples of Equipment Available as Grab-Sample Radiation Monitors*

All of the equipment described under continuous working level monitors can be used as grab sample working level monitors. The pumps are run for a five minute period instead of continuously during the sampling period.

## REFERENCES


1. The National Radon Measurement Proficiency (RMP) Program Applications and Participation Manual. December 1988. U.S. E.P.A., EPA-32011-88-056.
2. Indoor Radon and Radon Decay Product Measurement Protocols. February 1989. U.S. EPA Problem Assessment Branch, Radon Division.
3. Interim Protocols for Screening and Follow-up Radon and Decay Product Measurements. February 1987. U.S. E.P.A., EPA 520/1-86-014-1.
4. Cohen, B.L. and E.S. Cohen. 1983. Theory and Practice of Radon Monitoring with Charcoal Adsorption. Health Physics 45, 501-508.
5. Prichard, H.M. and K. Mariën. 1985. A Passive Diffusion $^{222}$Rn Samples Based on Activated Carbon Adsorption. Health Physics, 48, 6, 797-803.
6. George, A.C. 1984. Passive Integrated Measurement of Indoor Radon Using Activated Carbon. Health Physics 46, 867-872.
7. Jester, W.A., R.W. Granlund, M.E. Linsley, and R. Batschelet. 1989. An Improved Method for Determining Airborne Radon-222 Concentrations Using EPA Standard Activated Charcoal Canisters. Transactions of the American Nuclear Society, 59, pp. 49-50.
8. Kotrappa, P., J.C. Dempsey, J.R. Hickey and L.R. Stieff. 1988. An Electret Passive Environmental $^{222}$Rn Monitor Based on Ionization Measurement. Health Physics, 54, 1, 47-56.

9. Lovett, D.B. 1969. Track Etch Detectors for Alpha Exposure Estimation. Health Physics, 16, 623-628.
10. Randte, M.W., L.J. Tucker, and M.K. Robertson. 1987. Tuffak, A Practical Nuclear Track Radon Detector. Radiation Prot. Dosim. 19, 1, 29-33.
11. Lucas, H.F. 1957. Improved Low-Level Alpha Scintillation Counter for Radon. Review of Scientific Instruments, 28, 680-683.
12. Lehtimäki, M., and G. Graeffe. 1986. Measurement of Air Ions," Environment International, 12, pp. 109-113.
13. Sciaser, K.J. 1974. Integrating Radon Progeny Air Sampler. Amer. Ind. Assoc. J., 35, 165-174.
14. Bigu, J. 1988. Conversion Factors for Calculating Working Levels Using Continuous Radon Time-Integrating Monitors. Radiat. Prot. Dosim. 22, 1, 19-25.
15. Kusnetz, H.L. 1956. Radon Daughters in Mine Atmosphere - A Field Method for Determining Concentrations. American Industrial Hygiene Assoc. Quarterly, 17.
16. Tsivoglov, E.C., H.E. Ayer, and D.A. Huladay. 1953. Occurrence of Nonequilibrium Atmospheric Mixtures of Radon and Its Daughters. Nucleonics 1, 40.
17. Thomas, J.W. 1972. Measurement of Radon Daughters in Air. Health Physics, 23, 783.

*Environmental Radon: Occurrence, Control and Health Hazards.* Edited by S. K. Majumdar, R. F. Schmalz and E. Willard Miller © 1990, The Pennsylvania Academy of Science.

# *Chapter Eleven*

# DETECTION AND MITIGATION OF INDOOR RADON

**E. WILLARD MILLER[1] and ROBERT F. SCHMALZ[2]**

[1]Associate Dean for Resident Instruction
and Professor of Geography (Emeritus)
and
[2]Professor of Geology
College of Earth and Mineral Sciences
The Pennsylvania State University
University Park, PA 16802

## INTRODUCTION

The recognition that radon is a geographically widespread health hazard is rather recent. Concern about elevated indoor concentrations of the gas first arose in the 1960s when houses were built in the western United States with materials contaminated by waste from uranium mines. In eastern United States, the health danger from radon did not receive wide attention until the discovery of very high levels of radon in the Watras house near Reading, Pennsylvania, in 1984. Scientists now estimate that out of 130,000 annual lung cancer deaths in the United States, between 5,000 and 20,000 may be attributable to radon (Fabrikant, 1988). The Surgeon General has attributed about 85 percent of all lung cancer deaths to smoking; virtually all of the lung cancer deaths among non-smokers he attributes to radon exposure. It is the purpose of this chapter to discuss the means of detecting radon indoors and to review the means of reducing its concentration to acceptable levels.

## RADON DETECTION

It is important to understand that most houses and other buildings do not have a radon problem. At the same time, the concentration of radon may differ significantly from building to building in a small area, within a single neighborhood, or even between adjacent buildings. The fundamental problem is that no one knows which houses have a problem and which do not, for the gas has no color or odor and cannot be detected by ordinary means. Efforts are under way at local and state levels to delineate areas of high radon potential, but such areas are very broadly defined, and individual radon screening tests provide the only reliable estimate of the radon level inside a particular structure.

There are many reasons why a homeowner might be concerned about a possible radon hazard in his or her house. Elevated radon levels may have been reported in neighboring houses; bedrock in the region may be of a type which is sometime uranium-rich; the owner may simply wish to be reassured that there is no significant hazard; or he or she may be interested in buying or selling the property. In any case, the first step should be to carry out a preliminary screening test. Such a test need not be expensive, and can be carried out by the property owner. Two types of radon test kits are generally available in hardware and drug stores across the country. Charcoal canister detectors will provide screening measurements after exposures of three to fourteen days. They are available from reliable laboratories and usually cost $10 to $30. Alpha-track radon test kits are slightly more expensive and require an exposure period ranging from several weeks to several months to obtain reliable results. These kits come with complete instructions for their use, and include post-paid packages in which to return the detector to the parent laboratory for analysis. About a week after returning the kit the homeowner should receive a report from the laboratory advising him of the radon level observed (usually in "Working Levels" (WL) or pico-Curies per liter of air (pCi/L)). If the reported level exceeds the EPA guideline of 4 pCi/L (0.02 WL), the report may suggest that action be taken to reduce the concentration of indoor radon. Because radon levels fluctuate widely depending on the season, the time of day, the barometric pressure and a variety of other factors, no serious remediation program should be initiated before the elevated radon level has been confirmed by a follow-up test. In many states, including Pennsylvania, state health or environmental agencies will perform follow-up tests without charge if screening results indicate that a serious hazard exists (usually when radon levels exceed at least 20 pCi/L). If two or more tests indicate that remedial action is needed, then the owner may wish to consult a radon mitigation manual of the type published by the federal or state government, and he may want to consult a reliable radon mitigation contractor, or both (U.S. EPA, 1989).

Although suitable screening devices are simple to use and widely available at moderate cost, some property owners may prefer to seek the services of a professional radon consultant. The U.S. Environmental Protection Agency conducts a Radon Measurement Proficiency Program. This voluntary program allows staff members of laboratories and businesses to learn appropriate measurement and remediation protocols and to demonstrate their capabilities. In many states (including Pennsylvania) individuals and companies which provide radon measurement and remediation services must be licensed, and must demonstrate a familiarity with radiation measurements, with approved and effective remediation techniques, and must have a record of high ethical standards. A listing of licensed radon consultants and contractors can be obtained from the health or environmental protection agency in each state which has such a requirement.

## RADON HEALTH HAZARD

The National Research Council and the Environmental Protection Agency have attempted to estimate the number of lung cancer deaths expected over a period of time due to exposure to different radon concentrations. The actual risk is dependent upon the amount of time that an individual is exposed to a specific radon concentration. It also depends on the individual's age at the time when first exposed, and the time which has elapsed (if any) since his last exposure. Smoking and other factors related to life style will also affect the likelihood of contracting cancer. As a broad estimate, between one and five radon-induced lung cancer fatalities would be expected in a group of one hundred persons exposed for a lifetime (70 years) to a radon concentration of 4 pCi/L or 0.02 WL. A lifetime exposure to 20 pCi/L or 0.1 WL might be responsible for between 6 and 21 fatal cases of lung cancer in the same size population, and if the concentration is increased to 200 pCi/L or 1 WL, the probable number of lung cancer fatalities rises to between 44 and 77 per hundred persons exposed. It is estimated that among one hundred persons exposed at this level for only ten years, between 14 and 42 individuals would contract fatal lung cancer. For habitual cigarette smokers, the probability of contracting fatal lung cancer are approximately ten times greater at any level of radon concentration (Fabrikant, 1988).

Although the level of risk which one individual considers acceptable may be quite unacceptable to another, the U.S. Environmental Protection Agency has recommended the following radon "action levels" (U.S. EPA, 1986):

If screening measurement results exceed 1 Working Level or 200 pico-Curies per liter, a follow-up measurement should be carried out immediately. The detector should not be exposed for more than one week, and there should be as little ventilation of the building as possible. This is considered a dangerous radon level, and immediate action to lower the radon concentration is required.

If screening measurement results fall between 0.1 WL and 1.0 WL (or between 20 pCi/L and 200 pCi/L), follow-up measurements are recommended within several months. Detectors should not be exposed for more than three months and ventilation during the measurement period should be limited. If the follow-up measurements confirm the elevated radon levels, steps should be taken soon to improve ventilation or otherwise reduce the indoor radon level.

If screening measurement results show radon levels of 0.02 to 0.1 WL (or 4 to 20 pCi/L), follow-up measurements are recommended. Measurement devices should be exposed for periods of one week during each of the four seasons. At this concentration, remedial action is not urgent, but the Environmental Protection Agency recommends that action be taken to reduce the radon level within a few years.

In each of the above cases, measures should be taken to reduce the level of indoor radon. If the results of screening measurement show radon levels below 0.1 WL (4 pCi/L), radon reduction measures are probably not necessary. These values are below the level that can be related to increased incidence of lung cancer, and the cost and difficulty of achieving further reduction is probably unjustified (Fabrikant, 1988).

## RADON ENTRANCE TO BUILDINGS

Radon is produced by the radioactive decay of naturally occurring uranium and thorium in rocks and soil, and moves most rapidly through the earth where fissures and cracks in the rock or permeable soil allow swift gas movement. If a house is built in an area where uranium and thorium are unusually abundant, or where subsurface rock or soil properties are especially favorable, radon may penetrate into the building through dirt floors, cracks in concrete floors or walls, floor drains, sumps, pipe or cable penetrations and through pores in hollow block or concrete foundation walls.

Radon may also enter the building as a constituent in natural gas heating fuel or dissolved in domestic water. Natural gas is considered a very minor source of radon in most cases, but in some areas ground water may carry appreciable quantities of dissolved radon which is released inside the house when substantial quantities of water, especially heated water, are used. Radon is generally not a problem in large municipal water systems because the gas and its insoluble decay products are trapped or released to the atmosphere before the water enters the transmission mains. Private domestic wells which draw ground water from granitic terrains are the most likely sources of water-borne radon. Finally, in rare and unusual instances, radon may be introduced into a structure by the radioactive decay of radium contained in the materials of which the building is constructed.

## MITIGATION TECHNIQUES

Radon mitigation methods can be divided into three broad classes:

- *Ventilation methods* which lower indoor radon concentration by increasing the air exchange rate and thereby diluting the contaminant gas or exchanging contaminated for fresh air;

- *Barrier methods* which prevent radon from entering the structure by interposing barriers between the radon source (usually the soil) and the building; and
- *Pressurization methods* which prevent the movement of radon into the building by maintaining a slightly greater pressure inside the structure than in the soil beneath.

### *Ventilation Methods*

Radon concentration can be reduced by several different ventilation techniques. These include natural (passive) ventilation, forced circulation, heat recovery ventilation and block-wall ventilation (U.S. EPA, 1987-a; Henschel 1988).

Natural Ventilation: This is the simplest and least costly method to reduce radon concentrations by replacing contaminated indoor air with outdoor air. Natural ventilation occurs in every house as air is drawn through openings such as doors and windows, and through less obvious openings, cracks and unsealed leaks as well. In older, average American houses it is estimated that the air inside the house is replaced by outside air about once every hour. This exchange rate may be expressed as 1.0 ach (air change per hour) or simply in "reciprocal hours" ($h^{-1}$). Older poorly insulated houses typically show more rapid air exchange, often twice the above rate or 2.0 ach, while in newer well insulated buildings with a rating of 0.1 ach, ten hours will be required for complete exchange of indoor air.

When relying on natural ventilation to reduce indoor radon concentration, it is simplest to open windows and doors to allow free air circulation. Radon and other contaminants in the air will be reduced by dilution, the effectiveness of the method being determined by the relative rates of contaminant influx (or generation) and air replacement. Natural ventilation may also reduce the movement of soil gas into the building by eliminating any pressure differential between soil gas and the interior. Typically, natural ventilation can reduce indoor radon concentration by up to 90 percent (Table 1).

Although the use of natural ventilation involves no direct cost, its use will usually involve some increases in the cost of heating or air conditioning. If high radon levels occur in basement living spaces below grade, it may be difficult to achieve adequate ventilation without the use of fans. However, it must be remembered that radon is drawn into a building when the inside air pressure is less than that in the sub-soil. Thus, if it is to be effective, any ventilation system must not reduce the air pressure inside the house. In order to guard against this possibility, there must be a free flow of air throughout the structure.

Forced Ventilation: If natural ventilation is not effective in reducing radon concentration to acceptable levels, or if it proves to be too costly (in terms of increased heating or cooling costs) or too difficult to regulate, some form of forced ventilation may offer a practical solution. In the simplest case, one or more fans may supplement natural ventilation. Fans are particularly effective in maintaining continuous air flow when outside air is still and in restricted areas like crawl spaces.

Forced ventilation is especially helpful in newer well-insulated houses. In a typical house, to achieve a 90 percent reduction in radon concentration by whole-house ventilation, a fan capable of moving 500 to 1,000 cubic feet of air per minute is needed. The fan is usually most effective when mounted at the lowest level in the house. It is important, however, that an exhaust fan, if used, not reduce the air pressure inside the house relative to the soil gas pressure, and thus draw radon from the soil into the building. Experimentation will be required to determine the optimum placement of fans, and the advice of a competent heating and air conditioning engineer with sensitive pressure measuring instruments may be required.

Forced ventilation systems directed toward the radon source, the soil gas beneath the structure, rather than the contaminated air inside the building, are also frequently used. Two important pathways provide soil gas radon with access to the interior of a structure: cracks, joints and pipe penetrations in the basement floor or foundation slab, and channels formed by the hollow blocks of

TABLE 1

*Percentage Reduction in Radon Concentration Which May Be Achieved by Various Mitigation Procedures.*

| Method(s) | % Reduction* |
|---|---|
| Exterior foundation wall barrier | 30-40% |
| Crack/pipe penetration sealing | 50-75% |
| Sub-slab venting | |
| passive vent | 60-98% |
| forced draft** | 75-99% |
| Hollow wall venting | |
| passive vent | 50-80% |
| forced draft | 60-99% |
| Internal pressurization (HVAC) | |
| (estimated) | 40-99% |
| Sub-slab, hollow wall pressurization | no data |

(Sources: McKelvey 1989; Henschel, 1987)

*Reductions shown are estimates. Actual results will depend upon prevailing conditions in each individual case.

**Forced draft (wind turbines or duct fans in the vent stacks) usually results in a ten-fold improvement over passive venting.

foundation (and some interior) walls. Each of these pathways demands specialized ventilation methods.

In ordinary modern construction, foundation slabs or basement floors of poured concrete are usually laid over a sub-floor of packed and leveled crushed stone or gravel. This ballast layer provides drainage beneath the concrete and forms a uniform, stable base for the concrete. The porosity of the ballast allows the accumulation of soil gas beneath the slab, however, and it is from this gas reservoir that significant quantities of radon (and other soil gases) may diffuse into the building. It is relatively simple to reduce the concentration of gas in the ballast layer. During construction, a simple manifold of perforated 3″ or 4″ PVC drain pipe protected by a permeable geofabric membrane can be installed in the gravel ballast before pouring the basement floor or foundation slab. A stub collector stack located inside or outside of the building can be utilized at a later date if sub-slab ventilation or pressurization is deemed necessary. In an existing building, a slightly less efficient ventilation system can be built by drilling one (or more) holes, usually 2″ to 4″ in diameter, through the concrete floor slab and into the gravel below. A vent pipe of appropriate size is then sealed into the opening and connected to a stack, either inside or outside of the building, which is vented to the outside above the level of the roof. The stack may be fitted with a wind turbine or duct fan to increase the efficiency with which gases are drawn from the ballast. Experience has shown that such sub-slab ventilation can reduce indoor radon contamination by as much as forty percent (Table 1).

Block-wall Ventilation: Contemporary house construction commonly utilizes hollow pre-cast cement blocks for foundation walls and some interior basement walls. The cement used to manufacture the blocks is usually permeable to soil gases. Wherever such block walls will be in contact with soil they must be carefully sealed to prevent diffusion of soil gases into the openings within the blocks. If this is not done, the hollow block walls act as ducts through which soil gases, including radon, may enter the living space inside a building. To eliminate this problem in an existing structure without extensive excavation, the uppermost course of blocks may be infilled with cement wherever the openings are accessible. Holes may also be drilled into the hollow blocks to accommodate vent pipes connected to a manifold system with a roof stack. Forced air circulation with duct fans or wind turbines at roof level will increase the effectiveness of such a system. The efficiency of the system may be further enhanced by flushing the hollow blocks by blowing air into the wall openings through specially drilled holes and simultaneously exhausting air through an externally vented manifold. In an extreme case, the entire wall may be vented by drilling 1-2″ openings into the interior of each block at floor level, and exhausting or pressurizing the air through a metal "baseboard" manifold and stack.

Experience with sub-slab ventilation, vented hollow walls and forced circulation exhaust systems which combine these methods indicate that reduction of indoor radon levels in excess of 80% can be achieved at reasonable costs in existing structures. In new construction the costs are even lower (Table 1).

Heat Recovery Ventilation (HRV): Although high rates of air exchange may substantially reduce the concentration of indoor contaminants, under ordinary circumstances the benefits will be achieved only at the cost of substantially increased heating and/or cooling costs. These energy losses can be reduced substantially by installing a high-efficiency, large volume air-to-air heat recovery system or heat exchanger. In such a device, the heat in air being exhausted during cold periods is used to warm incoming air; during warm periods the process is reversed, and warm incoming air is cooled by exchange with the air being exhausted. The devices presently available are 50 to 60 percent efficient. When installed in conjunction with a large capacity air handling system, 50 to 75 percent reduction in indoor radon concentration can be achieved without increasing heating and cooling costs by more than about 50 percent. Such HRV systems are most effective when only a moderate reduction in radon concentration is required. The method is generally cost effective in lowering radon levels from approximately 10 or 15 pCi/L to a level of 4 pCi/L; at higher initial concentrations alternative methods will probably be more satisfactory. The HRV method is relatively expensive; in new construction, the cost of the heat exchanger and associated air handling equipment will range from $500 to $3,000 for an average house. Higher costs may be associated with remediation in an existing building.

## PRESSURIZED CONTROL SYSTEMS

Each of the foregoing radon reduction methods depends upon the extraction of radon from one or more sites of gas accumulation either inside or adjacent to the building. To the extent that forced air circulation is involved, the site of gas accumulation (sub-slab gravel ballast, hollow block wall, etc.) is subjected to reduced pressure. Pressure reduction inside a building even as small as a few Pascals caused by the draft of a fireplace or furnace is believed to significantly increase the rate of gas penetration into the building. Obviously, deliberate depressurization of a hollow wall or sub-slab gravel bed will tend to draw in additional soil gas. The potential benefits of exhausting gases from such reservoirs by forced draft must be carefully weighed against the risk that the depressurization will draw additional soil gases into the reservoirs. Although there is probably no general answer to this question, in most cases it may be more appropriate to use differential pressures to prevent soil gas penetration into the structure. To achieve this end, it is only necessary to maintain a slightly higher pressure (a few Pascals will be sufficient) inside the building than the prevailing pressure in the soil gas beneath the structure. In large modern office buildings and apartments equipped with central heating, ventilating and air conditioning (HVAC) systems this condition is generally met. To the extent that data are available, radon and radon decay product concentrations in such buildings are usually found to be low (cf: Silberstein and Grot, this volume). Although the low radon concentrations observed probably result from a combination of causes, including the large ratio of enclosed volume to ground contact area, the effect of positive internal pressure is quite surely a significant contributor.

This observation suggests that efforts to maintain even a very small positive internal pressure in a residential structure would be very helpful in reducing indoor radon levels. Where use patterns permit, basement spaces can be pressurized using unheated outside air. In cases where the basement is used regularly, fans could be used to force air from upper floors into the basement to maintain a slight positive pressure. It may also be beneficial to reverse the air flow in a forced-draft hollow wall or sub-slab ventilation system. Outside air forced into hollow walls and into the sub-slab gravel pack should effectively reduce diffusion of soil gas into the structure, and should have little effect on heating/cooling costs in the building. Few data are available at this time on the effectiveness and cost of operating a positive pressure system of the type envisioned, but under ordinary circumstances such a system might prove to be significantly more cost-effective than any other method of radon mitigation.

## SEALING AND BARRIER MITIGATION METHODS

Since the recognition of indoor radon contamination as a national problem, a number of physical mitigation techniques have been developed. These techniques focus upon the interposition of physical barriers to block the movement of radon gas from the soil into the protected structure. They include crack sealing methods, membrane and barrier technologies, pipe and cable penetration seals, improved trap designs for floor drains and sumps, and closures for sump wells. Many of these methods are described in detail in the Final Report of the Pennsylvania Radon Research and Demonstration Project (PA DER, 1988).

### *Sealing*

The entry of radon into the interior of a house can generally be prevented by the application of a sealing material to cracks and small holes in basement walls and floor. Sealing is commonly carried out in conjunction with other techniques since it is difficult to locate and seal all minute pores and fissures, and since normal settling will continue to open cracks in the future. Large visible cracks in floor and walls are usually widened and keyholed, roughened and cleaned thoroughly, then filled with a flowable sealant. The sealant must be compatible with the wall or floor material, gas-proof, flexible and non-shrinking. There are several appropriate sealants available, but the one most commonly used is a specially formulated flowable urethane. Complete specifications for both the procedure and materials to be used in sealing floor cracks, floor-wall joints, French and perimeter drains, and floor penetrations are presented in the Pennsylvania *Final Report* cited earlier (PA DER, 1988).

### *Membrane and Barrier Techniques*

A gas-proof membrane may be used to control or retard the flow of radon from an identified source such as rock or soil into the interior living space of a building. There are several types of membrane barrier and these may be applied in any of several ways. It can be applied to seal off and isolate an accessible area, or it may be tightly sealed around the perimeter and at all penetrations. Generally, high radon concentrations require sealing of membrane seams at joints in conjunction with depressurization to control soil gas. In some cases epoxy coatings can be applied to the inside surface of foundation walls or floor to form a gas-retardant membrane and prevent the entry of radon and other soil gases. In new construction, foundation walls can be sheathed with gas-proof plastic barriers of "Trocal" (registered trade mark) or a similar material prior to final grading. The installation of similar barriers can also be carried out at substantially greater cost after the building is finished (PA DER, 1988). Exposed earth, particularly in crawl spaces, can be sealed with concrete, with or without provision for sub-slab ventilation.

Membranes are effective barriers when exposed soil contributes to soil gas concentrations. Usually membranes are used to control radon if the protected space contains air handling equipment, duct work or water supply piping. Exposed duct work acts as a collector and distribution system for soil gases, radon and radon decay products. The type of membrane used will depend in each case upon the type of space, available access, and the size of the space to be sealed.

### *Sealing Drains and Sumps*

Many houses have drain holes drilled through the basement floor to receive small quantities of condensate from gas heaters, air conditioners and dehumidifiers. Such drain holes provide a pathway for soil gas entry. Radon entering through such a penetration may not be confined to the basement, but may be distributed throughout the building by the circulating fan of the heating or air conditioning system. If the drain hole has a condensation pump, the danger of radon entry is minimal. If there is no pump, the existing hole must be sealed to prevent the escape of radon.

Trapped and untrapped floor drains offer another common entry pathway for radon. Trapped drains utilize a water reservoir that acts as a barrier to upward gas flow through the drain, while allowing free downward flow of water to the residential drainage system. If radon is detected entering through a floor drain, the installation of a trap or improved flow control valve is indicated. In an extreme case it may be necessary to seal the drain opening.

Sump wells are widely used as a means of ground-water collection and control. Normally a sump consists of a hole through the basement floor in which rock and/or soil are exposed. In many instances, large diameter perforated pipes may be installed beneath the basement floor which enter the sump pit to facilitate ground-water collection and removal. Depending upon its physical condition, this system can be an active radon mitigation system if the water blocks the movement of radon. Under other circumstances the system may act as a radon collection manifold through which radon can migrate from the entire sub-slab area into the sump and from there into the building. In this circumstance, the sump must be sealed, usually with a concrete slab. The joint between the floor and the concrete cover should be sealed with flowable sealant, and if necessary a trapped floor drain can be installed to receive incidental water from the basement.

### *Water Treatment*

When radon enters a house in the domestic water supply, the radon may be absorbed by an activated charcoal filter or removed by aeration. The appropriate device for this purpose is a granular activated charcoal (GAC) unit. The device is placed directly in the water supply line between a pre-filter and a post-filter. The pre-filter removes particulate matter and some dissolved minerals and serves to extend the life of the GAC unit. The post-filter removes any charcoal fines from the treated water. The GAC unit is usually a fiber-glass cylinder about four feet high and fourteen inches in diameter. Radon gas is absorbed by the charcoal in the unit. Radon decay products ingrown on the charcoal medium emit gamma radiation, and some shielding may be required.

Radon may also be removed from domestic water by aeration. Although simple and inexpensive, this method requires a long aeration period for complete degassing. Radon removed from the water is vented to the open atmosphere. Installation of suitable aeration equipment is difficult, and the sound level during operation may be unacceptably high. Both aeration and charcoal filtration have been shown to be effective in reducing radon concentration in domestic water to acceptable levels (U.S. EPA, 1987-c).

## MITIGATION HANDBOOKS AND SOURCES

If radon screening tests confirm the necessity for remedial action, a property owner may wish to consider the procedures set forth in any of the radon mitigation manuals of the type published by the state and federal governments, or he may want to consult a reliable radon mitigation contractor, or both.

Two types of references and guides are available for homeowners who seek to reduce the level of radon in their homes. Methods of minimizing the radon hazard in new construction are described in detail in several publications distributed by the state and federal governments. The most comprehensive of these is the U.S. Environmental Protection Agency's publication, "Radon Reduction in New Construction" (U.S. EPA 1987-a). Similar handbooks are also available through the several state environmental protection departments (e.g., PA DER, 1985).

Radon mitigation or remediation in existing structures usually presents a more complex problem. The simple installation of one or more fans may suffice in one case; in another extensive reconstruction may be required. Detailed testing, evaluation and remediation protocols are described in the U.S. EPA publications, "Radon Reduction Techniques for Detached Houses" (Henschel, 1988) and "Radon Reduction Methods" (U.S. EPA, 1987-b). Working drawings and general specifications for use by contractors have been compiled in "General Remedial Action Details for Radon Gas Mitigation" (PA DER, 1985) and in the U.S. EPA Manual, "Reducing Radon in Structures" (U.S. EPA, undated). Sources of materials and specialized equipment for radon remediation are listed in the EPA volume cited above. Finally, many states have licensing requirements for radon remediation contractors, and homeowners who plan to invest in any significant radon reduction effort should first contact the appropriate state office or agency for advice on selecting a reliable contractor to test the radon level in the building and to make corrective modifications if such corrective measures are found to be necessary.

## REFERENCES

Fabrikant, J.I., et al. 1988. *Health Risks of Radon and Other Internally Deposited Alpha-Emitters: BEIR IV.* Report of the Committee on the Biological Effects of Ionizing Radiation of the National Research Council. National Academy Press, Washington, D.C. 576 pp. + glossary, index.

Henschel, D.B. 1988. *Radon Reduction Techniques for Detached Houses: Technical Guidance.* (2nd Edition), EPA/625/5-87/019. U.S. Environmental Protection Agency, Office of Research and Development, Research Triangle Park, North Carolina. 192 pp. + 2 appendices.

McKelvey, W.F. March 1989. *Close Out Evaluation and Report of the Radon Remedial Action Demonstration Project.* Performed for Philadelphia Electric Company by VERSAR, Inc. 12 pp.

PA. DER. May 1985. *General Remedial Action Details for Radon Gas Mitigation.* Department of Environmental Resources, Bureau of Radiation Protection, Harrisburg, PA. 36 pp.

PA DER. June 1988. *Final Report of the Pennsylvania Radon Research and Demonstration Project.* Prepared by R.F. Weston, Inc. and R.F. Simon Co., Inc. for the Pennsylvania Department of Environmental Resources, Harrisburg, PA. 78 pp. + 5 exhibits, appendix.

*U.S. EPA. August 1986. *A Citizen's Guide to Radon: What It Is And What To Do About It.* OPA-86-004. U.S. Environmental Protection Agency, Office of Air and Radiation, Washington, D.C. 13 pp.

U.S. EPA. 1987a. *Radon Reduction in New Construction; An Interim Guide.* OPA 87-009. U.S. Environmental Protection Agency, Office of Air and Radiation and Research and Development, Washington, D.C. 7 pp.

*U.S. EPA. 1987b. *Radon Reduction Methods: A Homeowner's Guide.* 2nd Edition. OPA-87-010. U.S. Environmental Protection Agency, Office of Research and Development, Washington, D.C. September 1987, 19 pp.

*U.S. EPA. 1987c. *Removal of Radon from Household Water.* OPA-87-011. Environmental Protection Agency, Office of Research and Development, Washington, D.C. September 1987, 8 pp.

U.S. EPA. 1989. *Indoor Radon and Radon Decay Product Measurement Protocols.* U.S. Environmental Protection Agency, Office of Radiation Programs, Washington, D.C. February 1989, 70 pp. + refs., 2 appendices.

U.S. EPA. Undated. *Reducing Radon in Structures; Manual.* 2nd Edition. U.S. Environmental Protection Agency, Office of Radiation Programs. Washington, D.C. 320 pp. + glossary, appendix.

*These public information leaflets are available without charge from the U.S. Environmental Protection Agency and many state environmental and health departments.

*Environmental Radon: Occurrence, Control and Health Hazards.* Edited by S. K. Majumdar, R. F. Schmalz and E. Willard Miller © 1990, The Pennsylvania Academy of Science.

# *Chapter Twelve*

# MODELING OF INDOOR RADON*

A. S. PASCHOA
Pontificia Universidade Católica do Rio de Janeiro (PUC/RJ)
Dept. de Fisica, C.P. 38071,
Rio de Janeiro, RJ 22453, Brazil

Human beings have been exposed to natural radiation since the beginning of their existence on Earth. However, only at the end of last century men became aware of natural radioactivity.[1-11] Promptly after the discovery of natural radioactivity, scientists started using this phenomenon as a tool for basic research as well as for a number of other applications; too numerous to be mentioned in a book chapter on modeling of indoor radon. However, it is worth mentioning that more recently, many symposia, congresses and workshops were solely dedicated to examine the results of studies undertaken by scientists in their attempt to understand the natural radiation environment.[12-20] Since 1963, when the first symposium on the Natural Radiation Environment took place in Houston, Texas, the number of published papers on the subject increased considerably. The United Nations Scientific Committee on the Effects of Atomic Radiation (UNSCEAR) has been summarizing the world knowledge on the natural radiation environment since its first report was published in 1956.[21-31]

Members of the public and the national authorities of many countries have more recently manifested a resurging interest on the subject of the natural radiation environment, in general, and in the monitoring of indoor radon (i.e. $^{222}Rn$, 3.82d) in particular. Indoors radon is today subject of regulations in several countries, because of the high exposure to radon in certain buildings. Well sealed energy efficiency houses and buildings became popular in cold climates after the oil crisis of the seventies. Indoors yearly average radon concentrations reach levels as high as 71 $Bq \cdot m^{-3}$ (1.9 $pCi \cdot L^{-1}$), as it can be seen from the values presented in Table 1 for selected countries.[32] However, radon concentrations can reach levels up to 1160 $Bq \cdot m^{-3}$ in certain houses built entirely with alum shale aerated concrete.[30] A list of publications concerning indoors radon published between 1980 and 1986 illustrates the increasing interest in this subject during this decade.[33]

---

*Work partially supported by FINEP and CNPq.

The average dose from indoors radon is believed to be the largest of all non accidental doses received from radiation sources, either natural or man made. The effective dose equivalent received from radon by an average person is about 2 mSv•y$^{-1}$ (200 mrem•y$^{-1}$), while the average dose received from all other natural sources is about 1 mSV•y$^{-1}$ (100 mrem•y$^{-1}$).[34]

Radon monitoring is important not only because of indoors radon exposure, but also because of variations in outdoors radon concentrations can indicate upcoming seismic activity or ongoing temperature inversions that create the necessary conditions for an air pollution episode to occur. Thus, sudden and/or extraordinary increases of well documented indoors and outdoors radon levels should be carefully examined to find the actual causes.

Natural gamma radiation emitted by $^{40}K$ (1.39 x $10^{9}$y) and radionuclides of the three distinct chains that originate with $^{238}U$ (4.5 x $10^{9}$y), $^{235}U$ (7.1 x $10^{8}$y) and $^{232}Th$ (1.41 x $10^{10}$y) irradiate human beings all the time, and particularly when indoors. When the radon concentration indoors is low, gamma radiation

TABLE 1

*Average* indoors radon concentrations in selected countries.*

| Country | Bq/m$^3$ | pCi/L | Ref. |
|---|---|---|---|
| U.K. | 18 b | 0.49 | 43 |
| Greece | 20 | 0.54 | 51 |
| Netherlands | 24 | 0.65 | 53 |
| U.K. | 25 l | 0.68 | 43 |
| Italy | 25 | 0.68 | 40 |
| China (Beijing) | 25.9 | 0.7 | 45 |
| Ireland | 37 | 1.0 | 52 |
| Germany (FRG) | 40 | 1.1 | 50 |
| Luxembourg | 40 | 1.1 | 54 |
| U.S.A. | 40 | 1.1 | 38 |
| Belgium | 41 | 1.1 | 41 |
| France | 44 | 1.2 | 44 |
| Canada | 51.4 | 1.39 | 39 |
| Denmark | 52 s | 1.4 | 42 |
| France | 58 | 1.6 | 56 |
| Switzerland | 60 | 1.6 | 55 |
| Brazil | 61** | 1.7** | 32 |
| Finland | 63 | 1.7 | 55 |
| Sweden | 71 | 1.9 | 37 |
| Denmark | 88 w | 2.4 | 42 |
| U.S.A.*** | 740 | 20 | 57 |

*The term average is used here loosely, having one of the following meaning: mean, geometric mean, arithmetic mean, median.

b = bedroom, l = living room, s = summer, and w = winter.

**Range from 24 Bq/m$^{-3}$ (0.65 pCi•L$^{-1}$) to 114 Bq/m$^{-3}$ (3.1 pCi•L$^{-1}$).

***Five all-electric homes in central eastern Pennsylvania.

may become the main contributor for indoors irradiation of humans.[32] In addition, cosmic rays are responsible for a radiation dose at sea level of about 0.26 $mSv \cdot y^{-1}$ (26 $mrem \cdot y^{-1}$).[34]

Models are developed not only to describe the behaviour of radon and daughters since the moment that radon is created in natural sources by the alpha decay of $^{226}Ra$ (1.6 x $10^3$y) up to the point that doses to humans are estimated based on the inhalation of radon and its progeny. The objective of a model should be determinant in defining the model structure and boundaries.

Modeling indoors radon is particularly useful when the $^{226}Ra$ concentration in building materials and soils can be known before a house will be built with such $^{226}Ra$ bearing materials and over $^{226}Ra$ rich soils. The reported concentrations of $^{226}Ra$ in building materials range from 0.3 $Bq \cdot kg^{-1}$ (8 $fCi \cdot kg^{-1}$) in wood to about 2.6 x $10^3$ $Bq \cdot kg^{-1}$ (70 $pCi \cdot kg^{-1}$) in aerated concrete based on alum shale.[30] In addition, when a house is built on a soil containing a high $^{226}Ra$ concentration, radon exhalation from the soil contributes to increase radon concentration indoors. The reported radon exhalation from soils range from 3.4 $Bq \cdot m^{-2} \cdot s^{-1}$ in latosolic soil from Osaka, Japan to about 53 $mBq \cdot m^{-2} \cdot s^{-1}$ in chernozemic soil from Illinois.[30]

## MODELING STRUCTURE

Modeling radon indoors and outdoors will be examined taking into account a model structure often used in environmental and dosimetric modeling. Figure 1 illustrates the basic elements for environmental and dosimetric modeling.[48] However, when the objective is only to establish whether or not the radon concentration indoors is below a pre-defined level that corresponds to a dose limit, there is no need to consider the box on direct internal contamination. As a matter of fact, the only two boxes that appear in Figure 1 that will be considered in this chapter of modeling of indoor radon are those relative to terms on source, and transport and interaction.

The classical study of gaseous diffusion consists in describing through which way gas molecules with well defined velocity vectors enter or leave a volume element located at a certain point in a system of interest. In the particular case of radon, a steady state situation for radon emanation in the source material containing its parent radionuclide $^{226}Ra$ can be reached. However, the radon concentration in the source material will not be uniformly distributed throughout the volume, even when the $^{226}Ra$ concentration is itself approximately uniformly distributed, as it is often the case. Radon gas tends to diffuse out from the place where it is produced to finally escape from the material through its external surfaces or cracks. Thus, for example, one can expect higher radon concentrations in the mid-section of a slab-like wall than near the external surfaces.

Although the radon concentration indoors is usually time dependent, the variation is slow in the very important case of energy efficient houses after a steady state radon and daughter products in air is reached. Thus, this chapter deals essentially with a steady state model of radon and daughters when the flux of radon from the source terms and the exchange rate with outdoor air are constant. Three aspects should then be considered in accordance with Wicke and Postendorfer:[49] (i) exhalation of radon from materials (source terms); (ii) air exchange with outdoor air (interaction and transport terms); and (iii) plateout and deposition of radon daughters (interaction terms). There are, however, interrelations among these three aspects that cannot be easily incorporated analytically in a model. As a consequence, the usefulness of any model of indoor radon and daughters ought to be considered in accordance with the adequacy of its assumptions to its objectives, rather than solely in relation to the aspects that it incorporates.

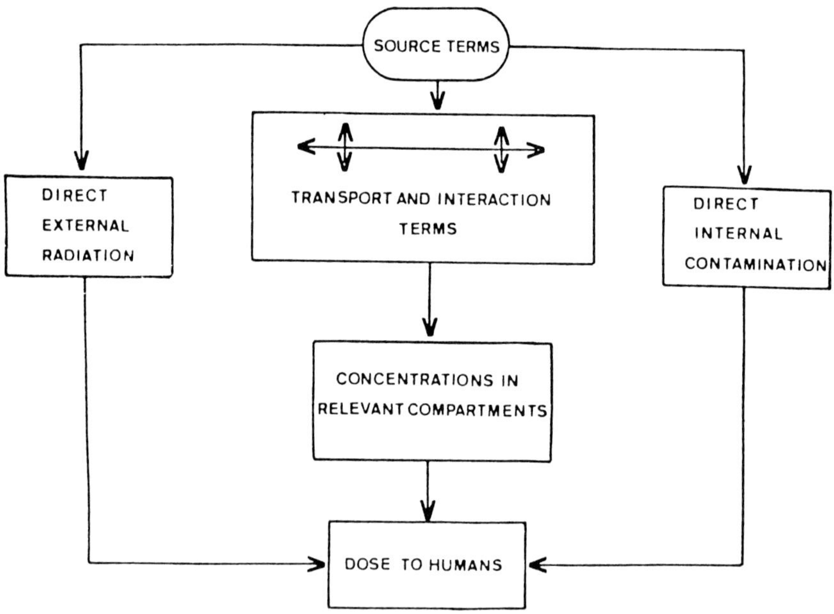


FIGURE 1. Basic elements for environmental and dosimetric modeling. (From ref. 48).

The boxes shown in Figure 1 can be filled once the objective of the model is established. In the case of radon modeling, the objective is usually human protection from excess radon inhalation. As a consequence, it is necessary to know whether or not the radon concentration indoors is beyond a pre-defined limit. The box indicating direct external radiation will not be considered in the model structure for indoors radon. Source terms are the main subject of modeling indoors radon, because the $^{226}$Ra, the radon parent, concentration in building materials and surrounding soils together with the porosity of these $^{226}$Ra containing media determine the overall radon concentration indoors. The box on direct internal contamination cannot be ignored when the model objective is to estimate the dose to humans.

## MATHEMATICAL MODEL

A large number of mathematical models including at least the first of the three aspects mentioned above has been developed to predict indoors radon and daughter concentrations.[47,49,58-78] There is an overall agreement in the literature that the Fick's law of diffusion theory can be applied successfully to predict the behaviour of radon gas molecules. This is the case because the radon gas flux from a source term is essentially the product of gas density and velocity. When one assumes constant velocity, the dependency becomes on position only, and the transport equation reduces to the diffusion theory approximation with radon gas molecules undergoing successive scattering processes. Such a model can be easily used to predict indoors radon and daughters concentrations at steady state as a function of ventilation rate, outdoor radon concentration, and radon source strength related with the $^{222}$Ra content in walls and soils.

### *Source terms*

It is well known that at steady state the number of radon gas molecules with a given velocity that enter a volume element of the source per unit time is equal to that number of molecules with similar characteristics that leave it. However, the one dimensional time-dependent transport equation for the radon concentration $C_i(x,t)$ throughout a source, at a point of x inside a slab-like wall or a semi-infinite slab-like soil (being i = w for wall or s for soil), can be conveniently described by:

$$\frac{\partial C_i(x,t)}{\partial t} = D_{ei}\frac{\partial^2 C_i(x,t)}{\partial x^2} - \lambda C_i(x,t) + \frac{S_i \eta_i \varrho_i \lambda}{\epsilon_i} \tag{1}$$

while, when the radon concentration in the source reaches a steady state (i.e., when the same number of gas molecules with a given velocity vector enters as leaves a volume element at any given time), equation (1) becomes:

$$D_{ei}\frac{d^2 C_i(x)}{dx^2} - \lambda C_i(x) = \frac{-S_i \eta_i \varrho_i \lambda}{\epsilon_i} \tag{2}$$

where, $\lambda$ is the radon decay constant (= 1.260 x $10^{-4}$min$^{-1}$), and for the source material i, $D_{ei} = D_i \epsilon_i$ is the effective diffusion coefficient of radon, $\epsilon_i$ the porosity, $S_i$ the $^{226}$Ra activity concentration, $\varrho_i$ the density, and $\eta_i$ the emanation coefficient defined as the activity of radon per activity of $^{226}$Ra.

Once there is a gradient at the interface between two radon diffusion media, the Fick's law of diffusion theory can be applied to determine the radon exhalation rate per unit area (or flux), $J_{i,j}$ (in pCi•m$^{-2}$min$^{-1}$ or Bq•m$^{-2}$•s$^{-1}$) from a medium i containing $^{226}$Ra, to a medium j (i.e., the indoor air). Fick's law can be written in this case as:

$$J_{i,j} = - D_{ei} \frac{dC_i(x_i)}{dx} \mid \text{at an interface} \tag{3}$$

where, $J_{i,j}$ is the radon flux from the medium i into medium j, commonly known as the exhalation rate.

The boundary conditions determine the solution of equation (2) that gives an expression for $C_i(x)$; as a consequence, the actual shape of $J_{i,j}$ depends on the source. Mathematical expressions for $J_{i,j}$ were derived for interfaces between wall-air, soil-air, and double media-air by Paschoa et al.[47]

## *Indoors radon and daughters*

Radon

Five terms combined in a differential equation constitute teh basis for the indoors radon model.[47,78]

$$\frac{dA_0}{dt} = S_0 + IA_0^{\circ} - IA_0 - \lambda_0 A_0 \tag{4}$$

where the five terms have the following meaning: (i) $dA_0/dt$ (in $Bq \cdot m^{-3} \cdot s^{-1}$ or $pCi \cdot m^{-3} \cdot min^{-1}$) is the rate of the indoors radon activity concentration $A_0(t)$ at a time t; (ii) $S_0 = J_{i,j}/H_j$ is the radon source strength, based on a radon flux $J_{i,j}$ and a linear dimension $H_j$ (in m) perpendicular to the interface between the media i and j; (iii) $IA_0^{\circ}$ is the rate through which radon is brought indoors form an outdoors concentration $A_0^{\circ}$ (in $Bq \cdot m^{-3}$ or $pCi \cdot m^{-3}$) at an air exchange rate I (in air-changes$\cdot s^{-1}$ - acs - or air-changes$\cdot min^{-1}$ acmin); $-IA_0$ is the rate of decrease of the indoor radon concentration $A_0(t)$ due to the air exchange rate I; and (v) $-\lambda_0 A_0$ is the decay rate of the indoor radon concentration because of the radon decay constant $\lambda_0$.

The time dependent solution for equation (4) is:

$$A_0(t) = A_0(oo) \left[1 - e^{-(\lambda_0 + I)t}\right] + A_0(O)\, e^{-(\lambda_0 + I)t} \tag{5}$$

where the indoor radon concentration at the steady state $A_0(oo)$ is given by:

$$A_0(oo) = \frac{\dfrac{S_0}{\lambda_0} + \dfrac{I}{\lambda_0} A_0^{\circ}}{1 + \dfrac{I}{\lambda_0}} \tag{6}$$

However, in the relevant case of steady state situation equation (5) reduces to equation (6), that is $\lim A_0(t)_{t \to oo} = A_0(oo)$. Figure 2 represents in a graph $A_0(oo)$ as a function of I and $A_0^{\circ}$ for radon source strength of 0.4, 4.0, and 40.0 $pCi \cdot m^{-3} \cdot min^{-1}$(0.24, 2.4, and 24 $mBq \cdot m^{-3} \cdot s^{-1}$), which correspond to $^{226}$Ra concentrations in a medium i of 50, $5.0 \times 10^2$, and $5.0 \times 10^3$ pCi•kg$^{-1}$ (1.9, 19, and $1.9 \times 10^2$ Bq•kg$^{-1}$), respectively. It is observable in Figure 2 that $A_0(oo)$ depends strongly on $S_0$ for I up to about 2 air-changes per hour (ach), but the same is not true for higher values of I.[47]

Radon Daughters

The differential equations for radon daughters concentrations up to $^{214}$Po in indoor Air $A_j$ can be synthetized as:

$$\frac{dA_j}{dt} = \lambda_j A_0 + IA_j^o - [\lambda_j + \lambda_{jR}(I) + I]\, A_j,\ j = 1 \cdots 4 \tag{7}$$

where j = 1 ($^{218}$Po), 2 ($^{214}$Pb), 3 ($^{214}$Bi), and 4 ($^{214}$Po); $\lambda_{jR}(I)$ are the rates of removal of radon daughters by deposition or plate out, which is an unknown function of the air exchange rate I;$\lambda_j$ are the radon daughters decay constants; and all other symbols have the same meaning as in the preceeding equations.

The steady state solutions for equations (7) are:

$$A_j(oo) = \frac{A_{j-1}(oo) + \dfrac{I}{\lambda_j} A_j^o}{1 + \dfrac{\lambda_{jR}(I)}{\lambda_j} + \dfrac{I}{\lambda_j}} \qquad j = 1 \cdots 4 \tag{8}$$

The function $\lambda_{jR}(I)$ is not usually known. However, for purposes of illustration, one can take from the literature that $\lambda_{jR} = 0.7\text{h}^{-1}$ ($1.17 \times 10^{-2}$ min$^{-1}$),[60,61] and that the fraction of unattached radon daughters is zero. Taking into account the above value of $\lambda_{jR}$, the ratios $\lambda_{jR}/\lambda_j$ are for j = 1 ••• to 4 the following:

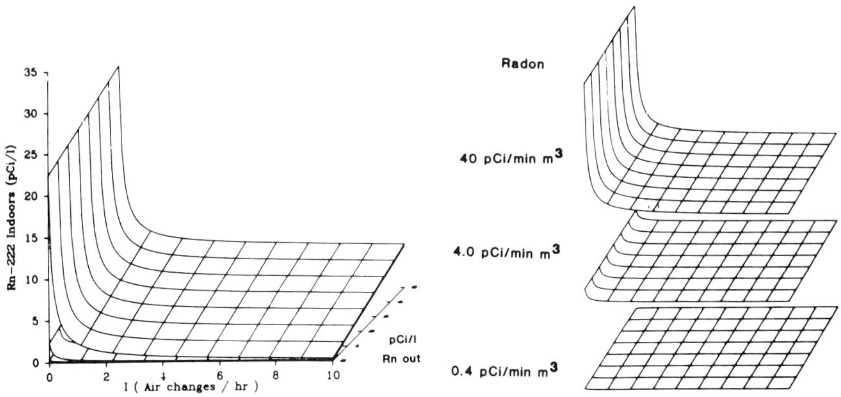


FIGURE 2. Steady state radon concentration $A_0(oo)$ as a function of air exchange rate I and outdoor radon concentration $A_0^o$ for three different radon source strengths. Left, actual graph; right, spatial representations of the surfaces for better observation. (From ref. 47).

j = 1 - 5.14 x $10^{-2}$; j = 2 - 4.50 x $10^{-1}$; j = 3 - 3.31 x $10^{-1}$; and j = 4 - 4.59 x $10^{-8}$. The ratios for $1 \leq j \leq 3$ are in the same order of magnitude, or even higher than the corresponding $I/\lambda_j$ for I < 1 ach. This means that for low and moderate air exchanges, typical of energy efficient houses, the plateout removal of $^{218}Po$, $^{214}Pb$, and $^{214}Bi$ affect significantly the radon daughter concentrations indoors.

Equilibrium Factor

Historically, the concentration of radon and daughters in uranium mines have been measured with a unit called Working Level (WL). This unit relates to the potential alpha energy of the short lived radon daughters present in 1 L of air. The potential alpha energy U (MeV) in a volume V ($m^3$) is given by:[72]

$$U = V \sum_{j=1}^{4} N_j E_j \tag{9}$$

where $N_j$ is the number of atoms of the radon daughter j per $m^3$, and $E_j$ is the potential alpha energy (MeV) for each j.

When the overall activity concentration is 3.7 x $10^3$ Bq•$m^{-3}$ (100 pCi•$L^{-1}$), the quantity U/V is about 1.3 x $10^5$ MeV•$L^{-1}$, which is used in the definition of the unit WL. To be more specific, 1 WL is defined as any combination of the short lived radon daughters ($^{218}Po$, $^{214}Pb$, $^{214}Bi$, and $^{214}Po$) per L of air which will result in a emission of a total alpha energy of 1.3 x $10^5$ meV. This is somewhat equivalent to state that an activity concentration of 3.7 x $10^3$ Bq•$m^{-3}$ of $^{222}Rn$ in equilibrium with its daughters corresponds to a potential alpha energy concentration of 1 WL.

Considering now that $A_j = \lambda_j N_j$, equation (9) can be re-written as:

$$\frac{U}{V} = V \sum_{j=1}^{4} \frac{A_j E_j}{\lambda_j} \tag{10}$$

When a non-equilibrium concentration of short lived radon daughters in air has the same potential alpha energy concentration as an equivalent concentration in air of radon daughters in equilibirum, the latter is called the equilibrium equivalent concentration of radon (EEC) of the non-equilibrium concentration in air of the mixture of short lived radon daughters.[72,79]

The EEC or $A_{eq}$ can be expressed, for convenience as:

$$A_{eq} = \frac{\sum_{j=1}^{4} A_j E_j / \lambda_j}{\sum_{j=1}^{4} E_j / \lambda_j} \tag{11}$$

The equilibrium factor F can then be defined in terms of the quantity $A_{eq}$ and the actual activity of radon concentration in air $A_0(oo)$, that is given by equation (6) as:

$$F = \frac{A_{eq}}{A_0(oo)} \qquad (12)$$

Taking into account equations (10) through (12) one can write:

$$F = \frac{\sum_{j=1}^{4} f_j A_j(oo)}{A_0(oo)} \qquad (13)$$

where each factor $f_j$ is given in Table 2, with other values for the constant values to be used in the preceding equations. Each factor $f_j$ is calculated as:

$$f_j = \frac{\frac{E_j}{\lambda_j}}{3.45 \times 10^4} \qquad (14)$$

Figure 3 shows graphs of the equilibrium factor as a function of I and $A_0^0$, assuming alternatives: a) no plateout removal of radon daughters (i.e., $\lambda_{jR} = 0$); or with plateout, making $\lambda_{jR} = 0.7\ h^{-1}$. As it can be seen from Figure 3, the equilibrium factor F given by equation (13) - with $f_j$ taken from Table 2, and $A_0(oo)$ and $A_{eq}$ calculated by equations (6) and (11), respectively - decreases by about 40% when the plateout effect is considered for low and moderate air exchange rates (i.e., for I < 1 ach). Such a result is in good agreement with reports published in the literature.[64,68,72] The overall effect of plateout removal, including the variation of $\lambda_{jR}$ with I and other parameters, needs further investigation because of its importance in the indoor concentration of radon daughters. Equations describing plateout removal rates have been proposed and tested in an air chamber with low levels of condensation nuclei.[80–82]

TABLE 2

*Data on $^{226}$Ra plus daughters up to $^{214}$Po**

| Nuclide | j | $T_{1/2}$ | $\lambda_j(s^{-1})$ | $E_j/\lambda_j$(MeV·Bq$^{-1}$) | $f_j$ |
|---|---|---|---|---|---|
| $^{226}$Ra(Ra) | −1 | $1.6 \times 10^3$y | $1.4 \times 10^{-11}$ | $(1.71 \times 10^{12})$** | |
| $^{222}$Rn(Rn) | 0 | 3.82 d | $2.1 \times 10^{-6}$ | $(9.10 \times 10^6)$ | |
| $^{218}$Po(RaA) | 1 | 3.05 min | $3.78 \times 10^{-3}$ | $3.61 \times 10^3$ | 0.105 |
| $^{214}$Pb(RaB) | 2 | 26.8 min | $4.32 \times 10^{-4}$ | $1.78 \times 10^4$ | 0.516 |
| $^{214}$Bi(RaC) | 3 | 19.7 min | $5.87 \times 10^{-4}$ | $1.31 \times 10^4$ | 0.380 |
| $^{214}$Po(RaC) | 4 | $1.64 \times 10^{-4}$s | $4.23 \times 10^3$ | $1.82 \times 10^{-3}$ | $5.28 \times 10^{-8}$ |

*Data taken from ref. (89).

**Numbers within parenthesis are not included in the summation $\sum_{j=1}^{4} E_j/\lambda_j = 3.45 \times 10^4$ that appears in equation (14).

Radon through cracks

Radon may also penetrate a room through cracks in walls and floor, affecting the indoor concentration.[83-87] This phenomenon has not been addressed in the model presented in this chapter, however, a mathematical model has been developed to examine the penetration of radon indoors through a vertical crack in a concrete soil.[87] Further development of such model is desirable to improve the prediction of indoors radon.

Using Ficks's laws as given by equation (3), a model was developed to represent the reduction of the radon flux from a $^{226}$Ra rich soil into air, by covering the soil surface with layers of material measured in units of their diffusion lengths.[88] Figure 4 illustrates the reduction of the radon flux as a function of

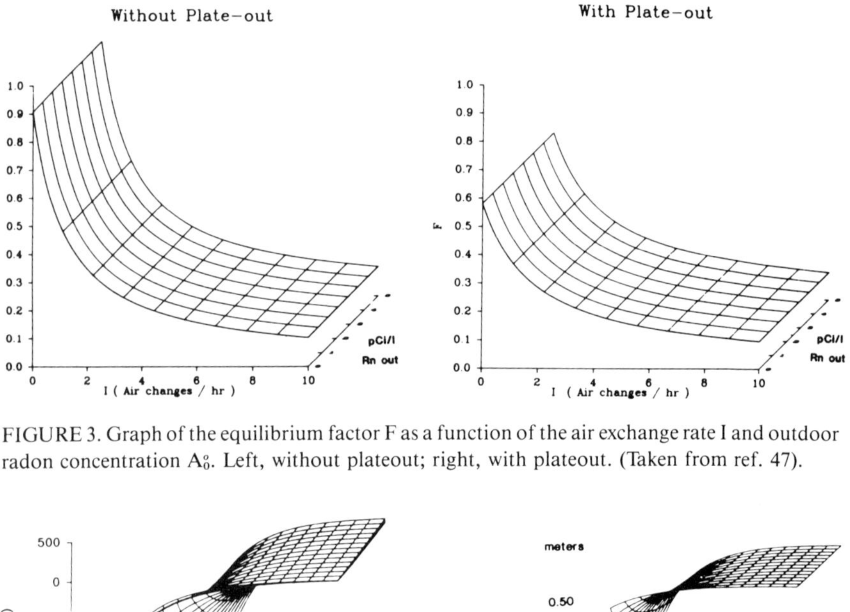


FIGURE 3. Graph of the equilibrium factor F as a function of the air exchange rate I and outdoor radon concentration $A_0^o$. Left, without plateout; right, with plateout. (Taken from ref. 47).

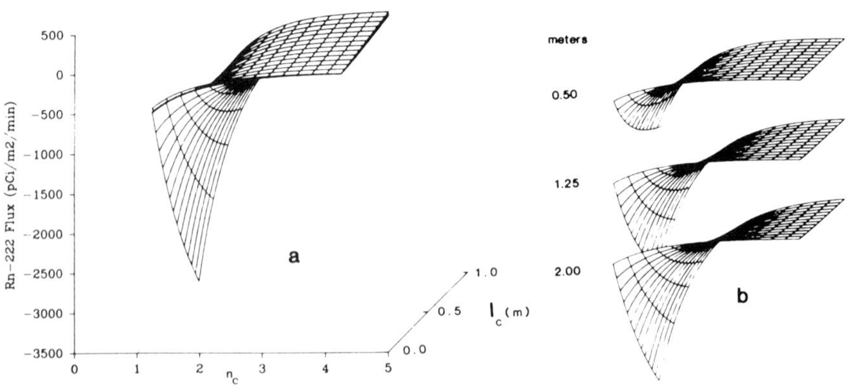


FIGURE 4. Radon flux as a function of the number $n_c$ of radon diffusion lengths $l_c$ of a cover, and the actual values of $l_c$; a) Graph of three surfaces superimposed; b) spatial representation of the three surfaces that appear superimposed in (a). (Taken from ref. 88).

the number $n_c$ of radon diffusion lengths $l_c$ of the cover, and the actual values of $l_c$. The flux appears with negative values in Figure 4 to facilitate the three dimensional representation. It can be interpreted from Figure 4 that a cover layer thicker than about three ($n_c$ = 3) radon diffusion layers (i.e. thickness $\geq$ 3 $l_c$), the radon flux from the soil becomes negligible. As a consequence, covering adequately radon sources can contribute to decrease significantly the radon concentration indoors.

## CONCLUDING REMARKS

A mathematical model can be used to predict the concentration of indoor radon and daughters, once the source strength and other relevant parameters are known.

It is still necessary to introduce into the model the removal and deposition rates of radon daughters as a function of the air exchange, however, the current model can be helpful in predictions whenever these removal and deposition rates are known.

The plateout removal rate of radon daughters plays an important role in reducing the equilibrium factor at low and moderate air exchange rates, but is not significant at air exchange rates higher than about 3 achr.

Although improvements are necessary to use the current model, computer codes can be used to make a theoretical assessment of the concentration of indoor radon even prior to build a house or a room.

## REFERENCES

1. Becquerel, H. 1896. Compt. Rend. 122:420-421.
2. Becquerel, H. 1896. Compt. Rend. 122:501-503.
3. Becquerel, H. 1896. Compt. Rend. 122:689-694.
4. Becquerel, H. 1896. Compt. Rend. 122:1086-1988.
5. Schmidt, G.C. 1898. Verhandl. Deut. Physik. Ges. 17:14-16.
6. Curie, M. 1898. Compt. Rend. 126:1101-1103.
7. Curie, P. and M. Curie. 1898. Compt. Rend. 127:175-178.
8. Curie, P., M. Curie and G. Bemont. 1898. Compt. Rend. 127:1215-1217.
9. Elster, J. and H. Geitel. 1898. Ann. Physik. 66:735-740.
10. Elster, J. and H. Geitel. 1899. Ann. Physik. 69:83-90.
11. Rutherford, E. 1899. Phil. Mag. 47:109-163.
12. The Natural Radiation Environment I. 1964. *In:* John A.S. Adams and Wayne M. Lowder, eds. The University of Chicago Press.
13. The Natural Radiation Environment II. 1972. *In:* John A.S. Adams, Wayne M. Lowder and Thomas F. Gesell, eds. CONF. 720805.

14. The Natural Radiation Environment III. 1980. *In:* Thomas F. Gesell and Wayne M. Lowder, eds. CONF. 780422.
15. The Natural Radiation Environment IV. 1982. *In:* K.G. Vohra, U.C. Mishra, K.C. Pillai, and S. Sadasivan, eds. Wiley Eastern Bombay/New Delhi.
16. Radon Workshop. 1977. *In:* A.J. Breslin, ed. HASL-325.
17. Areas of High Natural Radioactivity. 1977. *In:* T.L. Cullen and E. Penna Franca, eds. Academia Brasileira de Ciencias.
18. The Assessment of Radon and Daughter Exposure and related Biological Effects. 1980. *In:* G.F. Clemente, A.V. Nero, F. Steinhausler and M.E. Wrenn, eds. RD Press, Salt Lake City.
19. Indoor Exposure to Natural Radiation and Associated Risk Assessment. 1984. *In:* G.F. Clemente, H. Eriskat, M.C. O'Riordan and J. Sinnaeve, eds. CONF. 831049, Radiat. Prot. Dosim. 7.
20. Proceedings of the Fourth International Symposium on the Natural Radiation Environment. 1988. *In:* J. Sinnaeve, ed. Lisbon. To be published.
21. Yearly Progress Report of the United Nations Scientific Committee on the Effects of Atomic Radiation (UNSCEAR). 1986. Official Records of the General Assembly, Eleventh Session (A/3354). United Nations.
22. Yearly Progress Report of the United Nations Scientific Committee on the Effects of Atomic Radiation (UNSCEAR). 1957. Official Records of the General Assembly, Twelfth Session (A/3659). United Nations.
23. Report of the United Nations Scientific Committee on the Effects of Atomic Radiation (UNSCEAR). 1958. Official Records of the General Assembly, Thirteenth Session, Supplement No. 17 (A/3838). United Nations.
24. Report of the United Nations Scientific Committee on the Effects of Atomic Radiation (UNSCEAR). 1962. Official Records of the General Assembly, Seventeenth Session, Supplement No. 16 (A/5216. United Nations.
25. Report of the United Nations Scientific Committee on the Effects of Atomic Radiation (UNSCEAR). 1964. Official Records of the General Assembly, Nineteenth Session, Supplement No. 14 (A/5814). United Nations.
26. Report of the United Nations Scientific Committee on the Effects of Atomic Radiation (UNSCEAR). 1966. Official Records of the General Assembly, Twenty-first Session, Supplement No. 14 (A/6314 and Corr. 1). United Nations.
27. Report of the United Nations Scientific Committee on the Effects of Atomic Radiation (UNSCEAR). 1969. Official Records of the General Assembly, Twenty-fourth Session, Supplement No. 13 (A/7613 and Corr. 1). United Nations.
28. Report of the United Nations Scientific Committee on the Effects of Atomic Radiation (UNSCEAR). 1972. Official Records of the General Assembly, Twenty-seventh Session, Supplement No. 25 (a/8725 and Corr. 1). United Nations. also available as: Ionizing Radiation Levels and Effects, Vol. 1 and Effects Vol. II, E.72.IX. 17 and 18.

29. Report of the United Nations Scientific Committee on the Effects of Atomic Radiation (UNSCEAR). 1977. Official Records of the General Assembly, Third-second Session, Supplement No. 40 (A/32/40). United Nations. Also available as: Sources and Effects of Ionizing Radiation. E.77.IX.1. (1977).
30. Report of the United Nations Scientific Committee on the Effects of Atomic Radiation (UNSCEAR). 1982. Official Records of the General Assembly, Thirty-seventh Session, Supplement No. 45 (A/37/45). United Nations. Also available as: Ionizing Radiation: sources and biological effects, E.82.IX.8 (1982).
31. Report of the United Nations Scientific Committee on the Effects of Atomic Radiation (UNSCEAR). 1987. Official Records of the General Assembly. Report.
32. A.S. Paschoa, R.R. Pinho and D.A.C. Binns. 1989. Workshop on Radon Monitoring in Radioprotection, Environmental Radioactivity and Earth Sciences. Trieste, Italy, April 3-14. In Press.
33. Nero, Jr. A.V. 1987. Health Phys. 52, xvii-xix.
34. Eisenbud, M. and A.S. Paschoa. 1989. Nuclear Instruments and Methods for Physics Research A 280; 470-482.
37. Swedjemark, G.A. 1980. "Radon in Dwellings in Sweden". in CONF. 780422, 1237-1259.
38. Nero, Jr., A.V., R.G. Sextro, S.M. Doyle, B.A. Moed, W.W. Nazaroff, K.L. Revzan and M.B. Schwehr. 1985. The Science of the Total Environment. 45:233-244.
39. McGregor, R.G., W.B. Walker and E.G. Letourneau. 1985. The Science of the Total Environment. 45:271-278.
40. Sciocchetti, G., F. Scacco, P.G. Baldassini, C. Battella, M. Bovi and L. Monte. 1985. The Science of the Total Environment. 45:327-333.
41. Poffijn, A., R. Marijns, H. Vanmarcke and J. Uyttenhove. 1985. The Science of the Total Environment. 45:335-342.
42. Sorensen, A., L. Botter-Jensen, B. Majborn and S.P. Nielsen. 1985. The Science of the Total Environment. 45:352-356.
43. Green, B.M.R., L. Brown, K.D. Cliff, C.M.H. Driscoll, J.C.H. Miles and A.D. Drixon. 1985. The Science of the Total Environment. 45:459-466.
44. Rannou, A., C. Madelmont and H. Renouard. 1985. The Science of the Total Environment. 45:467-474.
45. Tianshan, R., L. Lianqing, C. Zhipeng, L. Guiyuan and C. Amin. 1987. Health Physics. 33:219-225.
46. Roser, F.X., G.H. Kegel, ZL. Hainberger, T.L. Cullen and R.T. Drew. 1964. *In:* A Study of Artificial and Natural Radioactive Contamination in Brazil. USAEC-AT (30-1)-2577, 54-59.
47. Paschoa, A.S., M.E. Wrenn and J.A. Torrey. 1984. Radiation Protection Dosimetry. 7:139-142.
48. Paschoa, A.S., N.P. Singh and M.E. Wrenn. 1990. To appear in Health

Physics.
49. Wicke, A. and J. Porstendörfer. 1982. *In:* The Assessment of Radon and Daughter Exposure and Related Biological Effects. (G.F. Clemente, A.V. Nero, F. Steinhausler, and M.E. Wrenn, Eds.). RD Press, Salt Lake City. 135-143.
50. Urban, M., D.A.C. Binns and J.J. Estrada. 1985. Radon Measurements in Dwellings. KfK 3866, CNEN 1101.
51. Papastefanou, C., M. Manolopoulou, E. Savvides and S. Charalambous. 1984. Proc. of 3rd Inst. Conf. on Indoor Air, Stockholm. 2:55.
52. McLaughlin, J.P. and P. Wasiolek. 1987. Proceedings of the Fourth International Symposium on the Natural Radiation Environment, Lisbon.
53. Put, L.W., R.J. deMeijer and Hogeweg. 1985. The Science of the Total Environment. 45:441-448.
54. Commission of the European Communities. 1987. Exposure to Natural Radiation in Dwellings of the European Communities, Report Euratom Treaty Article 31, Group of Experts, Luxembourg.
55. Commission of the European Communities. 1988. Report No. 1, Radon in Indoor Air, 1-15.
56. Rannou, A. 1989. Workshop on Radon Monitoring in Radioprotection. Environmental Radioactivity and Earth Sciences, Italy.
57. George, A.C., M. Duncan and H. Franklin. 1984. Radiation Protection Dosimetry. 7:291-294.
58. Jacobi, W. 1972. Health Phys. 22:441-450.
59. Porstendorfer, J., A. Wicke and A. Schraub. 1978. Health Phys. 34:465-473.
60. Jonassen, N. and J.P. McLaughlin. 1980. The Natural Radiation Environment III. (T.F. Gesell and W.M. Lowder, eds.). CONF. 780422 (Vol. 2) 1211-1236.
61. Moeller, D.W., D.W. Underhill and G.V. Gulezian. 1980. The Natural Radiation Environment III. (T.F. Gesell and W.M. Lowder, eds.). CONF. 780422 (Vol. 2) 1424-1443.
62. Stong, K.P., D.M. Levins and A.G. Fane. 1981. Radiation Hazards in Mining: Control, Measurement, and Medical Aspects (M. Gomez, ed.). Society of Mining Engineers, New York, 713-719.
63. Jonassen, N. 1981. Radiation Hazards in Mining: Control, Measurement, and Medical Aspects (M. Gomez, ed.). Society of Mining Engineers, New York, 927-930.
64. O'Riordan, M.C., A.C. James and K. Brown. 1982. Radiat. Prot. Dosim. 3:75-82.
65. Wicke, A. and J. Postendörfer. 1982. The Natural Radiation Environment IV (K.G. Vohra, U.C. Mishra, K.C. Pillai and S. Sadasivan, eds.). Wiley Eastern Limited, New Delhi. 481-488.
66. Landman, K.A. 1982. Health Phys. 43:67-71.
67. Bruno, R.C. 1982. J. Air Poll. Control Assoc. 34:105-108.

68. Cliff, K.D., A.D. Wrixon, B.M.R. Grenn and J.C.H. Miles. 1983. Health Phys. 45:323-330.
69. Jonassen, N. 1983. Health Phys. 45:369-376.
70. Zapalac, G.H. 1983. Health Phys. 45:377-383.
71. Knutson, E.O., A.C. George, J.J. Frey and B.R. Koh. 1983. Health Phys. 45:445-452.
72. Swedjemark. G.A. 1983. Health Phys. 45:453-462.
73. Rudnik, S.N., W.C. Hinds, E.F. Maher and M.W. First. 1983. Health Phys. 45:463-470.
74. Bruno, R.C. 1983. Health Phys. 45:471-480.
75. Jonassen, N. 1983. Health Phys. 45:487-491.
76. Nielson, K.K. and V.C. Rogers. 1982. A Mathematical Model for Radon Diffusion in Earthen Materials. NUREG/CR-2765, PNL-4301, RAE-18-2, RU 42pp.
77. Stranden, E. 1982. The Assessment of Radon and Daughter Exposure and Related Biological Effects. (G.F. Clemente, A.V. Nero, F. Steinhäusler and M.W. Wrenn, eds.). RD Press, Salt Lake City, 127-134.
78. Kusuda, T., S. Silberstein and P.E. Mcnall, Jr. 1980. J.Air Poll. Contr. Assoc. 30:1201-1207.
79. Jacobi, W. and K. Eisfeld. 1980. Dose to Tissues and Effective Dose Equivalent by Inhalation of Radon-222, Radon-220 and Their Short-lived Daughters. GSF Rep. S-626, Gesellschaft für Strahlen und Umweltforschung mbH München.
80. Okuyama, K., Y. Kousaka, Y. Kida and T. Yoshida. 1977. Journ. Chem. Eng. Japan, 10142-10147.
81. Postendörfer, J., G. Robig and A.J. Ahmed. 1979. Aerosol Sci. 10:21-28.
82. Holub, R.F. 1984. Radiat. Prot. Dosim. 7.
83. Landman, K.A. and D.S. Cohen. 1983. Health Phys. 44:249-257.
84. Nero, A.V. and W.W. Nazaroff. 1984. Radiat. Prot. Dosim. 7:23-39.
85. Akerblom, G., P. Andersson and B. Clavensjö. 1984. Radiat. Prot. Dosim. 7:49-54.
86. Janssens, A., F. Raes and A. Pffijn. 1984. 7:81.86.
87. Wilkinson, P. and P.J. Dimbylow. 1985. The Science of the Total Environment. 45:227-232.
88. Paschoa, A.S., J.A. Torrey and M.E. Wrenn. 1985. The Science of the Total Environment. 45:187-194.
89. Lederer, C.M. and V.S. Shirley. (eds.). 1978. Table of Isotopes, Seventh Edition, 1599 pp.

*Environmental Radon: Occurrence, Control and Health Hazards.* Edited by S. K. Majumdar, R. F. Schmalz and E. Willard Miller © 1990, The Pennsylvania Academy of Science.

# *Chapter Thirteen*

# INDOOR CONCENTRATIONS OF RADON DAUGHTERS AND REDUCTION METHODS

**NIELS JONASSEN[1] and J.P. McLAUGHLIN[2]**

[1]Laboratory of Applied Physics
Technical University of Denmark
2800 Lyngby, Denmark
and
[2]Physics Department
University College
Dublin, Ireland

## INTRODUCTION

As a result of the varied physical and chemical properties of radon and its daughters their behavior in the indoor environment presents us with a system of such complexity that quite often one can describe it with some degree of confidence only in steady or quasi-steady state conditions.

In order to obtain an understanding of this behavior it may be useful to consider the radon series as forming a type of microclimate within the indoor environment. For example phenomena such as the attachment of radon daughters to aerosols and their deposition onto room surfaces are somewhat analogous to the processes of condensation and precipitation which occur in the outdoor climatic system. Because of this microclimate-like behavior the control and reduction of radiation exposures from the radon series indoors present difficulties not usually encountered in more conventional and routine radiation protection situations.

In non-accident situations in the nuclear and industrial environment, in hospitals etc., radioactive sources, whether in solid, liquid or gaseous form are generally kept in a well designed and controlled configuration. To keep unsolicited doses within acceptable limits the sources are adequately shielded and regularly monitored.

Radon and its daughters in dwellings, on the other hand, migrate through the air spaces, undergo changes in configuration and other physical characteristics under the influence of highly variable meteorological and house parameters.

The radiation doses received from the radon series in a dwelling arise predominantly as a result of alpha particle irradiation of lung tissue following inhalation. Unlike external doses received from fixed beta or gamma ray sources such radon daughter alpha doses cannot be measured but must be estimated on the basis of dosimetry models which take into account such disparate and unrelated factors as the radon daughter aerosol size distribution, lung morphology, breathing rate cycle etc.

Even in terms of making a physical measurement of the concentrations of radon and its daughters in an air space there are special difficulties which are not just instrumental in origin. Because of the aforementioned temporal and spatial variability in the concentrations and characteristics of the radon series within the air spaces of a dwelling it is not an easy matter to establish a procedure to obtain measurements of these species which are sufficiently typical and representative of a dwelling on an annual average basis such that meaningful and useful dose estimates can be made.

A failure, unintentional or otherwise, to recognize this has often resulted in measurements in dwellings being made, which are less than useless.

In this chapter an attempt will be made to highlight what currently are considered to be the essential behavior patterns of the radon series in the indoor environment and also to indicate the effectiveness of a number of direct action approaches aimed at reducing indoor exposures and doses.

## RADON DAUGHTERS

The concentration of a given radon daughter in indoor air is determined by its rate of production and the rates with which it is being removed from the air. The only process with any significant rate of production is normally the radioactive decay of the airborne fraction of the immediate parent product. The contribution to indoor air radon daughter concentrations by outdoor air activity, brought in by ventilation, can normally be neglected in terms of radiological significance.

As far as the removal processes are concerned, the situation, however, is much more complex. Radon daughters may be removed from the air not only by the inevitable radioactive decay to the next daughter in the series but also by ventilation, plateout on room surfaces (possibly enhanced by electric fields) and filtration. In addition the rates of the various removal processes may be affected by the interaction of the radon daughters with the ambient aerosol.

Let us as an example consider the decay of a $^{222}$Rn-atom. The recoil resulting from the emission of an $\alpha$-particle causes the immediate decay product, $^{218}$Po, to be a doubly charged positive ion.[1]

This ion rapidly enters into interactions with trace gases, polar molecules, like water, or other ions in the air and in a time probably of the order of milliseconds gives rise to the formation of a small molecular cluster with a diameter of maybe a few nm. This cluster of molecules, which may be positively charged or neutral, is generally referred to as unattached or free $^{218}$Po.

The corresponding attached form arises when unattached $^{218}$Po attaches to ambient aerosol particles.

While radioactive decay, and to a large extent ventilation and filtration, affects equally unattached and attached $^{218}$Po (and the other short lived airborne decay products), this is not the case when we consider processes like diffusional and electric field—enhanced plateout. We shall return to this question later on in this chapter.

In Figure 1 are diagrammatically shown the various processes, which the individual short-lived radon daughters may partake in, during radioactive decay from $^{222}$Rn to the long lived $^{210}$Pb.

The direction of the processes are indicated by arrows and coded in the following way: (D): radioactive decay, (A): attachment to aerosols, (d): detachment

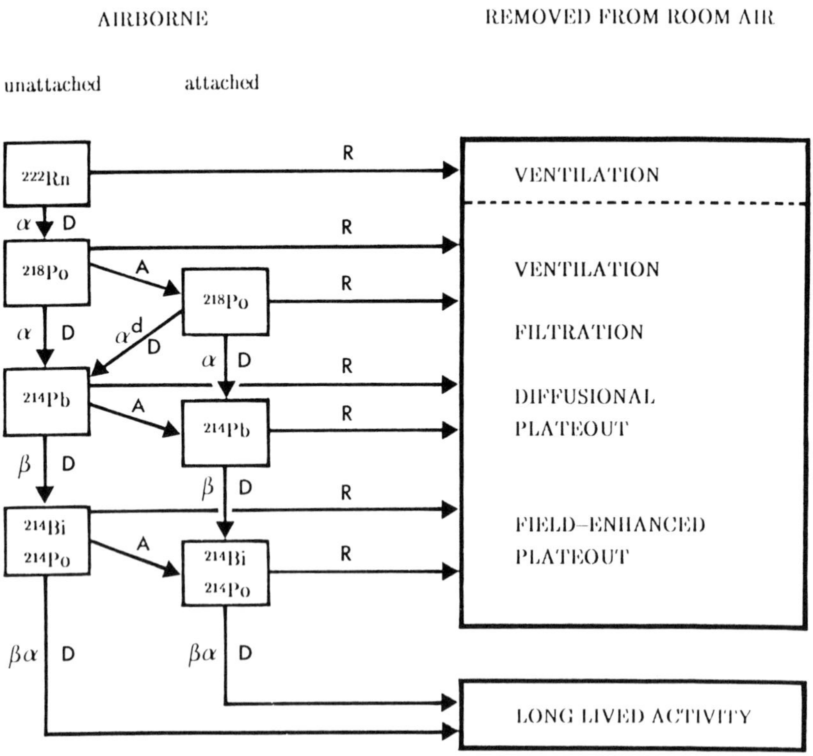


FIGURE 1. Processes affecting indoor radon and radon daughters

from aerosols (only relevant in the case of attached $^{218}$Po, where the decay, with a recoil energy of 110 keV, may result in the formation of $^{214}$Pb either in the unattached form or still attached to an aerosol particle)[2], (R): removal from the air space.

In this chapter a general overview of the behavior of the short-lived radon decay products in indoor air under the influence of the various processes will be given.

## RATE CONSTANTS

All the processes in which the airborne radon daughters participate can be characterized by rate constants, i.e., the rate of removal of a given species by a given process is equal to the appropriate (rate) constant, characteristic for the process, multiplied by the concentration of the species in question, and correspondingly for the rate of production. While the rate constants for the radioactive decay processes are known with great precision, only approximate values can be ascribed to the ventilation, filtration and plateout processes.

In Table 1 are shown the known values and likely range of values of the various time constants.

Obviously it is technically possible to create much higher ventilation and filtration rates than the maximum values quoted above. These values, however, represent what will normally be considered as acceptable from a comfort point of view.

The attachment rate constant $\lambda_a$ for unattached decay products attaching to aerosol particles is given by the product of the aerosol concentration $Z$ and the attachment coefficient $\beta$. The concentration $Z$ may vary from about $10^3$ $cm^{-3}$ in very clean air to maybe $10^5$ $cm^{-3}$ when cooking and/or smoking takes

TABLE 1

*Rate Constants for Radon Daughters in Indoor Air*

| decay constant | value $h^{-1}$ |
|---|---|
| $^{222}$Rn-decay, $\lambda_0$ | $7.6 \cdot 10^{-3}$ |
| $^{218}$Po-decay, $\lambda_1$ | 13.37 |
| $^{214}$Pb-decay, $\lambda_2$ | 1.55 |
| $^{214}$Bi-decay, $\lambda_3$ | 2.10 |
| $^{2214}$Po-decay, $\lambda_4$ | $1.5 \cdot 10^7$ |
| | |
| parameter | likely range |
| Ventilation, $\lambda_v$ | 0.3 - 1.5 |
| Filtration, $\lambda_f$ | 0.5 - 3 |
| Aerosol attachment, $\lambda_a$ | 5 - 500 |
| Unattached plateout, $\lambda_{u,p}$ | 5 - 100 |
| Attached plateout, $\lambda_{a,p}$ | 0.1 - 0.8 |

place. The attachment coefficient $\beta$ is a function of the size of the aerosol particles with a typical value of may be $5 \cdot 10^{-3}$ $cm^3 \cdot h^{-1}$, and the value of $\lambda_a$ for attachment to the normally polydisperse indoor air aerosol may consequently vary from as little as 5 to as much as 500 $h^{-1}$.

Both unattached and attached radon daughters are removed from the room air by plating out or depositing by diffusion on room surfaces like walls, floor, ceiling and furniture or, for that matter, people.

Since the unattached daughters are much smaller and consequently have a much higher coefficient of diffusion than the attached ones, the plateout rate constant $\lambda_{u,p}$ is much greater than $\lambda_{a,p}$ for radon daughters attached to aerosols.

The value of a plateout rate constant is given as the plateout velocity of the species in question multiplied by the surface to volume ratio of the room. The plateout velocities may be affected by air flow patterns or by unplanned electric fields due to local static charges.

As a consequence it is difficult for all indoor environments to predict with any accuracy expectable values of plateout rates. In addition very few direct measurements of plateout rates in real rooms have been made as distinct from laboratory chamber measurements. The ranges given in Table 1 represent largely the results of such laboratory based experiments.

## STEADY STATE MODEL

A series of mathematical models dealing with radon and its daughters have been developed over the last two decades.[3,4]

The models describe the interplay between the various processes affecting the radioactive elements and are very useful in predicting the radiological exposure and are also a useful tool in assessing the effectiveness of different types of remedial techniques.

In the present account the main features of steady state room models will be considered where a transient equilibrium between consecutive radioactive elements exists, i.e., where the rate of change of the activity concentration of the i'th element $\frac{dC_i}{dt} = 0.$

(As indicated in Table 1 the indices 0 - 4 refer to $^{222}Rn$, $^{218}Po$, $^{214}Pb$, $^{214}Bi$ and $^{214}Po$, respectively.)

The activity concentrations of the unattached and attached airborne radon daughters are then given by

$$C_{i,u} = \frac{\lambda_i(C_{i-1,u} + p_{i-1}C_{i-1,a})}{\lambda_i + \lambda_a + \lambda_{u,p} + \lambda_r} \quad (1)$$

and

$$C_{i,a} = \frac{(1-p_{i-1})\lambda_i C_{i-1,a} + \lambda_a C_{i,u}}{\lambda_i + \lambda_{\pi,\alpha} + \lambda_r} \tag{2}$$

and

$$C_i = C_{i,u} + C_{i,a} \tag{3}$$

Here $p_i$ is the alpha recoil factor ($p_0 = p_2 = p_3 = p_4 = 0$ and $0.8 \leq p_1 \leq 1.0$[2,5]), and $\lambda_4$ is the sum of the removal rates due to ventilation, filtration and electrical methods. Because of the very high value of $\lambda_r$, $C_3 = C_4$.

Equations (1), (2) and (3) express activity concentrations, i.e., number of disintegrations per unit time per unit volume. The SI-unit is bequerel per cubic meter ($Bq \cdot m^{-3}$). It should be noted that some regulatory and government agencies still use the old non-systematic unit picocurie per liter ($1\ pCi/\ell = 37\ Bq \cdot m^{-3}$).

In a normal indoor environment radioactive equilibrium between radon and its airborne daughter products (i.e., $C_0 = C_1 = C_2 = C_3$) never exists, due to plateout on room surfaces and the inevitable (however low) air exchange with the outdoor air. The degree of equilibrium can be expressed by the concentrations of the daughter products relative to radon $C_1/C_0$, $C_2/C_0$ and $C_3/C_0$, but when dealing with human exposure it is more relevant to characterize the air by what is known as the equilibrium factor $e_p$ with respect to Potential Alpha Energy Concentration (PAEC), $E_p$.

The PAEC is given as the sum of the potential alpha energy of all the daughter atoms present in a unit volume of air in their decay down to $^{210}Pb$.

Since only $^{218}Po$ and $^{214}Po$ emit alpha particles (with the energies $E_1 = 6.00$ MeV and $E_2 = 7.69$ MeV) the PAEC can be expressed as

$$E_p = E_1 \frac{C_1}{\lambda_1} + E_2\left(\frac{C_1}{\lambda_1} + \frac{C_2}{\lambda_2} + \frac{C_3}{\lambda_3}\right) \tag{4}$$

The product of potential alpha energy concentration and time is the radiological exposure of persons in a given atmosphere, and consequently the potential alpha energy concentration expresses the exposure rate for the atmosphere in question.

The unit for PAEC (or exposure rate) is joule per cubic meter ($J \cdot m^{-3}$). The old unit is working level ($1\ WL = 2.08 \cdot 10^{-5}\ J \cdot m^{-3}$ = the PAEC of 3700 $Bq \cdot m^{-3}$ EER (equivalent equilibrium radon)).

It should be stressed that working level is not a quantity but merely a unit for PAEC or exposure rate.

If radioactive equilibrium between radon and its daughters existed we would have

$$E_{p,max} = C_0\left(\frac{E_1}{\lambda_1} + \frac{E_2}{\lambda_1} + \frac{E_2}{\lambda_2} + \frac{E_2}{\lambda_3}\right) \tag{5}$$

The equilibrium factor for PAEC is then defined as

$$e_p = \frac{E_p}{E_{p,max}} \tag{6}$$

The equilibrium factor can also be considered as a normalized exposure rate, i.e., the exposure rate per unit of radon concentration.

Because of the large radiological dose considered to be associated with the unattached decay products as compared to the attached form[6], it is useful to characterize the radiological state of the atmosphere in terms of the attached fraction $f_a$ and the unattached fraction $f_u$ of PAEC, defined as

$$f_a = \frac{E_{p,a}}{E_p} \tag{7}$$

and

$$f_u = \frac{E_{p,u}}{E_p} \tag{8}$$

or by the attached and unattached equilibrium factor $e_{p,a}$ and $e_{p,u}$

$$e_{p,a} = \frac{E_{p,a}}{E_{p,\ max}} = (1 - f_u)\ e_p \tag{9}$$

and

$$e_{p,u} = \frac{E_{p,u}}{E_{p,\ max}} = f_u e_p \tag{10}$$

The expressions in equations (9) and (10) are especially useful when evaluating the effect of daughter removal processes.

## AEROSOL ATTACHMENT AND PLATEOUT

In a room with a given influx of radon and a given ventilation rate the concentration of the individual daughter products as well as the PAEC and the equilibrium factor $e_p$ (eqs. (1) - (6)) depend upon the aerosol attachment rate $\lambda_a$ and consequently upon the aerosol concentration and size distribution.

The total daughter activity in the room appears as airborne, (unattached or aerosol—attached), or as deposited on surfaces in the room, and of these only the airborne activity is available for inhalation.

Knutson[7] has recently, using the following "baseline" values of the various parameters: $\lambda_v = 0.55\ h^{-1}$, $\lambda_a = 50\ h^{-1}$, $\lambda_{u,p} = 20\ h^{-1}$, $\lambda_{a,p} = 0.20\ h^{-1}$ and d = 0.83, predicted about 2% of the total daughter potential alpha energy to be unattached, 35% attached to aerosols, 45% to be surface deposited and 18% to be carried away by ventilation.

In Figure 2 is shown the expected partitioning between the various states as a function of the aerosol attachment rate, with the remainder of the parameters having the same values as quoted above.

At an (only theoretically possible) aerosol concentration of zero only about 6% of the available potential alpha energy is airborne, and, since no aerosol particles are present, all in the unattached state. About 2% will be ventilated out and about 92% will be surface deposited.

As the aerosol concentration is increased the effect of the airborne radon daughters attaching to the aerosol particles is to suppress both the unattached airborne and the surface deposited activities until at high aerosol concentrations ($Z \cong 10^5$ cm$^{-3}$) less than 1% is present as unattached airborne radon daughters, about 50 - 60% as attached, 16% is surface deposited and the rest, some 25%, is carried away by ventilation.

It should be stressed that as, with increasing aerosol concentration, a larger fraction of the daughter activity is airborne and thus available for ventilation,

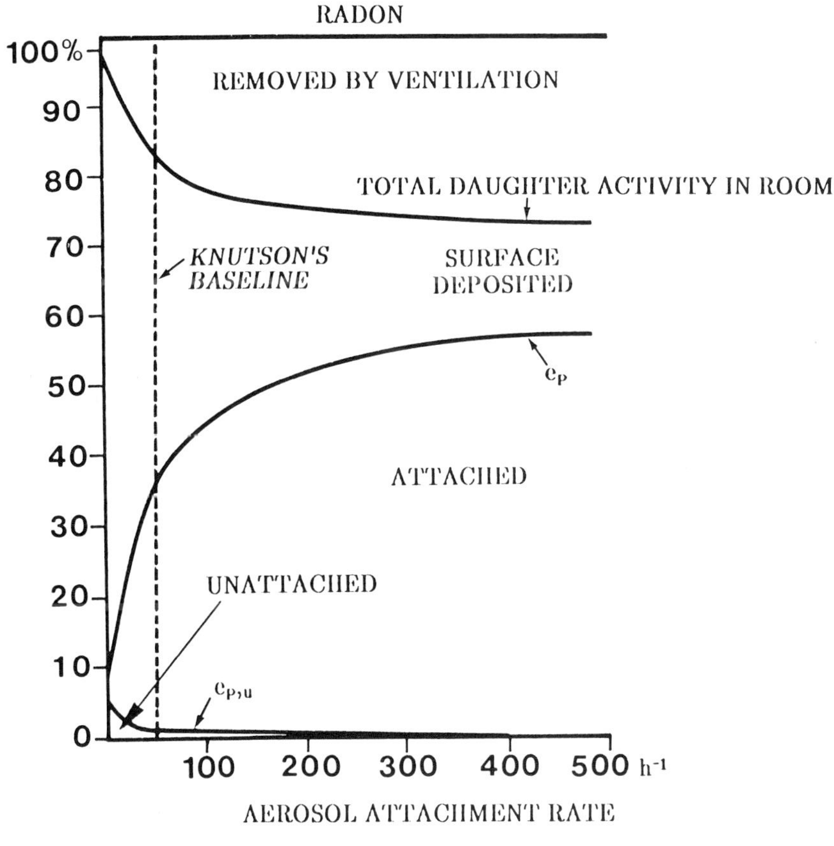


FIGURE 2. Partitioning of radon daughters as a function of the aerosol attachment rate.

the total daughter activity in the room will decrease with increasing aerosol activity.

The total PAEC and the equilibrium factor $e_p$ will, however, in all cases increase with increasing aerosol concentration, but the relative increase will be smaller with higher ventilation rates.[8]

The predictions of the models and the values of the parameters have been checked with several series of measurements in experimental chambers.

There have been very few investigations in real houses with sufficient measurements of ventilation and aerosol properties to permit checking the model. Knutson[7] has summarized these results which show that the equilibrium factor $e_p$ normally falls in the range 0.3 - 0.5 and the unattached equilibrium factor $e_{p,u}$ in the range 0.02 - 0.05.

In a series of measurements in Swedish houses, Jonassen and Jensen[9,10] found the values $e_p = 0.51$ and $e_{p,u} = 0.01$ in a house inhabited by smokers and $e_p = 0.46$ and $e_{p,u} = 0.04$ in non-smokers houses. No measurements of aerosol concentrations were made.

## REMOVAL OF RADON DAUGHTERS

### *Ventilation*

Some of the techniques aimed at reducing the radiological dose due to inhaled radon daughters work on the parent nuclide, the radon gas, as well as on the daughters. This is especially true in the case of applying increased ventilation.

For a given influx of radon into a room the equilibrium concentration $C_0$ will be inversely proportional to the sum of the radiological decay rate $\lambda_o << \lambda_v$, and the ventilation rate $\lambda_v$ (Eqs. (1) or (2)), and since in practice $\lambda_o << \lambda_v$, an increase in $\lambda_v$ from, say, 0.3 $h^{-1}$ to 1.5 $h^{-1}$ will cause a reduction in the radon concentration by a factor of 5. The PAEC will be reduced even more, because the production rate is reduced (by the factor of 5) and the removal rates are increased, however, not as strongly, because of the higher decay constants of the daughter products, see eqs. (1) and (2).

### *Filtration*

In regard to methods aimed at directly removing radon daughters from the air a number of possible approaches are available. The most commonly used technique is to filter the air using either mechanical or electrofilters.

When the air enters an electrofilter it passes through a region of high monopolar ionization by a corona discharge, usually from a thin wire kept at a high potential (5 - 20 kV). The ions move, by the field between the corona electrode and grounded surroundings transversely to the air flow. On their way the (small) air ions will attach to airborne particles, both inactive and radioactive, and these charged particles (or large ions) are brought by the air flow into a region, where an electric field will eventually deposit the particles on collection electrodes, with an efficiency that depends on the type of filter and the size distribution of the particles.

Most filters remove unattached and attached radon daughters and neutral aerosol particles from the air passing the filters with about the same efficiency, maybe 80 - 90%, but the effect of the filtration on the steady state concentration of the species in question depends strongly upon the interplay between the various removal and attachment processes.

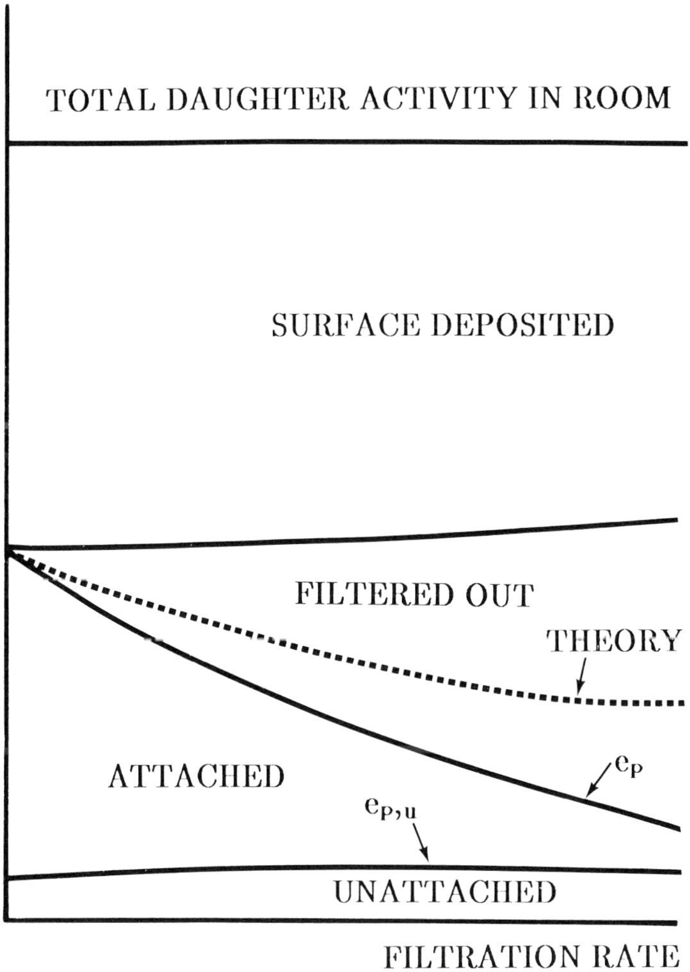


FIGURE 3. Effect of filtration at low aerosol concentration.

A given rate of filtration should, according to the models, lead to a certain decrease in the total (unattached and attached) concentration of a given radon daughter. If, however, the aerosol concentration is also substantially decreased by the filtration the attachment rate will be smaller and the fraction of the radon daughters remaining in the unattached state may increase.

In Figures 3 and 4 are shown the characteristic features of the effect of filtration on the partitioning of the radon daughters in a room at low and high aerosol concentrations. For the sake of simplicity the ventilation rate is assumed to be zero.

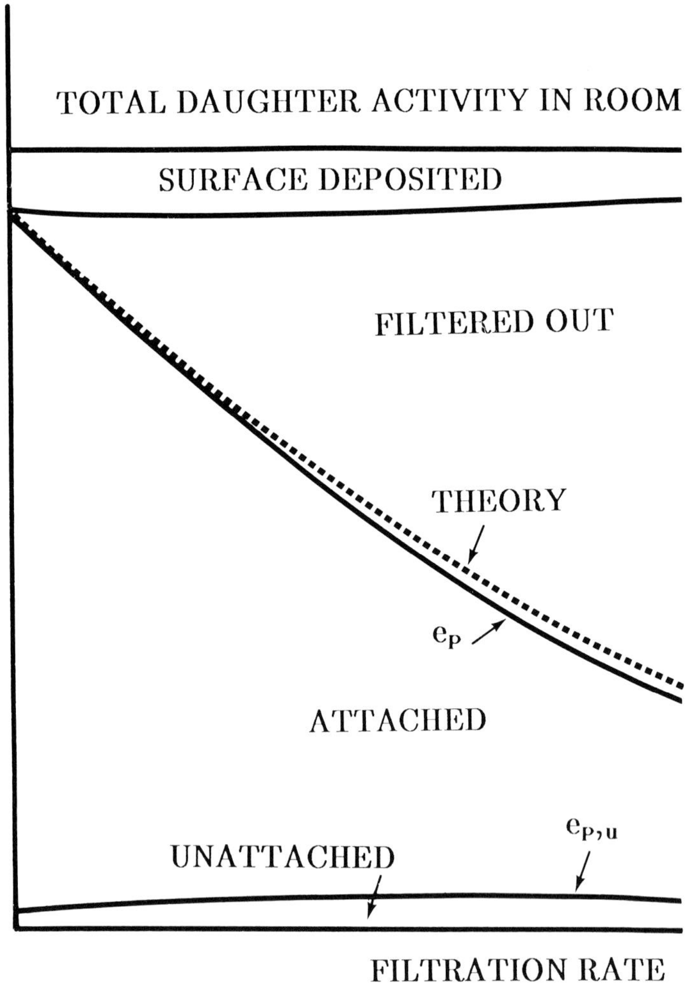


FIGURE 4. Effect of filtration at high aerosol concentration.

In the low aerosol situation increasing the filtration rate will reduce the concentration of attached radon daughters partly because of the direct effect of the filtration and also because of the reduced aerosol concentration. Also the unattached daughters will be filtered out, but the reduction in aerosol particles will furthermore reduce the attachment rate and thus (partly) counteract the filtration-induced removal of unattached daughters, and the result will normally be an almost constant (or slightly increasing) concentration of unattached daughters, with an unattached equilibrium factor $e_{p,u} \cong 0.02$—$0.04$.

The actual measurable reduction in equilibrium factor is always greater than predicted by the theory, which seems to indicate an increase in the plateout rate constants probably caused by the circulation of the air.[11]

In a situation, where the aerosol production rate and concentration is high, Figure 4, the concentration of unattached daughters remains small and the effect of circulation-increased plateout is therefore smaller. As a consequence the measured reduction in the equilibrium factor will follow the theoretical predictions rather closely.

Measurements in laboratory rooms[12] as well as in actual houses[9] show that it is possible by use of household electrofilters to lower the equilibrium factor (and thus the PAEC) to about 30% of the unfiltered value.

As most current lung dose models[6] ascribe a much higher (maybe 10-20 times) dose per unit of inhaled unattached PAEC than to the attached variety the filtration-induced reduction in the estimated dose is somewhat smaller than the actual measured reduction in PAEC, maybe to about 50% of the value for untreated air.

## *Field-Enhanced Plateout*

It is of historic interest to note that Elster and Geitel[13] in 1901 reported the original discovery of airborne radioactive substances by using a wire at a negative potential as collector for the so-called "active deposit", which probably for the major part consisted of positively charged unattached $^{210}Po$.

As mentioned earlier the unattached radon daughters may be neutral or positively charged (maybe up to 25%). The attached radon daughters will be charged to the same extent as the type of aerosol particles to which they attach. In the typical 0.1 - 0.2 $\mu$m range most particles are charged with approximately the same number carrying positive and negative charges.[14,15]

If a suitable electric field is established in a room, the charged radon daughters, and radon daughters attached to polarizable particles, will more towards and tend to plate out on the electrodes one of which is (normally) the room surfaces, which are essentially at ground potential.[16]

The field may be established by placing a non-ionizing electrode at a high potential (with respect to ground) for instance near the ceiling.

If the dimensions of the electrode, however, are not impractically large the voltage will drop off rapidly over the first centimeter or so from the electrode, and the field strength will be very low in the rest of the room and especially at the surfaces, where the plateout takes place, and since the plateout rate is proportional to the field strength at the plateout surface, the effect of the field on removing the airborne radon daughters will be small.

If on the other hand the electrode is made to ionize the air in its immediate vicinity ions (with the same polarity as the electrode) will move away from the electrode and set up a space charge in the room and thus modify the field distribution in such a way that the field strength is lowered in the vicinity of the electrode and increased in the rest of the room, and therefore also at the boundaries, and as a result the electrical plateout rate may increase substantially.

The use of field-enhanced plateout of radon daughters has been studied in several investigations.[17,18,19,20]

From the experimental investigations it appears that fields (and ions) from electrodes at a positive potential with respect to ground are considerably more effective in removing radon daughter ions from the air than are those from negative electrodes. This is probably caused by the fact that charged unattached radon daughters always seem to be positively charged. It is not quite clear if neutral radon daughters, to any major extent, are being charged by the space charge and made to move by the field.

The effect of applying electric fields to the air is in many ways equivalent to filtering the air. With both techniques primarily the activity of attached radon daughters is being reduced while the activity connected with unattached radon daughters stays more or less constant, see Figure 3 and 4.

As a consequence the measured reduction in PAEC will be higher than that of radiological dose as estimated on basis of current long dose models.

It has been demonstrated[19] that it is possible by use of a single electrode at a potential of + 30 kV to remove 94% of the PAEC in an experimental room. In another investigation[20] a line-operated ionizer (+ 18 kV), the size of half a carton of cigarettes, was mounted under the ceiling in a 120 $m^3$ room. At a typical indoor aerosol concentration of 20.000 - 30.000 $cm^{-3}$ the ionizer reduced the PAEC to about 30% and the estimated dose to about 50% of the value in the untreated state of air.

## COMPARISONS OF REMEDIAL METHODS

Of the three methods, discussed above, for removing radon daughters from the air ventilation is in principle by far the most effective, primarily because of its effect on the relatively long lived radon gas.

It can be assumed that the application of ventilation will reduce the PAEC and the dose to more or less the same extent, in practice down to maybe 20%

of the unventilated value.

Some caution should, however, be noted in respect to using ventilation as a radon control technique.

On a human level it may conflict with household comfort and energy budgeting and at a physical level it may, if incorrectly applied, cause an increase in indoor radon due to a depressurization induced influx of radon-rich soil gas. In terms of overcoming these problems successful heat recovery systems are available.[21,22]

In a well-designed system of this type highly efficient heat exchangers are used and the problem with depressurization may be minimized by using a balanced indoor-outdoor airflow system.

A major disadvantage with the use of ventilation is that it often requires a rather complicated and costly installation, and especially in existing houses this may be the most important deterrent for its use.

As far as the removal methods, filtration and field-enhanced plateout, are concerned, they are in many respects equivalent to ventilation.

It should, however, be kept in mind, that neither filtration nor field-enhanced plateout, on the one hand, either increases or decreases the radon concentration, but, on the other hand, do have the advantage of not involving the energy-consuming side effect of introducing cold outdoor air.

The maximum effect of both processes seems under practical conditions to be a reduction in the PAEC to about 30 - 40% of the value in untreated air and a reduction in the estimated dose to maybe 40 - 60%.

The advantage of these methods over ventilation is primarily a question of cost. For both methods installation is extremely simple, and, especially the ionization/field devices are also inexpensive in comparison with conventional ventilation appliances.

A possible disadvantage with using ionizers with an external field in living area might be an electrostatic charging of persons in the room. Measurements of such chargings have shown lower values than what is often encountered by walking on ordinary floor coverings.

It should be mentioned that most of the tests of the use of filtration and field-enhanced plateout have been laboratory experiments, but they have, on the other hand, (also) been performed under conditions closely resembling normal living and working environments.

## SUMMARY AND CONCLUSION

Several models have been developed which by use of mass balance equations predict the behavior and interrelations of radon and its daughters in indoor environments. The parameters of the models are the ventilation or air exchange rate, the aerosol attachment rate, the plateout rates for unattached as well as attached airborne radon daughters, the alpha recoil factor and the removal rates for control techniques where such are used.

Although the behavior of the individual daughter products is rather complex, under certain circumstances the influence of one of the parameters, such as ventilation or filtration rate, may be quite dominating resulting in a more accurately predictable situation. In most houses where radon in the soil gas is the main source, remedial measures, such as the use of barriers, aimed at preventing radon entry are to be preferred.

It should, however, be cautioned, that the long term reliability of many radon barrier techniques now being used have yet to be confirmed and verified experimentally.

Even where an effective barrier technique is properly installed the settling of a house over a period of a few years may cause new entry points to be formed. It is important to appreciate, that where conditions are favorable for radon entry (i.e., high soil permeability, an underpressure across the house foundation slab etc.) only a small entry point being available will allow significant radon flux into a house.

Such realistic considerations argue strongly in favor of having indoor air treatment systems installed as a necessary back-up to the primary radon barrier techniques. Also considerations of economic cost, technical feasibility and maybe even human acceptability may in some cases suggest that a direct treatment of the indoor is preferable. This may especially be the case where the main or essentially contributing source of indoor radon is the building materials of the house. In such situations direct removal of radon daughters from the air by filtration and electrically enhanced plateout may have advantages over traditional ventilation.

Some methods of direct treatment of the indoor air, may appear less effective in reducing the radiological doses, calculated on basis of the current lung dose models, than might be suggested by their high efficiency in reducing the overall exposure.

It should, however, be kept in mind that the lung doses are theoretical predictions based on models still under development, and it is thus prudent and common sense to assume that a reduction in the exposure will always be of beneficial nature.

## REFERENCES


1. Wilkening, M. 1987. Effect of Radon on Some Electrical Properties of Indoor Air. *In:* P. Hopke (ed.). *Radon and Its Decay Products, Occurrence, Properties and Health Effects.* ACS Symposium Series 331, p. 252-263.
2. McLaughlin, J.P. 1972. The Attachment of Radon Daughter Products to Condensation Nuclei. Proc. Roy. Irish Acad., Sect. A, 72:51.

3. Jacobi, W. 1972. Activity and Potential Alpha Energy of $^{222}$Rn and $^{220}$Rn Daughters in Different Air Atmospheres. Health Phys., 22:441.
4. Porstendorfer, J., A. Wicke and A. Schraub. 1978. The Influence of Exhalation, Ventilation and Deposition Processes Upon Radon ($^{222}$Rn) and Thoron ($^{220}$Rn) and Their Decay Products in Room Air. Health Phys., 34:465.
5. Mercer, T.T. 1976. The Effect of Particle Size on the Escape of Recoiling RaB Atoms from Particulate Surfaces. Health Phys., 31:173.
6. James, A.C. 1988. Lung Dosimetry. *In:* W.W. Nazaroff and A.V. Nero (eds.). *Radon and Its Decay Products in Indoor Air.* Wiley, New York, NY. 259.
7. Knutsson, E. 1988. Modeling Indoor Concentrations of Radon's Decay Products. *In:* W.W. Nazaroff and A.V. Nero (eds.). *Radon and Its Decay Products In Indoor Air.* Wiley, New York, NY. 161.
8. Jonassen, N. and J.P. McLaughlin. 1988. Removal of Radon and Radon Progeny from Indoor Air. *In:* W.W. Nazaroff and A.V. Nero (eds.). *Radon and Its Decay Products in Indoor Air.* Wiley, New York, NY. 435.
9. Jonassen, N. and B. Jensen. April 1987. The Effect of Filtration on Radon Daughter Atmospheres. Laboratory and Field Experiments. Proc. Second Int. Spec. Conf. on Indoor Radon. APCA, Cherry Hill.
10. Jonassen N., and B. Jensen. February 1987. Radon Daughter Levels in Indoor Air, Effect of Filtration and Circulation. Prog. Report XI, Lab. Appl. Physics, Tech. University of Denmark.
11. Jonassen, N., and J.P. McLaughlin. 1982. Air Filtration and Radon Daughter Levels. Environ. Int., 8:71.
12. Jonassen, N., and B. Jensen. February 1986. Radon Daughter Levels in Indoor Air, Effect of Filtration and Circulation. Prog. Report X, Lab. Appl. Physics, Tech. University of Denmark.
13. Elster, J., und H. Geitel. 1901. Über eine fernere Analogie in den elektrischen Verhalten der natürlichen und der durch Bequerelstrahlen abnorm leitend gemachten Luft, Phys. Z. 2:590.
14. Jonassen, N., and J.P. McLaughlin. 1985. The Reduction of Indoor Air Concentration of Radon Daughters Without the Use of Ventilation. Sci. Tot. Environ., 45:485.
15. Jonassen, N. 1985. Electrical Properties of Radon Daughters. *In:* H. Stocker (ed). *Occup. Rad. Safety in Mining.* Proc. Int. Conf., Vol. 2, Can. Nucl. Assoc., 561.
16. Jonassen, N., and B. Jensen. 1988. Modification of Electric Fields by Space Charges, Effects on Airborne Radon Daughters. Rad. Prot. Dos. 24: No.1/4, 497.
17. Jonassen, N., and B. Jensen. October 1988. Removal of Radon Daughters by Filtration and Electrostatic Plateout. Proc. EPA. The 1988 Symp. on Radon and Radon Red. Tech. Denver.
18. Moeller, D.W., S.N. Rudnick, and E.F. Maher. 1988. Laboratory and Field

Tests of a Hassock Fan Ion Generator Radon Decay Product Removal Unit. Rad. Prot. Dos., 24: No. 1/4, 503.

19. McLaughlin, J.P., and P. Wasiolek. 1988. Experimental Room Studies on the Reduction of Radon Progeny Concentrations Using Unipolar Ions and Electric Fields. Proc. EPA, The 1988 Symp. on Radon and Radon Red. Techn. Denver.
20. Jonassen, N., and B. Jensen. July 1989. Radon Daughter Levels in Indoor Air, Effect of Filtration and Circulation. Prog. Report XVI, Lab. Appl. Physics, Tech. University of Denmark.
21. U.S. Environmental Protection Agency. September 1987. Radon Reduction Methods. (2nd Edition). OPA-87-010.
22. Nero, A.V. April 1989. Earth, Air, Radon and Home. Phys. Today, 32.

*Environmental Radon: Occurrence, Control and Health Hazards.* Edited by S. K. Majumdar, R. F. Schmalz and E. Willard Miller 

# *Chapter Fourteen*

# WILKES UNIVERSITY RADON CHAMBER: A REVIEW OF THE RADON CHAMBER, ITS CHARACTERIZATION, AND ITS USE TO CALIBRATE DIFFUSION BARRIER CHARCOAL ADSORPTION DETECTORS

**ROGER L. MAXWELL,[1] BRIAN REDMOND[2] and JOSEPH GALLETTA[3]**

[1]Director Radon Laboratory
[2]Director Center for Environmental Quality
[3]A.S.H.R.A.E. Member, Retired

Wilkes University
Wilkes-Barre, PA 18766

## INTRODUCTION

Since May 22, 1989 Wilkes University (formerly Wilkes College) has had an operational, large controlled radon chamber. The chamber, measuring 3.87m wide by 6.22m long by 2.09m high, encloses a volume of 50.4 cubic meters. Currently the chamber is controlled in the sense that the worst case standard deviation of the measured radon concentration is 3.4% of the average radon concentration. The chamber, installed in early November, 1988, became operational with the installation of electric power. Since becoming operational, we have continued to characterize the chamber and have used it to calibrate diffusion barrier activated charcoal adsorption detectors manufactured by TEQ Corporation of Wilkes Barre, PA. This characterization and calibration have been supported by a Ben Franklin Partnership Grant.

## BACKGROUND

In 1985 Wilkes College and the Northeastern Pennsylvania Environmental Council (NEPEC) participated in a program to provide radon gas detectors to the people of northeastern Pennsylvania in order to help determine whether the indoor radon gas problem was significant in northeastern Pennsylvania. Using data obtained in this program the Earth and Environmental Science Senior Seminar classes of 1986 and 1987 attempted to correlate radon gas levels with geology and soil type.

Through the first few years of the program passive detectors, alpha track detectors, were made available. In April 1986 Wilkes College purchased a commercial working level monitor to participate in active testing of radon progeny. Later on, diffusion barrier charcoal canister detectors from the Radon Project at the University of Pittsburgh were distributed as well. In 1987 NEPEC convinced Wilkes College of the need to become a site for the manufacture and testing of charcoal canisters. Dr. B. Cohen of the University of Pittsburgh provided parts lists, including vendors, and manufacturing instructions for his diffusion barrier charcoal canisters as well as a tour of his testing laboratory at the Radon Project.

In April 1987 Wilkes College manufactured its first diffusion barrier charcoal canister detectors and established a scintillation/multichannel analyzer test system in order to evaluate those detectors for exposure to radon gas. For this evaluation the test system/detector combination needed to be calibrated. At this time detectors were calibrated by placing them in a region of known radon concentration for various exposure times and then testing the detectors in the scintillation/multichannel analyzer test system to determine the detector calibration constant. The problem with this process was that the "region of known concentration" was a room in the basement of Stark Learning Center at Wilkes College, that had tested high in radon gas, and although its radon gas concentration was continuously monitored, the values obtained were not constant but varied with normal variations that one sees in indoor radon gas measurements, i.e. diurnal and weather front variations. Typical maximum, minimum, and mean concentration values from the 1/89 calibration were 34 pCi/l, 13 pCi/l, and 23 pCi/l. These radon gas concentrations were measured with a commercial passive type scintillation cell/photomultiplier tube counting system.

To test the calibration some Wilkes College charcoal canisters and some purchased from the University of Pittsburgh were exposed in the same regions for the same times and then evaluated. The test results were similar. This calibration was therefore considered effective, indeed, effective enough for Wilkes College to pass the Environmental Protection Agency's Radon Measurement Proficiency Program, Round Four in May 1987 and again in Round Five in July 1988.

Calibrations of the detector/test system combination are performed every six months, or whenever the source of charcoal canisters changes a calibration is performed using the new detectors. But each calibration involved taking averages of the changing concentration over the exposure periods. To overcome this problem of concentration variation the construction of a controlled radon chamber was planned. While there were no specifically formulated plans regarding the chamber, the original idea was to build a small chamber, perhaps similar to that described by Cohen and Nason (Cohen B.L. and Nason R., "Radon Chamber for Calibration of Charcoal Detectors," Health Phys. 51:135-137, 1986).

The problem of developing a plan to construct a controlled radon chamber was given to several seniors in the School of Engineering and Physical Sciences as their senior project. In November 1987 in the course of working on that project, the seniors and one of the authors (R.L.M.) visited Andreas George and the Department of Energy's radon chamber located in the Environmental Measurements Laboratory in New York City. George described the construction and operation of his chamber. His advice to the Wilkes College seniors was to build as large a chamber as possible, since there will always be a need for large chambers where many pieces of test equipment could be tested at one time, and a large chamber would be less subject to radon gas concentration variations caused by entry to and exit from the chamber as well as by charcoal canister exposures in the chamber. In their report the seniors concluded "that a large walk in type chamber is the best choice for Wilkes" and presented preliminary plans for construction and location of a walk-in type radon chamber and considered sources, control features, and test equipment for the chamber. While their first plan called for the construction of a wooden walled room, one of the seniors investigated the cost of a commercial walk-in cooler as the basic unit. It was determined that this could be a viable alternative. Negotiations between the School of Engineering and Physical Sciences of Wilkes College and Penn Refrigeration Service Corporation of Wilkes Barre, PA produced a favorable quotation for a 50 cubic meter insulated room.

## RADON CHAMBER

Figure 1 presents a schematic view of the main chamber and the antechamber showing the window, the doors, the exit port, the positions of the source, fan, radon detector (PRD), and selected test points (A through E) in the chamber, as well as the power outlet boxes. The shelf is a four level (28.5cm, 74.5cm, 120cm, and 166cm) moveable wire shelf which allows the passage of gas in all directions. A Pylon Rn-1025, Flow Through Radon Gas Source, rated at an equilibrium activity of 96.0 kBq and as having continuously available 12,080 mBq of $Rn^{222}$ is located on the third shelf up from the floor, while a MSA Flow Life Portable Pump, whose exhaust, at one liter per minute, flows through the source, is located on the top shelf. (This latter is called source fan later in this report). A Pylon PRD-1, Passive Radon Detector, attached to a Pylon AB-5, Portable Radiation Monitor, is located on the second shelf up from the floor. A second Pylon AB-5, Portable Radiation Monitor, with Pylon LCA-2, Lucas Cell Adaptor, and Pylon Type 300 Scintillation Cell is attached to the chamber via the exit port and is used as a continuous radon gas monitor. A fixed length of Tygon tubing, connected at one end to the exit port via a filter assembly, has its open end connected to a movable stand that can be located anywhere in the chamber. The fan is a BF Model C-4, normally used to cool computers and/or electronics circuits, having a capacity of approximately 40 cfm.

Except for the window panel, the chamber was constructed of standard size prefabricated 4 inch thick urethane insulated, embossed aluminum finished panels. The panels were sealed with Dow Corning 999-A, a silicon caulk, along their tongue and groove joint during construction. Possible leak paths connecting the outside to the interior of the chamber 1) around the exit port, 2) around the window, and 3) around and inside of the electric conduits were caulked with Sikaflex MultiCaulk, 50 year polyurethane. The chamber was constructed in SLC 46, a basement room of Stark Learning Center. Stark Learning Center is the main academic building at Wilkes College.

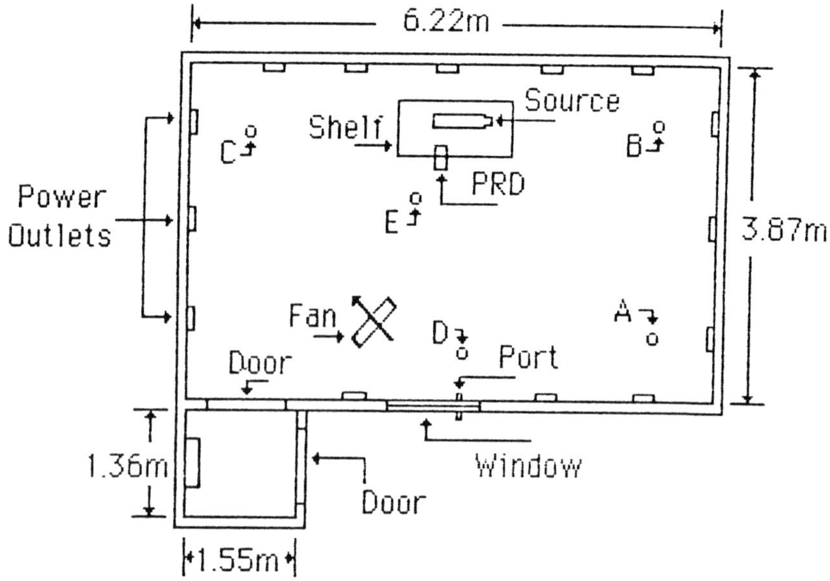


FIGURE 1. Sketch of Radon Chamber.

## CHAMBER CHARACTERIZATION

Immediately after construction and before any new holes were made in the chamber the first two experiments performed were efforts to determine how leaky the chamber was. In the first experiment a Pylon PRD-1, Passive Radon Detector, attached to a Pylon AB-5, Portable Radiation Monitor, (the combination herein referred to as PRD) was placed in the opened chamber and the normal room background for that area of the basement of Stark Learning Center was monitored. The diurnal variation in the radon concentration was clearly evident. The range of the readings was from about .6 pCi/l to 7.5 pCi/l/l. When the chamber doors were closed the radon concentration in the chamber was seen to decay with a half life measured to be 4.86 days. Since the half life of radon-222 is about 3.82 days, radon gas was leaking into the chamber.

The second experiment consisted of discharging the 96 kBq secular-equilibriated source into the chamber and again measuring the decay time of the gas in the chamber. Using the conversion factor 1 Bq = 27.0 pCi, one finds that 96 kBq of Rn gas is equivalent to $2.6 \times 10^{-6}$ Ci. Since this gas is released in a volume of $5 \times 10^{4}$ liters, the concentration expected is $(2.6 \times 10^{-6})/(5 \times 10^{4}) = 52$ pCi/l. The radon gas concentration was again monitored using the PRD. First the closed chamber was monitored and then the source gas was blown out of the source in the chamber using the MSA pump at one liter per minute. The measured radon concentration peaked at 54.6 pCi/l and was determined to decay with a half life of 3.52 days. This indicated that radon gas was leaking out of the chamber. A working level monitor in SLC 46 just exterior to the chamber door showed no increased in readings during this test. The Sikaflex caulking around the window and in the original electrical conduit was done after these tests.

Major design features yet to be decided were what kind of source to use, how "strong" should it be, where to place it, and how to bring the radon gas into the room. Review of other chambers regarding these matters, as well as other operating characteristics, through personal visits at the EPA facility in Montgomery, Alabama (in connection with the RMP program) and the DOE Environmental Measurements Laboratory in New York City and through articles about the DOE facility in Grand Junction, Colorado (Kuennen R. W. and Roth R.C., "Reduction of Radon Working Level by a Room Air Cleaner," American Waste Management Association, 82nd Annual Meeting and Exposition, Anaheim, CA, Paper Number 89-79.6, June 1989) and the Elliot Lake Laboratory in Ontario, Canada (Bigu J., "A Walk-In Radon/Thoron Test Facility," Am. Ind. Hyg. Assoc. J. 45(8): 424-532, 1984) showed that for chambers ranging from 20 cubic meters to 44 cubic meters, source strengths varied from 3.4 microCuries to one MilliCurie. Thus other factors besides room size were important in determining source strength. Since Wilkes had only one controllable source, the Pylon Rn-1025, Flow Through Radon Gas Source, rated at 96kBq (2.6 microCuries) the third experiment performed was to determine the equilibrium concentration obtainable with this source. Assuming no leaks the chamber concentration would be given by

$$dC/dt = -\lambda C + S \tag{1}$$

where C is the concentration, S the source input rate, and $\lambda$ is the decay constant. In equilibrium, $dC/dt = 0$, so that

$$C_0 = S/\lambda. \tag{2}$$

With $S = (12{,}080 \times 10^{-3}$ Bq/m)/(50,000 liters) = .391 pCi/liter/hr and $\lambda = (\ln 2)/(\tau) = .00755$ (Hours)$^{-1}$, where $\tau = 3.825$ days = 91.8 hours, then $C_0 = (.391)/(.00755)$ pCi/l = 51.8 pCi/l.

Initial PRD tests at the end of March, with no power in the chamber to operate the mixing or source flow fans, gave a five day average concentration of 45.8 pCi/l with a standard deviation of 1.5 pCi/l. Later, in full operation, the continuous radon monitor consistently indicates concentrations of about 49 pCi/l. Measured concentrations with standard deviations for the mixing fan and the source fan on and the source full open are given in Table 1. The PRD is located on shelf three as indicated above. The position of the Tygon tubing input to the scintillation cell is location E of Figure 1.

These figures indicate about a 6% loss of radioactive material from the chamber. This corresponds to about 326 pCi/minute into a room (SLC 46) at least 4 times the volume of the chamber, i.e. 200,000 liters, for an increase in radon concentration of $1.6 \times 10^{-3}$ pCi/l/minute or about 2.3 pCi/l/day if that room were sealed. The room isn't sealed and we do not measure any increase in the radon concentration outside the chamber. We plan, however, to provide fresh air intake to SLC 46 and an emergency exhaust vent to the roof of Stark Learning Center.

The antechamber was added to the original design to minimize chamber concentration changes due to entrance to and exit from the chamber. The PRD

TABLE 1

*Maximum Chamber Radon Concentrations Measured with Pylon PRD and with Pylon Type 300 Scintillation Cell (Serial Number 238). [Concentrations are in pCi/l. Standard Deviations are indicated in parentheses.]*

| PRD Concentration | SN238 Concentration | Time Interval |
|---|---|---|
| 48.2(1.5) | 49.9(1.2) | 7/21 - 7/24 |
| | 48.9(1.5) | 8.4 - 8/9 |
| 47.0(0.8) | 48.5(1.0) | 8/9 - 8/11 |
| 47.5(1.0) | 48.9(1.4) | 8/11 - 8/15 |

was placed in the antechamber in order to try to determine more about radon leaks from the chamber, since a possible leak position is through the door seal. This was done after the wiring installation. Preliminary tests with a decay half life of 4.4 days indicate that radon is leaking into the antechamber from the chamber. We did see discrete jumps in the concentration levels in the antechamber as measured by the PRD when the external door was opened and closed. We do not see jumps in the chamber concentration levels, again as measured by the PRD, when the door between the chamber and the antechamber is opened and closed. Thus we believe the antechamber to be an effective addition to the design of a large chamber for the stabilization of the radon concentration levels.

After becoming operational and after the placement of the wire shelf and the fan, radon concentration measurements made with the PRD were compared with those made the continuous radon monitor using scintillation cell, serial number 238. These comparisons proved the identity of the measured concentration values when the Tygon tubing input to the scintillation cell was at the position of the PRD, i.e. location E of Figure 1. To measure the uniformity of the chamber concentration the Tygon tubing input to the scintillation cell was moved to the other locations indicated in Figure 1 and it was also move to the height corresponding to the lowest shelf. The measured concentrations with the PRD and the scintillation cell were again compared and were again identical. All of these comparisons are shown in Table 2.

The last of these initial characterizations was to determine if the concentration could be easily controlled by removing the source fan and by partially closing

TABLE 2

*Chamber Radon Concentrations Measured with Pylon PRD and with Pylon Type 300 Scintillation Cell (Serial Number 238). [Concentrations are in pCi/l. Standard Deviations are indicated in parentheses. The input location of the tubing to SN238 is listed under Location. Low means that the height of the tubing input was the position of the lowest shelf. No indication means that the tubing input was at the position of the PRD input.]*

| PRD Conc | SN238 Conc | Location | Time Interval |
|---|---|---|---|
| 46.2(1.2) | 46.9(1.4) | E | 6/5 - 6/6 |
| 45.9(1.5) | 46.5(1.1) | E | 6/6 - 6/7 |
| 45.1(1.2) | 45.6(1.3) | A | 6/7 - 6/8 |
| 44.8(1.5) | 45.7(1.3) | B | 6/8 - 6/9 |
| 43.7(1.3) | 44.6(1.4) | C | 6/9 - 6/10 |
| 43.8(1.2) | 44.4(1.2) | D | 6/10 - 6/11 |
| 43.1(1.3) | 44.4(1.3) | E, Low | 6/11 - 6/12 |
| 43.3(1.1) | 44.4(1.3) | A, Low | 6/12 - 6/13 |
| 43.7(1.2) | 45.1(1.5) | A, Low | 6/13 - 6/14 |
| 43.9(1.5) | 44.9(1.0) | A, Low | 6/14 - 6/15 |
| 44.3(1.4) | 45.4(1.2) | A, Low | 6/15 - 6/16 |
| 44.3(1.0) | 45.3(1.2) | A, Low | 6/16 - 6/17 |
| 44.7(1.4) | 46.1(1.4) | C, Low | 6/17 - 6/19 |

the output valve of the source. The input valve was closed completely. The Tygon tubing input to the scintillation cell was relocated at position E. With the mixing fan on, the source fan off, and the source output open two turns, the scintillation cell measured concentration varied monotonically from its initial value at 47.8 pCi/l to 41.8 pCi/l and then stayed low with 42.5 pCi/l as it mean value. With the same fan conditions, but with the source output open one turn, the mean value was measured to be 43.3 pCi/l. Thus changing the opening to a lower value did not lower the concentration further and other effects must dominate in determining the concentration. Closing the source input and removing the source fan flow did lower the concentration by about 15%.

Several working level measurements were made in the chamber with and without the mixing fan to determine the range of the working level. With the fan on the average working level for a three day measurement period was .032 W.L. plus or minus 0.65%. During this interval the PRD reading was 48.2 plus or minus 1.5 pCi/l and the Scintillation cell reading was 49.6 plus or minus 1.2 pCi/l. With the mixing fan off the average working level for a three day measurement period was .209 W.L. plus or minus 0.25%. For this interval the PRD measurement average was about 40 pCi/l.

## CHARCOAL CANISTER CALIBRATION

Before the chamber was operational a quick check was made on the calibration of the detectors that used to be manufactured by Wilkes. For this test four detectors were exposed in the chamber for seven days. They were located in the chamber adjacent to the PRD. After the chamber was operational a complete calibration of the diffusion barrier charcoal canisters now manufactured by TEQ Corporation of Wilkes Barre was performed. For this calibration thirty six detectors were exposed in the chamber, four for one day, four for two days, etc. up to four for nine days. These detectors were placed on the first and second level of the wire shelf. While these detectors were being exposed the radon concentration in the chamber was measured using the PRD. The average concentration during each exposure period was determined. These concentrations were then used to calibrate the detectors as indicated below.

The expression used at Wilkes to determine the radon concentration of the region in which the diffusion barrier charcoal detectors are exposed is

$$\text{Conc} = k(T)\,[((\text{Counts})/(\text{Interval})) - \text{Bkgdrate}]/(\exp(-\lambda t))(K)(\text{Eff}) \quad (3)$$

where

*Conc* is the concentration to be determined,

*k(T)* is a temperature correction factor which takes into account the changing adsorption of radon gas with temperature, T,

*Counts* is the gross gamma counts for a region of interest in a multichannel analyzer/NaI(Tl)/photomultiplier gamma spectrum that includes the 242keV, 295keV, and 352keV peaks from Pb-214 and the 609 keV peak from Bi-214,

*Interval* is the test time interval of counting (minutes),

*Bkgdrate* is background counting rate in counts per minute (CPM) of the same region of interest measured by the same test system as in counts (above),

$\lambda$ is the decay constant of Radon-222,

*t* is the time interval between the close of the detector after exposure and the midtime of the counting interval,

*K* is the factor that relates the activity of the detector at the time of closing to the concentration of the region where the detector was exposed, i.e. (activity) = K(Concentration) and is measured in units disintegrations per minute per picoCurie per liter (dpm/(pCi/l), and

*Eff* is is the counting efficiency of the counting system.

The factor K depends on the exposure time and can be represented by an expression

$$K = K_0 \, [1 - \exp(-T/\tau)] \tag{4}$$

where

*T* is the exposure time for the detector,

$\tau$ is the integration time constant of the detector,

$K_0$ is the infinite exposure time constant that relates region concentration and detector activity. Substituting Equation 4 in Equation 3 gives an expression equivalent to the algorithm presented by Strom (Strom D.J., Radiation Health Department, Graduate School of Public Health, University of Pittsburgh, private communication, 1988).

As noted by Strom this algorithm is not the same as that used by EPA in analyzing the open type charcoal canisters (Gray D.J. and Windham S.T., "EERF Standard Operating Procedures for Rn-222 Measurements Using Charcoal Canisters," EPA 520/5-87-005, June 1987). The main difference between the two calculation methods is in the timing of the decay term. The timing used in the decay factor in the EPA treatment starts at the midtime of the exposure interval and continues to the beginning of the test interval. Again the timing above is from the end of the exposure interval to the midpoint of the test interval. (Another apparent difference is not real. The calibration factor, CF, times the exposure time, Ts, in the EPA method corresponds to our K. Thus again K depends on exposure time and if the K is determined for the same exposure time as that of the detectors in whose analysis it is used, the exposure time will cancel out of the EPA equations.) During a previous calibration we determined

K using both timing methods and then calculated the exposure concentrations of a series of diffusion barrier charcoal canisters using both algorithms. The greatest percentage difference in the results was less that 2.4%.

To calibrate the counting system one must determine the counting efficiency using a calibrated source with the same geometry as the diffusion barrier charcoal adsorption detector to be tested. To calibrate the diffusion barrier charcoal adsorption detectors one must determine K, i.e. $K_0$ and $\tau$.

In the quick check calibration mentioned first above K is the seven day value of Equation 4, i.e. just one number. Equation 3 was solved for K and once the average measured concentration was used for"Conc' all the factors were known and K could be determined. It was determined to be 51.9 [(dpm/(pCi/l)]. The old calibration for those detectors was k(7day) = 51 [dpm/(pCi/l)].

For the full calibration of currently manufactured detectors the graph of K vs Time given in Figure 2 represents the calculated K points and the least squares fit for Equation 4. The least square values for $K_0$ and $\tau$ are 49.2 dpm/(pCi/l) and 3.0 days. The value of $\lambda$ determined by Cohen and Gromicko (Cohen B.L. and Gromicko N., "Adequacy of Time Averaging with Diffusion Barrier Charcoal Adsorption Collectors for $^{222}$Rn Measurements in Homes," Health Phys. 54:195-202, 1988) for their detectors was 3.5 days. The percent change in the detector calibration constant K from the last calibration is about −5%

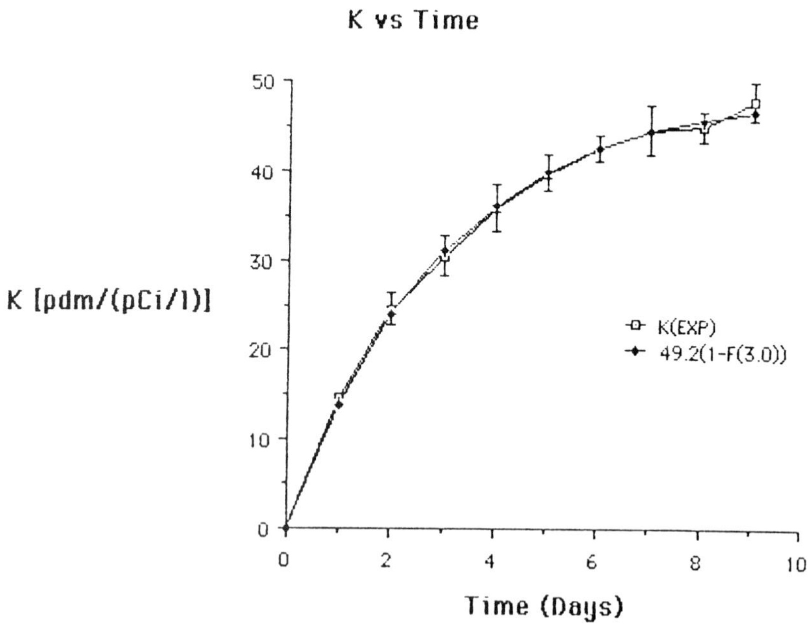


FIGURE 2. K vs Time for June 1989 Calibration of TEQ Charcoal Canisters.

for seven day exposure detectors, about 0% for three day exposure detectors, and +2% for two day exposure detectors.

## MEASUREMENTS RELIABILITY

We have tried to calibrate our test equipment with radiation standards traceable to those of the National Institute of Standards and Technology (NIST).

In order to calibrate our continuous radon monitor's scintillation cells, we use a Pylon RN-150 Calibrated Radon Gas Source. Using the manufacturer's instructions a known quantity of radon gas plus or minus 4% is dispensed to the scintillation cells. With moderate care the percentage error in the efficiency should be less that 5%. With the same moderate care this should mean a scintillation cell measured concentration with a percentage error of less than 6% neglecting any humidity or temperate corrections.

The source used to determine the efficiency of the NaI(Tl)/photomultiplier/MCA counting system is a Nucleus RN-226-SC, Radon Standard, with the same source geometry as a diffusion barrier charcoal canister. The source accuracy is 5%. Thus the counting system efficiency could be determined to less than 6% with moderate care. This percentage error and that of the scintillation cell measured concentration would give K values with percentage errors on the order of 8.5% or less and charcoal canister measured concentrations with percentage errors on the order of 10% or less.

We have also obtained a Pylon Rn-190-47, 47 mm Radon Daughter Standard, to determine the efficiency of Pylon working level counting system.

## FUTURE PLANS

Our original plans were for a temperature-, humidity-, particulate-, and radon gas concentration-controlled chamber. While we have a constant radon gas concentration chamber, we really can not set the radon gas concentration to any value below our maximum at this time. We need to work out a method to do this. Perhaps a manifold system that feeds part of the source emanation to the chamber and part to an outside vent will work. For humidity and temperature control we plan to install a low profile evaporator coil with a disconnectable propeller type fan, controlled by a variable speed switch, inside the chamber and connect it to an air cooled condensing unit outside the chamber. Dehumidification of the chamber can be accomplished with the cooling coil. Humidification can be accomplished by allowing the moisture removed by the cooling coil to accumulate in and evaporate from a pan in the room. A thermostatically controlled electrical baseboard heater in the room will introduce heat when needed. One idea directed toward particulate control is not to continuously exchange the air with an outside conditioner but to allow us to test under various conditions by accomplishing the air conditioning inside the chamber.

We will have to make more penetrations through the chamber walls to do the above conditioning. We propose to create a positive pressure inside the chamber and then to directly measure the pressure changes inside the chamber to determine the leak rate. Then the leak rate after a penetration can be compared to that before to assure proper sealing around the penetration. Or the leak rate at any time can be compared to previous values to assure the continued integrity of the chamber.

Future use of the chamber will obviously include 1) continued characterization of methods to achieve the desired control, 2) continued use of the chamber to calibrate charcoal canisters, 3) continued use of the chamber to calibrate radon gas and radon gas progeny concentration measuring equipment, and 4) research involving materials and methods of affecting and reducing radon and radon progeny concentrations.

## REFERENCES

Cohen B. L. and Nason R., Health Phys. *51*:135 (1986).

Kuennen R. W and Roth R. C., presented at the 82nd Annual Meeting and Exposition of the American Waste Management Association, Anaheim, CA., Paper Number 89-79.6, (June 1989).

Bigu J., Am. Ind. Hyg. Assoc. J. *45(8)* 424 (1984).

Strom D. J. Radiation Health Department, Graduate School of Public Health, University of Pittsburgh, (private communication).

Gray D. J. and Windham S. T., EPA 520/5-87-005 (June 1987).

Cohen B. L. and Gromicko N., Health Phys. *54*:195 (1988).

*Environmental Radon: Occurrence, Control and Health Hazards.* Edited by S. K. Majumdar, R. F. Schmalz and E. Willard Miller © 1990, The Pennsylvania Academy of Science.

# *Chapter Fifteen*

# PRELIMINARY RADON PROGENY MEASUREMENTS IN THREE FEDERAL OFFICE BUILDINGS

**SAMUEL SILBERSTEIN and RICHARD A. GROT**

U.S. Department of Commerce
National Institute of Standards and Technology
National Engineering Laboratory
Center for Building Technology
Building Environment Division
Gaithersburg, MD 20899

## ABSTRACT

The National Institute of Standards and Technology (NIST) conducted preliminary radon progeny measurements of three Federal buildings as part of an indoor air quality study. Progeny levels in two buildings in the Washington metropolitan area were extremely low compared to levels commonly found in houses. The progeny levels of a Portland building were closer to typical house levels. These results suggest that measurements of additional buildings would be promising in finding construction and operating methods that could be applied to lower radon progeny levels of residences.

## INTRODUCTION

There have been extensive surveys of radon ($^{222}$Rn) or radon progeny (the short-lived $^{222}$Rn decay products $^{218}$Po, $^{214}$Pb, $^{214}$Bi, and $^{214}$Po) concentrations in residences, and to a lesser extent in schools.[1,2,3,4,5] (Only three nationwide surveys of houses are cited; there are too many regional surveys to cite.) However, there are few measurements of either radon or radon progeny levels in commercial and office buildings.[6,7] Occupants may spend 40 hours per week in these types of buildings, a sizeable fraction of the total time spent indoors. Exposure to radon progeny in offices and commercial buildings, therefore, is important in estimating occupants' total exposure.

Office buildings have many characteristics that are different from residences, for example the foundations, the ratio of volume to surface in contact with the ground, and the heating, ventilating and air-conditioning (HVAC) systems. HVAC systems in commercial buildings are generally operated so that the interior of the building is slightly pressurized with respect to outdoors, and outdoor air is mixed with recirculated air. Knowledge of how these factors influence radon progeny levels may suggest ways to lower levels in residences.

The National Institute of Standards and Technology (NIST) has been conducting indoor air quality surveys of Federal office buildings for several years. As part of these surveys, radon progeny measurements were made in the Portland East Federal Building in Oregon[8] and in the U.S. Geological Survey Building in the Washington metropolitan area.[9] Radon progeny levels were also measured in the Building Research Building at NIST, which is also located in the Washington metropolitan area.

## PROCEDURES

### *Buildings*

Radon measurements were made in the Portland East Federal Building (Fig. 1) in July 1987; progeny measurements were made in November 1987. Both radon and progeny measurements were made at the U.S. Geological Survey (USGS) Building in Reston, Virginia in January 1988, but not simultaneously. Progeny measurements were made at the Building Research Building in Gaithersburg, Maryland in April 1986.

The Portland East Federal Building and USGS Building are described elsewhere in detail.[8,9] The Portland East Federal Building, completed in 1987, is an office building with seven stories above ground and four stories below ground, including a parking garage on the lowest three. The USGS Building, constructed in 1974, is an office and laboratory building with seven stories above ground and a basement. The Building Research Building, built in 1963, is an office and laboratory building with three stories above ground and a basement. The Portland East Federal and USGS Buildings have pressurized HVAC systems. The Building Research Building has radiator heating and cooling with exhaust air ventilation.

### *Radon Progeny Measurements*

A portable working level monitor was used to record hourly radon progeny concentrations in working levels (WL). Air is pumped through an 80-$\mu$m membrane filter at approximately 0.15 L/min. The radioactive disintegrations originating from the radon progeny contained on the particulates retained by the filter are detected by a silicon diffused junction alpha detector, and counted by a microcomputer, which stores hourly counts.[10] The counts are converted into progeny concentrations using a calibration factor supplied by the instrument manufacturer. Approximately yearly calibrations showed calibration factor to be stable.

At each measurement site, the working level monitor was usually placed on a table or desk and left in place for approximately 12-24 hours without being disturbed. It was sometimes placed at ground level near sump holes.

*Radon Measurements*

Radon measurements were made by exposing charcoal canisters to air for three days, and sending them to an outside laboratory for analysis.[1]

*Air Exchange Rate Measurements*

The air exchange rate of the Portland East Federal and USGS Buildings was determined by a computer-controlled, automated tracer decay method with sulfur hexafluoride as tracer gas.[9,12] The system injects $SF_6$ into the building supply fans every 2-3 hours, allows the tracer gas to mix, and then monitors the tracer concentration. Two computer-based systems were used in the Portland

FIGURE 1. The Portland East Federal Building in Oregon

East Federal Building permitting tracer gas sampling at 20 locations including offices and return-air ducts.[8] Tracer gas was sampled in thirteen return-air ducts in the USGS Building.[9] The tracer concentrations did not vary appreciably at different locations in the two buildings, showing that the air was well mixed.[8,9]

*Results*

Average radon progeny and radon measurements in the Portland East Federal Building are summarized in Figs. 2 and 3, respectively. The highest progeny levels occurred in the lowest two sub-basement levels. The measurement in the lowest level of the building, B4, was taken on the floor next to a sump hole in a small, unventilated room. The highest level in the building, occurring next to the sump hole, was 0.009 WL (Fig. 4). The maximum level in the occupied areas of the building (levels B1 and higher) was 0.006 WL on the fifth floor. Radon progeny peaks seemed to occur at different times during the day. The air exchange rate was about 1 $h^{-1}$ during the progeny measurements.[8] Over the course of the building study, the air exchange rate ranged between 0.5 and 1.5 $h^{-1}$. It was 0.02 and 0.04 $h^{-1}$ when the HVAC system was turned off.

Radon concentrations in the Portland East Federal Building measured by charcoal canisters were reported to be 1.2 pCi/L (Fig. 3). The highest and lowest floors appeared to have higher radon concentrations than the intermediate

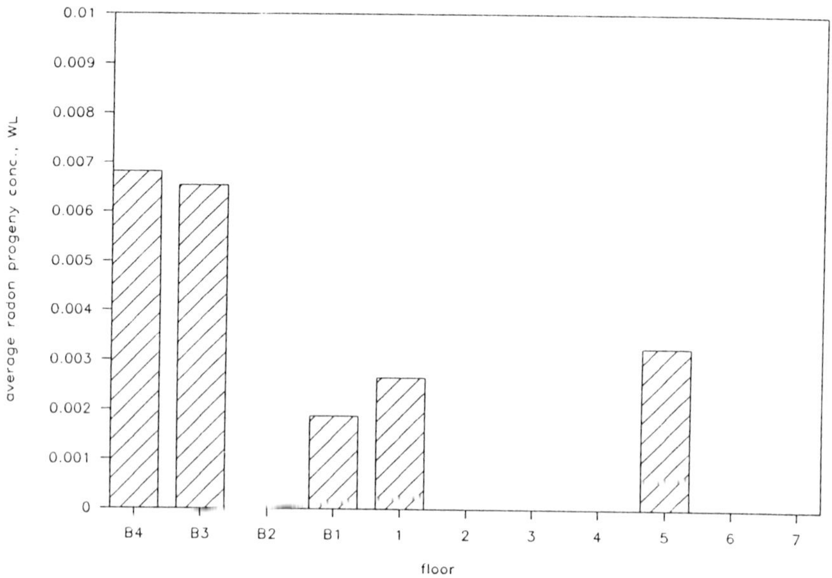


FIGURE 2. Average radon progeny measurements by floor in the Portland East Federal Building. (Measurements were taken on floors B4, B3, B1, 1, and 5.)

floors. The concentrations were 0.7-1.2 pCi/L on floors B1, 1, and 5-7, and 0.4-0.5 pCi/L on floors 2-4.

The average radon progeny concentration, 0.0007 WL, and the maximum concentration, 0.0012 WL, in the basement of the USGS Building, were nearly identical to those on the first floor, 0.0009 and 0.0013 WL, respectively. Radon concentrations, measured just before the progeny concentrations, were at or below 0.4 pCi/L, the lower detection limit of the charcoal canisters, at 15 locations from the basement to the seventh floor. The air exchange rate was approximately 1.5-2 $h^{-1}$. The average progeny concentration, 0.0004 WL, and the maximum concentration, 0.0006 WL, on the second floor of the Building Research Building, were also similar to those in the basement, 0.0005 and 0.001 WL, respectively. The second floor levels are indistinguishable from outdoor background levels; the basement levels are slightly higher.

## DISCUSSION

The radon progeny levels of all three Federal buildings studied were below the American Society of Heating, Refrigerating, and Air-Conditioning Engineers, Inc. (ASHRAE) guideline of 0.01 WL.[13] The progeny concentrations of the two Washington metropolitan area office buildings were especially low, at most twice as high as outdoor background levels. It is possible that the two

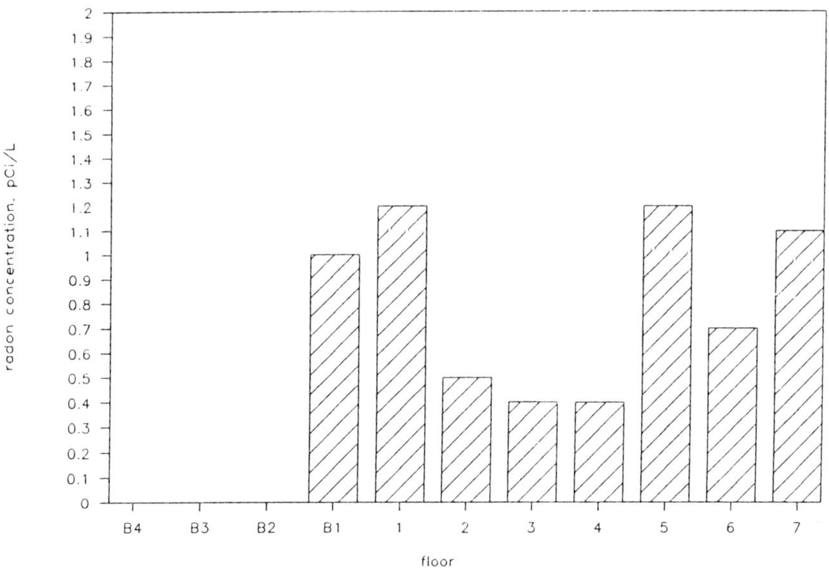


FIGURE 3. Radon measurements by floor in the Portland East Federal Building. (Measurements were taken on floors B1-7.)

Washington area Federal buildings were built on exceptionally low radon emitting sites. However, the maximum progeny concentration of the two buildings was lower than that of four Washington area schools,[5] and two dozen Washington area single family houses (unpublished results). A high ratio of volume to area of the surface in contact with the ground, or pressurization of the building by the HVAC system could also result in low progeny levels. However, the Building Research Building has approximately the same geometry as area schools, which have higher progeny levels, and is not pressurized. Likely causes of the low radon progeny concentration seem to be construction practices and high air exchange rates.

The radon levels in the Portland East Federal Building, 0.4-1.2 pCi/L, were similar to the average radon level, 0.7 pCi/L, reported by Lawrence Berkeley Laboratory (LBL),[4] for Portland commercial buildings. (LBL measured long-term average radon concentrations using a track-etch detector, not short-term concentrations as in the present study.) The radon levels of the lowest and highest floors of the Portland East Federal Building were higher than those of intermediate floors. A similar result was seen by LBL[4] in many commercial buildings in the state of Washington, but not in Oregon, where radon concentrations tended to be higher on the upper than on the lower floors. Abu-arad and Fremlin[7] also saw the same kind of pattern in a study of radon progeny concentrations of an 11- and a 17-story building in Great Britain. They explained that this pattern arose from differences in the air exchange rate of the rooms

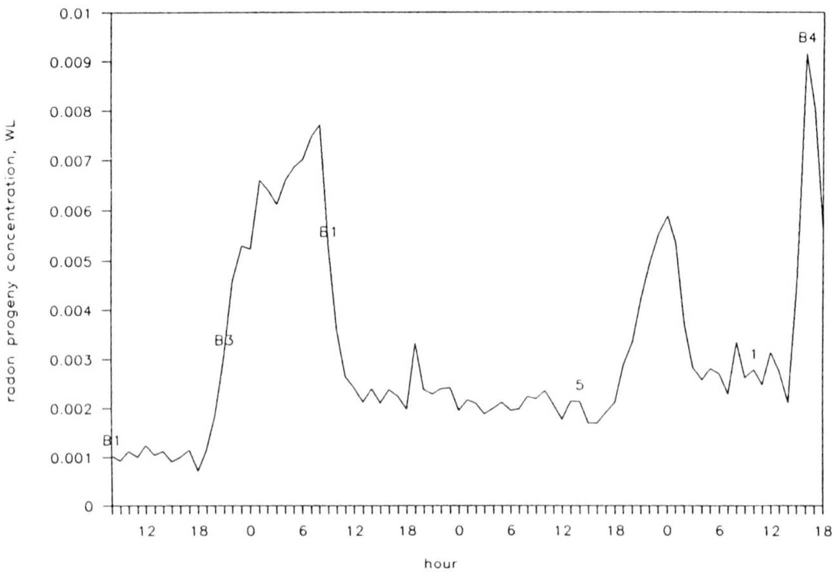


FIGURE 4. Hourly radon progeny measurements in the Portland East Federal Building. The floor numbers above the graph indicate the start of a measurement on that floor.

of the building, not their height above the ground. However, a more probable explanation for the Portland Building is that air enters the elevator shaft and stairways at the garage floors, which contain the radon sources, and exits at the upper floors. Radon is introduced into the lower occupied floors by recirculation throughout the HVAC system.

Measurement of additional government and private buildings nationwide would help determine whether low progeny concentrations generally occur in office buildings. Working level monitors giving hourly readings are desirable for surveys because they can distinguish between levels during occupancy and vacancy. Only levels during occupancy are useful in estimating exposure. Detailed study of buildings having low levels would help ascertain the responsible factors. Some of the construction or operating practices of buildings having extremely low progeny concentrations may be applicable to new residential and school construction.

## ACKNOWLEDGMENT

The project was supported by the U.S. Department of Energy, the General Services Administration, the U.S. Geological Survey, and the National Institute of Standards and Technology. The authors would like to thank Stuart Dols for taking measurements in the office buildings. We would also like to acknowledge the cooperation of the staff of the Portland East Federal Building and the USGS Building.

## REFERENCES

1. Cohen, B.L. and N. Gromicko. 1988. Variation of radon levels in U.S. homes with various factors. *JAPCA* 38:129-143.
2. Alter, H.W. and R.A. Oswald. 1987. Nationwide distribution of indoor radon measurements: a preliminary data base. *JAPCA* 37:227-231.
3. Nero, A.V., M.B. Schwehr, W.W. Nazaroff and K.L. Revzan. 1986. Distribution of airborne radon-222 concentrations in U.S. Homes. *Science* 234:992-997.
4. Turk, B.H., J.T. Brown, K. Geisling-Sobotka, D.A. Froehlich, D.T. Grimsrud, J. Harrison, J.F. Koonce, R.J. Prill and K.L. Revzan. 1987. Indoor air quality and ventilation measurements in 38 Pacific Northwest commercial buildings (2 volumes). Lawrence Berkeley Laboratory Report 22315, Berkeley, CA.
5. Peake, P.T., J.T. MacWaters, H. Chmelynski, G. Mollyn and K. Maconaughey. 1989. Radon and radon progeny measurements in five schools. U.S. Environmental Protection Agency, Washington, D.C.

6. U.S. Environmental Protection Agency. 1989. Radon measurements in schools. Interim report, Washington, D.C.
7. Abu-Jarad, F. and J.H. Fremlin. 1982. The activity of radon daughters in high-rise buildings and the influence of soil emanation. *Environ. Intl.* 8:37-43.
8. Grot, R.A. and A. Persily. 1989. Environmental evaluation of the Portland East Federal Office Building preoccupancy and early occupancy results. National Institute of Standards and Technology Interagency Report 89-4066, Gaithersburg, MD.
9. Dols, W.S., and A. Persily. Ventilation and Air Quality Investigation of the U.S. Geological Survey Building. National Institute of Standards and Technology Interagency Report 89-4126, Gaithersburg, MD.
10. Eberline. 1984. Working Level Monitor Technical Manual. Santa Fe.
11. Grot, R.A., C.M. Hunt, and D.T. Harrje. 1980. Automated air infiltration measurements in large buildings. *In: Proceedings, First Air Infiltration Centre Conference,* Bracknell, England.
12. American Society for Testing and Materials. 1987. Standard Test Method for Determining Air Leakage Rate by Tracer Dilution. Standard E741, pp. 768-779 in 1987 Annual Book of ASTM Standards, Volume 4.07, Philadelphia.
13. American Society of Heating, Refrigerating, and Air-Conditioning Engineers, Inc. 1989. Ventilation for Acceptable Indoor Air Quality. Standard 62, Atlanta.

*Environmental Radon: Occurrence, Control and Health Hazards.* Edited by S. K. Majumdar, R. F. Schmalz and E. Willard Miller 

# *Chapter Sixteen*

# RADON AND HEALTH CONSEQUENCES: AN OVERVIEW

**ANIRUDDHA GANGOPADHYAY[1,3] and SHYAMAL K. MAJUMDAR[2]**

[1]Department of Pathobiology
Auburn University
Auburn, AL 36849
and
[2]Department of Biology
Lafayette College
Easton, PA 18042

## INTRODUCTION

Radon and its daughter products have created enormous interest to the scientists all over the world because of their possible health related consequences.[1] An exceptionally high level of indoor radon concentration was detected in a home in Pottsville, Pennsylvania in 1984 that created a major concern for the residents of the United States (see Chapters 1, 2 and 3 in this book). The "acceptable" radon level is 4 pico Curies per liter of air (pCi/L), however, according to the United States Environmental Protection Agency (EPA) as many as 10% of U.S. homes may have average radon levels greater than 4 pCi/L (see Chapter 17). The distribution of radon levels in residences in Pennsylvania and adjoining states is discussed by Schnell and Monmonier in Chapter 4. Radon is one of the chain members of Uranium 238, that generates the radon isotope, Rn-222. This isotope escapes from soil and rocks and accumulates in the enclosed spaces of the atmosphere. One such space is the basement of houses.

The dose of radiation to the lung due to inhalation of radon gas is contributed by the decay of its particulate progeny which accumulate in the mucous lining of the bronchial air passages. The damage induced by radon is primarily due to the alpha particles produced by the short-lived daughters. There are three

[3]Present address: In Vitro Research and Development, Centocor Inc., 244 Great Valley Parkway, Malvern, PA 19355.

isotopes of radon that occur normally in the decay of uranium and thorium, namely $^{219}Rn$, $^{220}Rn$ and $^{222}Rn$ (see Chapters 5,6,7 and 9 in this book regarding radon properties, geology and geochemistry, occurrence of radon precursors and measurements of radon and radon decay products, respectively). Due to the short half-life of $^{219}Rn$ and $^{220}Rn$ compared to half-life of $^{222}Rn$, only the latter isotope is considered to contribute significantly to the radiation dose to the lung. Radon 222 is a noble gas that has virtually no physical or chemical reactions. It thus encounters few barriers to diffuse into our respiratory system and can reasonably be considered as a source of radiation due to alpha and beta particle emitting daughters. Although gamma rays are also generated, they contribute a small fraction to the total exposure. The possibility of the induction of cancer as a result of indoor exposure to radon gas has created increasing attention in the scientific community during the past decade. It is now widely accepted that indoor radon is the largest single source of exposure to ionizing radiation in our environment.[2] Exposure to outdoor radon concentrations is generally too low to be considered as significant risk. In view of the current concern about the health risks associated with indoor radon exposure, this chapter attempts to review the current understanding relative to radon-associated health-risk based on the studies discussed in Chapters 17 - 24 in this book.

## *Biological Effects Of Radon*

The biological effects of radon on living systems are complex. It may be noted that the presence of radon in an environment does not mean exposure to radiation, and similarly, exposure to radon does not necessarily involve adsorption of radiation. Radiation damage to a biological system may be caused by interaction with DNA molecules. In this condition, mutations may be induced by alteration in the chemical structure of the DNA (see Chapters 18 and 22 of this book for biological effects). At the cellular level, radiation may induce damage to the DNA or cytoplasm. Radiation that induces damage to the DNA molecule may contribute significantly to the development of neoplastic transformation. Irreversible damage to the DNA molecules may be lethal to the cell, whereas damage to one strand of the DNA double helix may be repaired. Nonlethal mutations due to radiation may contribute to the development of malignancy, since damage to the genetic material is propagated.

The mechanism of bio-injury induced by radiation involves several complex series of physiochemical events. It is initiated with the deposition of energy by radiation in the form of ionization and subsequent excitation of atoms or molecules in living organisms. This is followed either by the intermolecular or intramolecular transfer of energy to produce various chemically active and short-lived species, the free radicals. These free radicals may react among themselves or with cellular DNA and RNA of the biological system to produce alterations in biological expression.

The degree of biological damage depends on several factors, such as, dose of radiation applied, type of radiation (excitation or ionization) and type of organism involved. The energy that is adsorbed from radiation in a biological system may lead to excitation or ionization. The raising of an electron in an atom or molecule to a high energy level, without actual ejection of the electron from that atom or molecule, is called *excitation.* Radiation with sufficient energy to eject one or more orbital electrons from the atom or molecule is termed *ionization,* and the radiation is called *ionizing radiation.* Although different biological effects are produced by different types of radiation,[3,4] the cellular damage that occurs with different types of radiation has more similarities than differences. The response of organisms to radiation may vary even among closely related species due to different tolerance. Certain chemicals are known to modify the response of radiation in biological systems. Such chemicals are, for example, radioprotectors like the amino acid cysteine and radio-sensitizers like molecular oxygen.

*Experimental Studies With Animals*

For over fifty years, experimental animals have been involved in the studies to investigate effects of radon exposure on respiratory tissues and other organs. Initial studies demonstrated short-term pathological changes[5,6,7] and recommended measurements of radon gas concentrations rather than radon decay product exposures. Valuable information has been made available by more recent studies conducted primarily in two U.S. laboratories: the University of Rochester (UR), the Pacific Northwest Laboratory (PNL), and the Compagnie Generale des Matieres Nucleaires (COGEMA), a laboratory in France.[8] The UR investigators used rats and dogs in their studies and exposed them to radon and radon progeny attached to ambient aerosols. The study also included cigarette smoke, cerium hydroxide, and uranium ore dust as associated exposures. The researchers at PNL employed dogs, rats and hamsters as experimental animals and exposed them to mixtures of radon, radon progeny attached to uranium ore dust, cigarette smoke, and diesel engine exhaust to create the underground uranium mine environment. More recent PNL studies included females, younger animals, and animals bearing fetus to supplement previous studies conducted primarily on adult male animals in order to provide a comprehensive understanding relative to underground miner data to population exposures.

The UR studies documented extensive pulmonary injury and subsequent shortening of life span in experimental animals due to inhalation of radon and radon progeny. However, it did not seem to impact significantly on the incidence of lung cancer. The reports from PNL and COGEMA studies show remarkable similarity relative to pathological changes. These findings and other relevant investigations are carefully addressed in Chapters 18 and 24 of this book.

Teratologic data showed hazardous effects on rat fetus due to 10,000 times typical annual level exposure of radon progeny compared to indoor radon exposure (Chapter 18). It is therefore inconceivable that indoor radon-exposure levels will produce these effects in humans. There are several factors that appear to influence the tumorigenic potential of radon progeny in studies with experimental animals. These factors are: cumulative exposure, exposure rate, unattached fraction, disequilibrium, and concomitant exposures to cigarette smoke.[8] It appears that exposure to radon progeny and uranium ore dust, diesel-engine exhaust, or cigarette smoke increases the risk of premalignant lesions, but with the exception of cigarette smoke, it does not appear to impact on the incidence of radon-induced pulmonary tumors.

### *Studies With Humans*

Previous studies on the effects of radon and its daughter products were not adequately controlled and it was difficult to accurately measure the rates of exposure, the existence of cofactors, and compare the results of other studies. In recent years, however, several carefully controlled studies have been published[9,10,11,12] that suggest an increased occurrence rate of lung cancer because of radon exposure above levels of about 300 working level month (WLM). Recently, the Biological Effects of Ionizing Radiation Committee (BEIR-IV) demonstrated that the higher risk of lung cancer correlated directly with cumulative exposure. The greatest single identifiable risk factor for radon-induced lung cancer appears to be tobacco use.[13] There exist reasonable differences in the populations exposed to radiation. The miners, for example, are almost exclusively adult males receiving their exposures at a younger age than the population in a typical domestic environment. Breathing patterns are different in miners due to extensive exertion which certainly contributes to the distribution of dust and aerosols in the lung. In general, miners are more exposed compared to other populations. The exposure of the miners can be inferred from Equilibrium Equivalent Radon Concentration, $EEC_{Rn}$, which is dependent on the relative concentration of the daughter products. The accurate measurement of daughter concentration is difficult as it varies with time.

Recent analyses of the underground miner data have modeled age-specific lung cancer death rates as a function of total exposure. The factors that modify the effects are exposure rate, age at risk, and time since exposure have also been evaluated. Studies with humans suggest that health-related consequences are inversely related to exposure rate, and relative excess risks increase with increasing age at first exposure, whereas the relative excess risk declined with time since exposure. Similar experimental procedures have been used to analyze data on radon-exposed rats in studies conducted at COGEMA and PNL.[14]

Dose of radiation is dependent on the deposition of particles, and the time of their clearance. Clearance of radioactive particles from the lung before it decays should not contribute to the dose of radiation. It has been reported[15] that in estimating the dose of basal cells in the bronchi (of lung), the unattached fraction of the total alpha particle energy determines the dose conversion factor (dose per unit exposure to radon progeny) independently of the factor of equilibrium. Rate of breathing and lung dimensions are considered as major factors influencing the dose-dependence of the lung. Doses will be higher in the newborn, primarily because of the smaller dimensions of the bronchi. (See Chapters 18-20 and 22-24 for effects of radon and their daughter products on general health, on the lung, and lung cancer mortality).

## DISCUSSION

At this point it may be considered that radon appears to be a major radiation hazard and may contribute to a significant fraction of pulmonary mortality. The EPA suggests that inhalation of indoor radon daughter products may lead to 5000 to 20,000 lung cancer mortality per year alone in the United States.[16] Recent human epidemiological studies have provided the basis for assessment of risks associated with radon exposure, while a better understanding of the nature of exposure-response relationships has been made available by studies with animals and domestic determinations. Surveys (Ne 36, Co 86, Chapter 20) of homes in the United States indicate an average radon concentration of about 55 $Bq/m^3$ (1.5 pCi/L). It appears that exposure to radon and its decay products adds to the total dose of radiation to the bronchial epithelium that goes beyond the total contribution from all other natural sources. Further efforts to define the relationships among various factors contributing to radiation dose and also to improve the procedure of measuring radon in the air would certainly contribute to the better management of radon-related health consequences where a population is at risk. We are still uncertain about the amount of radiation received by underground miners and other populations. It will be appropriate to include surrogate animals to evaluate components of radon-exposure. Studies with experimental animals have provided valuable information relative to the dosimetry of radon daughters and the incidence of disease due to exposure.

It is estimated from the mortality of miners that in the United States, indoor radon exposure causes a substantial amount of lung cancer.[17] However, there is no confirmative evidence available to establish the validity of a linear extrapolation from miners' data to the doses and dose rates of typical homes. The relative impact of radon and radon daughter products on non-smokers and smokers is not fully understood. It is also not apparent how the risk varies with age at exposure or how the risk from past exposures increases with age. Existing information does not clearly prove that typical indoor exposure rates produce more cancer. Knowledge gained from human epidemiological and animal studies is certainly important enough to cause concern about health-related consequences due to radon exposure. At the same time, existing information may not be adequate enough to establish radon-related consequences with certainty. Additional information is needed from residential epidemiological studies, and from *in vitro* and *in vivo* experimental studies involving surrogate animals.

## REFERENCES


1. Schmalz, R.F. 1989. Radon and health. J. PA. Acad. Sci. 63(3):189-190.
2. National Council on Radiation Protection and Measures (NCRP). 1987. Ionizing radiation exposure of the population of the United States, NCRP Report No. 98, National Council on Radiation Protection and Measurements, Bethesda, MD.
3. Ord, M.J. 1973. Radiation studies. *In:* The Biology of Amoeba. Academic Press, New York, NY. pp. 401-422.
4. Gangopadhyay, A., and S. Chatterjee. 1987. Effects of industrial chemicals and radioactive materials in biological systems. *In:* Environmental Consequences of Energy Production: Problems and Prospects. Eds. S.K. Majumdar, F.J. Brenner, and E.W. Miller. Pennsylvania Academy of Science, Easton, PA. pp. 445-463.
5. Jansen, H. and P. Schultzer. 1926. Experimental investigations into the internal radium emanation therapy. I. Emanatorium experiments with rats. Acta. Radiol. 6:631.
6. Read, J. and J.C. Mottram. 1939. The "tolerance concentration" of radon in the atmosphere. Br. J. Radiol. 12:54.
7. Jackson, M.L. 1940. The biological effects of inhaled radon, Master's Thesis. Massachusetts Institute of Technology, Cambridge, MA.
8. Cross, F.T. 1988. Radon inhalation studies in animals. DOE/ER-0396. Report prepared for the U.S. Department of Energy, Office of Health and Environmental Research, Washington, D.C. NTIS, Springfield, VA.
9. Muller, J., W.C. Wheeler, and J.F. Gentleman, et al. 1983. Study of mortality of Ontario miners, 1955-77. Pt. 1. Ontario Ministry of Labour, Ontario Workmen's Compensation Board.
10. Radford, E.P., K.G. St. Clair-Renard. 1984. Lung cancer in Swedish iron miners exposed to low doses of radon daughters. N. Engl. J. Med. 310:1485-1494.
11. Howe, G.R., R.C. Nair, and H.B. Newcombe, et al. 1986. Lung cancer mortality (1950-80) in relation to radon daughter exposure in a cohort of workers at the Eldorado Beaverlodge uranium mine. J. Natl. Cancer Inst. 77:357-362.
12. Hornung, R.W. and T.J. Meinhardt. 1986. Quantitative risk assessment of lung cancer in U.S. uranium mines. National Institute for Occupational

Safety and Health. Center for Disease Control, Cincinnati.

13. Committee on Biological Effects of Ionizing Radiation. 1988. Health risks of radon and other internally deposited alpha emitters (BEIR-IV). National Academy Press, Washington, D.C.
14. Gilbert, E.S. 1988. Lung cancer risk models from experimental animals. *In:* Proceedings of Twenty-Fourth Annual Meeting of the National Council on Radiation Protection and Measurements, Washington, D.C. March 30-31.
15. James, A.C. 1988. Lung dosimetry. *In:* Radon and its decay products in indoor air. W.W. Nazaroff and A.V. Nero, Jr. Eds. John Wiley and Sons, New York, NY.
16. A Citizen's Guide to Radon, what it is and what to do about it. 1986. Report OPA-86-004. Environmental Protection Agency, Washington, D.C.
17. Nero, A.V., Jr. 1988. Radon and its decay products in indoor air. An overview. pp. 1-53. Eds. W.W. Nazaroff and A.V. Nero, Jr. Wiley-Interscience, New York, NY. pp. 518.

*Environmental Radon: Occurrence, Control and Health Hazards.* Edited by S. K. Majumdar, R. F. Schmalz and E. Willard Miller © 1990, The Pennsylvania Academy of Science.

# *Chapter Seventeen*

# PUBLIC HEALTH IMPLICATIONS OF RADON EXPOSURES IN THE UNITED STATES

**VERNON N. HOUK, DANIEL A. HOFFMAN and CHRISTIE EHEMAN**

Center for Environmental Health and Injury Control
Centers for Disease Control
U.S. Public Health Service
Atlanta, GA 30333

## INTRODUCTION

Although the biological effects of radon exposure have been known for some time, it has only been within the last few years that widespread public health concern over the extent of radon exposure in the United States has become evident. This chapter will focus on the distribution of radon levels in the United States, the variability inherent in various radon survey methodologies, and a brief summary of the most recent estimates of health risks from radon exposures.

## ESTIMATING THE EXPOSURE TO THE UNITED STATES POPULATION

Since most rocks and soil contain uranium in varying degrees, radon is ubiquitous in nature. Normally the radon released from the soil quickly dissipates into the atmosphere. However, radon entering a dwelling from the soil underneath the structure or through other routes may concentrate to hazardous levels. Other sources of radon such as building materials and water, while important in some areas, are secondary to soil in determining indoor radon concentrations. Because soil and rock are the primary source of indoor radon, information on the soil content of uranium and radium has been used to identify areas with the potential for elevated levels of radon. Geological regions high in uranium such as the Reading Prong in Pennsylvania and New Jersey have been associated with elevated indoor radon levels.[1] However, factors such as the permeability and moisture content of the soil, localized geology, and building construction greatly affect the flux of radon into a building so that only the potential for elevated indoor radon levels can be defined.

Because indoor radon levels in a particular residence cannot be predicted, actual measurements of radon are necessary to assess the risk to an individual or to the population. A number of different techniques have been used to measure both radon and radon progeny in homes.[2] Even though the radon decay products are the source of the health risk, radon is easier and less expensive to detect so that most measurements are of radon gas. Two radon detectors, charcoal canisters and alpha track detectors, have been widely used for large scale surveys by states and federal agencies and by individuals measuring radon in their own home. Charcoal canisters consist of a container filled with 25 to 100 grams of activated charcoal which is opened to room air for a period of 2 to 5 days. The canister is then analyzed in a laboratory using a gamma detector. Alpha track detectors integrate radon levels over an extended period of time—typically one to twelve months. These devices consist of a high quality piece of plastic inside a cup equipped with a filter which allows only radon inside. Alpha particle produced by radon marks the plastic. The number of these marks or tracks per unit area can then be converted to units of radon concentration in air.

Radon measurements in buildings vary significantly with the season in which the measurement is taken, the length of time over which it is made, and the location of the measurement device within the structure. Variations between indoor radon concentrations in the summer and winter have been well documented. Typically radon levels are highest in the winter. However, the ratio between winter to summer levels is not consistent and in one survey of 70 homes in Maine ranged from 0.5 to 7 with an arithmetic mean of 1.8.[3] Even in the same homes, the ratio between winter and summer radon concentrations varied from year to year. Air exchange rates and meteorological conditions all affect the level of radon inside a particular structure on a continual basis. Radon levels have been shown to fluctuate dramatically during a 24 hour cycle, with weather changes and with the comings and goings of the occupants. In addition, the floor nearest the soil surface will usually have the highest radon levels in a structure. For example, basement concentrations are usually higher than first floor readings.[4,5] For these reasons, the longer the measurement the better the chance of obtaining a good estimate of the annual average indoor radon levels in a particular residence. With all of these variations in the average radon level in a single residence, the task of estimating the average exposure level to the U.S. population is quite challenging.

A number of indoor radon measurement surveys have been designed by researchers to estimate the levels of radon to which the public is exposed. However, the goal of determining the average exposure to the public has often been overridden by the desire to locate homes with particularly high levels so that the risk to occupants can be reduced through mitigation. Often the underlying geology of a region and the type of building construction have been key factors in targeting regions of a state or county for indoor radon surveys so that measurements are concentrated in areas with elevated levels of radon. For these reasons, the data from surveys currently complete may be biased in estimating the average indoor radon concentration.

Results from these surveys has been used to estimate the distribution of indoor radon in the U.S. The U.S. Environmental Protection Agency assisted in developing and conducting 14 state-wide surveys comprised of short-term measurements in the winter season of 1986 and 1987.[6] These surveys were designed to locate areas with the potential for high levels of indoor radon and to provide an estimate of the distribution of radon levels by state. Estimates for the geometric mean and percent of homes above 5 pci/l are given in Table 1 for the first floor results alone and for the overall survey results including basement measurements for each state. Results based on the combined data would indicate a majority of homes in some states exceed 4 pci/l. However, since these are short-term winter measurements, many taken in basements, the results probably over estimate the annual average indoor levels. Results based on first floor measurements are more likely to represent radon levels in living areas and are more consistent with other survey results. Based on first floor results, between 3 and 27% of homes exceed 4 pci/l, depending on the particular state. The geometric mean radon concentrations on the first floors ranges from 0.8 to 2.3.

An analysis of 22 data sets available in 1985 had a range of geometric means from 0.3 to 5.7 pci/l. These data sets when combined had a geometric mean of 0.9 pci/l with a geometric standard deviation of 2.8. Based on these results about 7% of U.S. Homes or 4 million single family homes have average radon levels above 4.0 pci/l.[7,8] A database of over 90,000 alpha track measurements from all 50 states was analyzed by Tech/Ops Landauer, Inc. Results from these measurements predict that 21.7% of the single family homes in the U.S. could be above 4 pci/l with a national geometric mean of 1.91 pci/l. Most of these measurements were completed by homeowners who are more likely to make measurements when they have heard that radon is likely to be a problem in their area. In fact, many measurements may be from areas which tend to be high in radon. However, a sizable number of measurements are available from many states and only 3% of the measurements were taken over a period of less than 40 days.

## VARIABILITY IN MEASUREMENT OF RADON LEVELS

While charcoal adsorption and alpha track detectors remain the most widespread and least expensive methods of measuring radon levels in the home, other methods are available. These include: 1) Electret Passive Environmental Radon Monitors (E-Perms); 2) grab samples measuring both radon and working levels; 3) continuous radon monitors; and 4) continuous working level monitors. This section will focus on the variability inherent in charcoal adsorption and alpha track measurements.

One of the most important problems facing any homeowner who wishes to measure radon levels is the natural variability of the levels in the home over time. Radon concentrations are known to fluctuate widely with changes of 100 percent between "high" and "low" levels not unusual. These variations are due to occupant activities, temperature and ventilation changes, and other external environmental changes such as wind speed and temperature variations. Considering these diurnal and long-term variations in radon levels, the principal question for the homeowner is "how well do short-term radon measurements predict the actual annual radon levels?" A recent study by Mose et al.[10] measured the degree of variability for both charcoal adsorption and alpha-track detectors between short-term and annual radon levels. The study was conducted by comparing a single two to three day charcoal measurement to the estimate of annual radon levels (estimated by using the average of four seasonal three month alpha-track measurements) in 152 homes. Table 2 shows the results for the charcoal cannisters. The estimates of uncertainty for a *single* charcoal measurement were substantial. For example, 67% of the homes yielded a charcoal-to-annual deviation of up to plus-or-minus 40%. This means that a charcoal reading of 4 pCi/l could be as low as 2.4 or as high as 5.6 pCi/l. For a single three-month alpha-track measurement, the level of uncertainty was less

TABLE 1

*Results from the EPA Indoor Radon Survey*

| | | First Floor | | Overall Results | |
|---|---|---|---|---|---|
| State | N | % homes >4 pci/l | geo mean | geo mean | % homes >4 pci/l |
| AL | 1180 | 3.0 | 0.8 | 0.9 | 6.4 |
| AZ | 1507 | 6.3 | 1.0 | 1.0 | 6.5 |
| IN | 1217 | 14.2 | 1.4 | 2.0 | 25.1 |
| KS | 2009 | 5.9 | 1.1 | 1.9 | 22.5 |
| KY | 879 | 8.0 | 1.0 | 1.4 | 17.1 |
| MA | 1659 | 14.9 | 1.0 | 1.9 | 22.7 |
| MI | 1989 | 4.1 | 0.6 | 1.5 | 11.7 |
| MN | 919 | 24.2 | 2.1 | 3.5 | 45.4 |
| MD | 1859 | 9.2 | 1.1 | 1.7 | 17.0 |
| ND | 1596 | 27.8 | 2.3 | 4.8 | 60.7 |
| RI | 376 | 7.3 | 0.9 | 1.9 | 20.6 |
| TN | 1773 | 10.3 | 1.3 | 1.5 | 15.8 |
| WI | 1191 | 6.7 | 0.9 | 2.2 | 26.6 |
| WY | 777 | 11.6 | 1.5 | 2.3 | 26.2 |

(Table 2). For example, in 60% of the homes, the level of uncertainty was on the order of 20% compared with 40% for charcoal cannisters. For both charcoal and alpha-track detectors, the deviation of a single measurement from the annual indoor radon concentration was not a function of the radon levels but rather the result of natural variations in radon, not measurement inaccuracies, as deviations of about the same magnitude occurred for both high and low radon concentrations and for all seasonal intervals.

A primary interpretation of the results of this study is that charcoal monitors are useful as a screen or sample of the radon level for a particular short-term interval. A more useful measurement would be a series of charcoal measurements taken over a 6-10 week period. The level of uncertainty of a three month alpha-track measurement is substantially less than that for the charcoal monitor. Consequently, caution is advised when using the result of a single charcoal measurement to predict the annual radon concentration in a home.

## CURRENT ESTIMATES OF LUNG CANCER RISK FROM RADON EXPOSURE

In assessing the risks from radon exposure to the general public, the Environmental Protection Agency (EPA) has employed a relative risk model, which assumes that the risk from radon is proportional to the baseline or "natural" risk of lung cancer. For example, a relative risk of two indicates that the risk is twice that of the baseline rate. An absolute risk model assumes that the risk acts in an additive fashion, that is, the number of additional cases attributed to radon exposure are added to the number of baseline cases. Based on data derived from studies of lung cancer risk from underground (both uranium and hard-rock mining), the relative risk coefficients range from 1 to 4% increase in risk per Working Level Month (WLM). Extrapolating these data to the general public and making adjustments for differences in breathing rate and, for children, lung volume, which effectively reduces the risk by about 40%, the EPA

TABLE 2

*Estimates of Uncertainty for Charcoal and Alpha-Track Radon Detectors*

| % Deviation from Annual Level | Estimates of Uncertainty for Charcoal and Alpha-Track Detectors | |
|---|---|---|
| | Charcoal Canisters | Alpha-Track |
| < +/– 10% | 16% of Homes | 30% of Homes |
| 20% | 35% | 60% |
| 30% | 53% | 75% |
| 40% | 67% | 83% |
| 50% | 75% | 90% |
| >50% | 77 - 90% | — |

TABLE 3

*Comparison of Risk Estimates for Constant Exposure to Radon Decay Products*

| Source | Lung Cancer Risk Per $10^6$ WLM |
|---|---|
| EPA | 460 |
| BEIR IV | 350 |
| ICRP 50 | 350 |
| NCRP 78 | 130 |

estimated approximately 230-920 fatal lung cancers per million WLM's.[11] Table 3 compares this estimate with those for three expert committees, the National Academy of Sciences Committee (BEIR IV),[12] the National Council on Radiation Protection and Measurement (NCRP 78),[13] and the International Commission on Radiological Protection (ICRP 50).[14] With the exception of NCRP 78, which used an absolute risk model, each of the other estimates used the relative risk model to calculate lung cancer risk. This also assumes that the risks from radon and cigarette smoking act in a submultiplicative or multiplicative fashion.

Taking the EPA risk estimate of 460 fatal lung cancer cases per $10^6$ WLM for *lifetime* radon exposure and assuming approximately 59 million single family homes occupied by about 165 million persons, with a 75% occupancy factor and a 50% equilibrium factor for radon decay products, at an average radon exposure of 0.3 WLM/year, an estimated 17,000 lung cancer deaths per year could be attributed to radon exposure. This includes *both* smokers and non-smokers. About 85% of these deaths would occur among smokers.

## CONCLUSIONS

1. Based on the results of EPA surveys since 1985, about 7% or four million single family homes in the United States have average radon levels greater than 4 pCi/l.
2. There is a great deal of uncertainty in a single charcoal canister measurement as a predictor of the annual radon concentration in a home. There is less uncertainty attached to a single three-month alpha-track measurement.
3. The estimated number of lung cancers attributed in part to radon exposure ranges between 13,000 to 17,000 lung cancer deaths per year.

## REFERENCES

1. Michel, Jacqueline. 1987. Sources, pp. 81-129. *In:* C. Richard Cothern and James E. Smith, Jr. (Ed.). *Environmental Radon.* Plenum Press, New York, NY.

2. U.S. Environmental Protection Agency. 1986. *Interim Indoor Radon and Radon Decay Product Management Protocols.* EPA 520/1-86-04. Washington, D.C.
3. Hess, C.T., R.L. Fleisher and L.G. Turner. 1985. Field and Laboratory Tests of Etched Track Detectors for Rn-222: Summer vs. Winter Variations and Tightness Effects in Maine Houses. *Health Phys.* 49:65-79.
4. Fleisher, R.L. and L.G. Turner. 1984. Indoor Radon Measurements in the New York Capitol District. *Health Phys.* 46:999-1011.
5. Gessell, Thomas F. 1983. Background Atmospheric Rn-222 Concentrations Outdoors and Indoors: A Review. *Health Phys.* 45:289-302.
6. Dziuban, Jacolyn A., S.B. White, Jane W. Bersten, and Barbara V. Alexander. In Press. Residential Radon Survey of Fourteen States. *In: Proceedings of the 1988 Symposium on Radon and Radon Reduction Technology.* NTIS, Springfield, VA.
7. Nero, A.V., M.B. Schwehr, W.W. Nazaroff, and K.L. Revzan. 1986. Distribution of Airborne Radon-222 Concentrations in U.S. Homes. *Science.* 234:992-997.
8. Nero, Anthony, B., Jr. 1988. An Overview. *In:* William W. Nazaroff and Anthony V. Nero (Ed.). *Radon and Its Decay Products in Indoor Air.* John Wiley and Sons, New York, NY.
9. Alter, Ward H. and Richard A. Oswald. In Press. Nationwide Distribution of Indoor Radon Measurements. *In: Proceedings of the 1988 Symposium on Radon and Radon Reduction Technology.* NTIS, Springfield, VA.
10. Mose, Douglas G., George W. Mushrush, and Stephen Kline. 1988. Realistic Uncertainties for Charcoal and Alpha-Track Radon Monitors. *In: Proceedings of the 1988 Symposium on Radon and Radon Reduction Technology.* NTIS, Springfield, VA.
11. U.S. Environmental Protection Agency. 1986. *Final Rule for Radon-222 Emissions from Licensed Uranium Mill Tailings: Background Information Document.* EPA 510/1-86-009, Washington, D.C.
12. National Academy of Sciences. 1988. *Health Risks of Radon and Other Internally Deposited Alpha-Emitters: BEIR IV.,* National Academy Press, Washington, D.C.
13. International Commission on Radiological Protection. 1987. *Lung Cancer Risk from Indoor Exposure to Radon Daughters.* ICRP Publication 50, Oxford, United Kingdom.
14. National Council on Radiation Protection and Measurements. 1984. *Evaluation of Occupation and Environmental Exposures to Radon and Radon Daughters in the United States.* NCRP Report No. 78, National Council on Radiation Protection and Measurements, Bethesda, MD.

*Environmental Radon: Occurrence, Control and Health Hazards.* Edited by S. K. Majumdar, R. F. Schmalz and E. Willard Miller © 1990, The Pennsylvania Academy of Science.

# *Chapter Eighteen*

# HEALTH EFFECTS AND RISKS OF RADON EXPOSURE[1]

**F.T. CROSS**

Senior Staff Scientist
Pacific Northwest Laboratory
Richland, WA 99352

## INTRODUCTION

Humans have always been exposed to radon and radon decay products. The first documented evidence of health risks associated with radon[a] exposure dates back to medical observations by Paracelsus and Agricola in the sixteenth century concerning "lung disease" in the underground miners of Central Europe.[1] It was not until the nineteenth century that lung cancer was identified as the primary cause of death in well over half of these miners. In 1944, Lorenz[2] reported the growing conviction in the United States and abroad that the radiation emitted by radon and its decay products might cause lung cancer in humans even though it was well known that miners were also exposed to other potential carcinogens in the mines. In the 1950s, the significance of exposure to radon decay products rather than inhalation of radon gas itself was established on theoretical grounds,[3] and epidemiological studies in humans and extensive experimental studies in animals were begun. Owing to this historical development, risks associated with radon exposures are based on occupationally exposed mining populations with high radon decay product exposures. Studies on nonmining populations were not begun until the late 1970s, although as early as 1956 the importance of exposure to indoor radon was pointed out by Hultquist[4] in Sweden.

[1]Work supported by the U.S. Department of Energy under Contract DE-AC06-76RLO 1820.

[a] For convenience, general reference is made to radon-222 exposures even though health effects are attributable primarily to radon-222 decay product exposures and, to a lesser extent, associated radon-220 decay product exposures.

The possibility of cancer induction as a result of indoor exposure to radon gas has attracted increasing attention in the scientific community during the past decade. This resulted from general recognition of the importance of the lung dose from indoor radon and the existence of very high levels in some houses. It is now widely recognized that indoor radon is the largest single source of exposure to ionizing radiation in the environment, accounting for 50% or more of the total average effective dose equivalent to the human population.[5] Exposures to outdoor radon concentrations are generally too low to impose a significant risk.

Scientific interest in radon has been paralleled by increased activity on the part of governmental agencies, particularly the U.S. Environmental Protection Agency (EPA) and the Department of Energy (DOE). In 1986, EPA issued two widely publicized pamphlets: *A Citizen's Guide to Radon: What It Is and What To Do About It,*[6] and *Radon Reduction Methods: A Homeowner's Guide.*[7] In the former pamphlet, EPA, in conjunction with the Centers for Disease Control (CDC), estimated that between 5000 and 20,000 lung cancer deaths per year might be attributable to residential radon exposure. These estimates account for a substantial fraction of lung cancers that are not directly attributable to cigarette smoking.

The true seriousness of the indoor radon problem is not known at this time. While the impact could conceivably be as high as the EPA/CDC upper estimate of 20,000 annual deaths, it could also be substantially lower than the lower estimate of 5000 annual deaths. Understanding the radon problem may lead to various actions that could decrease radon exposure levels. At a minimum, radon exposure mitigation and energy conservation measures, as well as building construction practices, will be influenced by our understanding of the indoor radon problem.

In view of the current widespread concern about the risks associated with indoor radon exposures, this chapter explores the evidence on risk from human epidemiology studies, particularly in reference to EPA's 4-pCi/L action level value and from animal studies. While the EPA 4-pCi/L level has no legal force and is not a standard *per se*, it is becoming a *de facto* standard as several states are considering the level in pending legislation. Although risk can also be related to radiation dose from radon exposure, this perspective on risk is not treated in this chapter.

## EVIDENCE OF RISK FROM HUMAN STUDIES

The projected risk of lung cancer associated with exposure to average concentrations of indoor radon is very much higher than that considered acceptable for other environmental pollutants. The risks of exposure to indoor radon can be estimated by extrapolating risk estimates from studies of miners, or by conducting epidemiological studies of lung cancer in the general population. Risk estimates based on miner studies are subject to uncertainties because of limitations in estimates of exposure to radon and to cigarette smoke. It is also difficult to adjust for qualitative differences between mines and home environments. The interaction of cigarette smoking with radon exposure and the uncertainties related to the choice of risk projection model and control populations are other sources of uncertainty.

There have been several reviews of retrospective studies of underground miners by various scientific organizations.[8,9,10,11,12,13] There is also a prospective study in progress on the New Mexico uranium miners.[14] While radon and cigarette-smoke exposures are more accurately determined in the latter type of study, assessment of long-term effects is delayed by a long wait for results.

Several epidemiological and ecological studies on radon are also underway in the general population. Reviews appear in *Radon Epidemiology: A Guide to the Literature,*[15] *Estimating the Risk of Lung Cancer from Inhalation of Radon Daughters Indoors: Review and Evaluation,*[16] and in a review being prepared for DOE by Neuberger (School of Medicine, University of Kansas). These studies report some evidence of a possible link between presumed radon exposure and lung cancer. However, the data are insufficient to derive quantitative risk estimates for indoor radon or to validate the risk estimates derived from miner studies. Such studies suffer from many flaws, as do the miner studies; also, they must deal with generally lower exposures and, consequently, fewer health effects. Statistical power, in many cases, is inadequate; eventual pooling of data from several studies could increase statistical power. In addition, logistical issues regarding recruitment and subject participation add to the difficulties.

A feasibility study described by Steinhausler[17] was conducted to investigate the minimum detectable risk for 10,000 nonoccupationally exposed individuals to 1 WLM $y^{-1}$ radon decay products.[b] This exposure rate is comparable to continuous exposure at the EPA 4-pCi/L radon action level. The results indicated that for a follow-up period of 40 years, the smallest detectable risk at the 95% confidence level is $4 \times 10^{-6}$ $y^{-1}$ $WLM^{-1}$. Thus, even with known exposures, the threshold for detecting excess lung cancer may be too high for adequate results in many studies. Apart from careful case-control studies in nonsmokers, many residential epidemiology studies may provide only an upper limit to the risk. On the other hand, this would be direct evidence of a risk associated with indoor radon and would remove any doubt that highly protracted exposures in

[b] Historically, inhalation exposure is defined in terms of the air concentration of radon progeny in working level (WL) units. A working level is defined as a concentration of short-lived radon progeny (through $214_{Po}$) totaling $1.3 \times 10^5$ MeV of potential *a* energy per liter of air. A working level month (WLM) is an exposure equivalent to 1 WL for 170 hours. These definitions avoid the dosimetric problems of disequilibrium of the progeny and also the question of whether the progeny are attached (or unattached) to a carrier aerosol.

homes, in contrast to mining exposures, are associated with an increased risk of lung cancer.

Residential epidemiological case-control studies of lung cancer include populations from several countries. Some are investigations of only non-smoking women. Conducting non-smoking studies, or matching subjects for smoking status, allows an improved estimate of the risk coefficient for radon.

The National Center Institute (NCI) initiated three residential radon studies—in New Jersey, Sweden, and China—in 1985 and a fourth study in Missouri in 1987.[18] Only women are included in the four studies because they are more likely to spend time at home than men, making residential measurements among women more relevant to the question of lung-cancer risk. In addition to the NCI studies, DOE has initiated epidemiological investigations of radon-exposed populations in mines and houses. These investigations are part of a much broader radon health-effects program sponsored by that agency.[19]

Some risk estimates based on miner epidemiology, modified for population exposures, are shown in Table 1 for exposure at the 4-pCi/L EPA action level.

A rounded central risk value for lifetime indoor radon exposure at 4 pCi/L is 2%, with a reasonable lower bound of 1%. No attempt was made to place the various projections on a common basis with regard to population mix and age distribution, treatment of smoking effect, time spent indoors, radon progeny disequilibrium, etc. These projected risks for a 4-pCi/L exposure may be compared with the observed lifetime lung-cancer risks for the U.S. population of 0.9% for nonsmokers and 9% for smokers.[10] The EPA estimates that radon concentrations in eight million homes in the United States may exceed 4 pCi/L.[18] Other estimated values, in percent of U.S. houses exceeding 4 pCi/L, range between about 2 and 20%;[21,22] the upper value represents about 16 million homes.

The EPA choice of a 4-pCi/L radon concentration level, therefore, has a presumed risk which cannot be rigorously defended. The "noise" in the EPA risk associated with 4 pCi/L can be roughly determined by the range in risks shown in Table 1. The highest risk value in the table is that of EPA, which exceeds

TABLE 1

*Percent Lifetime Risk to Populations for Indoor Radon Exposure at 4 pCi/L and 50% Disequilibrium*

| Percent Risk | Source |
|---|---|
| 0.9 | NCRP, 1984a[13] |
| 1.3-5.0[1] | EPA, 1986a[6] |
| 1.1-1.6[2] | ICRP, 1987[12] |
| 2.5 | NRC, 1988[10] |

[1] Puskin and Yang[20] have subsequently reduced the EPA values by about 40%, to 0.8-3.

[2] Lower and higher values are based, respectively, on the ICRP absolute and relative risk projection models.

all other estimates (including their own lower-bound estimate) by at least a factor of two. The subsequent revision of the EPA estimates by Puskin and Yang[20] came after the 4-pCi/L value was recommended by EPA. In view of the other risks given in Table 1, it would seem reasonable to conclude that, if they are equally plausible values, those of the EPA are higher by a factor of at least two. Therefore, by implication, the 4-pCi/L pivotal action-level value may also be conservative by a similar factor. Factors other than risk are, of course, required in determining an "acceptable" exposure level.

## EVIDENCE OF RISK FROM ANIMAL STUDIES

### *Introduction*

Animal studies have been conducted for more than 50 years to examine the effects of radon exposure on respiratory tissue and other organs. Initial studies emphasized early effects or short-term pathologic changes[23,24,25] and measurements of radon gas concentrations rather than radon decay product exposures. The more modern data relate health effects to radon decay product exposures as in the human studies. They originated primarily in two U.S. centers—the University of Rochester (UR) and the Pacific Northwest Laboratory (PNL)—and in the Compagnie Générale des Matières Nucléaires (COGEMA) laboratory in France.[26]

Researchers at UR employed approximately 2000 mice, 100 rats, and 80 dogs in their completed studies; exposures were to radon and radon progeny attached to room aerosols. Approximately 10,000 rats have been employed in the ongoing COGEMA studies; exposures were to radon and radon progeny attached to ambient aerosols. Associated exposures included cigarette smoke, cerium hydroxide, and uranium ore dust. The ongoing PNL studies have included approximately 800 hamsters, 6000 rats, and 100 dogs exposed to mixtures of radon, radon progeny attached primarily to uranium ore dust, cigarette smoke, and diesel engine exhaust. The PNL studies, in particular, attempted to simulate the underground uranium mine environment, but they now include radon exposures more typical of indoor air. New PNL studies include adult female and younger animals, including the fetus, to supplement previous data on adult male animals and to help extrapolate underground miner data to population exposures.

While pulmonary injury was extensive in the UR studies, the lesions did not lead to a significant incidence of lung cancer. High-level pulmonary irradiation produced acute toxicity and resulting lifespan shortening in the experiments at all three laboratories. The generally high-level UR exposures with their consequent lifespan shortening, along with early termination of experiments to study developing pathology, precluded development of exposure-response relationships at that laboratory. Except to use their dosimetry data, the UR studies are therefore not discussed further. The data from PNL and COGEMA, which show remarkable similarity, are discussed together in a general way. Laboratory-specific data can be found in the recent review of radon inhalation studies in animals by Cross.[26]

### *Biological Effects*

Preliminary teratologic data indicate no apparent hazard to the rat fetus when cumulative radon progeny exposures to the dam are about 10,000 times typical annual levels associated with indoor radon.[27] It is unlikely, therefore, that indoor radon-exposure levels will produce these effects in humans. The potential carcinogenicity from fetal exposures in unknown.

The major biological effects produced in animal inhalation exposures to radon and radon progeny are lifespan shortening, pulmonary emphysema, pulmonary fibrosis, and respiratory tract carcinoma. The damage is primarily limited to the lung, although other organ systems such as kidney and skin are implicated.[28] The significance of nonpulmonary tumors, and tumors of the neck and head airways (which are produced following exposure to high unattached radon decay products), to human exposures is unclear at this time and is the subject of current review. The incidence of excess nonpulmonary tumors is, however, low compared to that of respiratory tract carcinomas and can probably be neglected at the low radon exposure levels typical of indoor air. Pulmonary fibrosis and emphysema, like lifespan shortening, do not occur to any appreciable extent until exposure levels approach a few thousand WLM. Therefore, it is unlikely that indoor radon exposure levels will produce these effects. Respiratory tract carcinoma, on the other hand, is produced in rats at exposure levels as low as 20 WLM, levels associated with lifetime exposures to radon gas in typical (but not high-level) homes. The carcinoma data in rats, and their significance to indoor radon exposures, will be the focus of the remainder of this section. While carcinomas were produced in PNL hamster and dog radon exposures, the data are too limited for development of exposure-response relationships in these species.

### *Carcinoma Data*

Five variables appear to influence the tumorigenic potential of radon progeny in studies with experimental animals: (1) cumulative exposure, (2) exposure rate, (3) unattached fraction, (4) disequilibrium, and (5) concomitant exposures to cigarette smoke.[26] Tumorigenic potential increases with increase in WLM exposures, and increase in radon progeny unattached fraction and disequilibrium, and with decrease in radon-progeny exposure rate. The influence of cigarette smoke depends on the temporal sequence of radon progeny and cigarette-smoke exposures. Exposures to radon decay products alone are capable of producing lung carcinomas in animals.

The lifetime lung-tumor risk coefficient for both benign and malignant tumors between approximately 100- and 5000-WLM exposures ranges from about 1 to 5 x $10^{-4}$ /WLM, based on raw tumor incidence data in rats (i.e., data uncorrected for lifespan differences with respect to control animals). Based on data from COGEMA, at exposures as low as 20 WLM, the risk is about 7.5 x $10^{-4}$/WLM.[29,30] No evidence of a threshold is evident in these exposures. Comparable risk values for human exposures are shown in Table 2.

With a decrease from 500 WLM/wk (high exposure rate) to 50 WLM/wk (moderate exposure rate), radon progeny exposure rate data show an increase in tumor incidence, at identical cumulative exposure levels, of about 200 to 300%.[26] Exposure rates as low as 5 WLM/wk (a low rate, comparable to underground miner exposure rates) reveal a tapering off of the exposure rate effect. At about 300 WLM, and presumably at lower cumulative exposure levels, tumor incidence data are not significantly different, regardless of exposure rate. The implications for indoor radon, where exposure rates are as much as three orders-of-magnitude lower, are not known at this time and cannot be directly tested in a short-lived animal species such as the rat.

Radon progeny unattached fraction and disequilibrium data show respective lung-tumor risk increases of approximately 50% per WLM exposure from 2 to 10% $f_a$[c] and 30% per WLM exposure from 0.4 to 0.1F.[d] The data suggest

TABLE 2

*Summary of Lifetime Lung-Cancer Risk Coefficients per WLM Exposure of Humans*

| Risk Coefficient | Reference |
|---|---|
| 2 to 4.5 x $10^{-4}$ (a) | UNSCEAR, 1977[8] |
| 2 x $10^{-4}$ (b) | Cliff et al. 1979[30] |
| 7.3 x $10^{-4}$ (b) | NRC, 1980[9] |
| 1.5 to 4.5 x $10^{-4}$ (a) | ICRP, 1981[11] |
| $\leq$1 x $10^{-4}$ (b) | Evans et al. 1981[31] |
| 1.3 x $10^{-4}$ (b) | NCRP, 1984a[13] |
| 1.6 to 2.3 x $10^{-4}$ (b,c,d) | ICRP, 1987[12] |
| 3.5 x $10^{-4}$ (b) | NRC, 1988[10] |

(a) Values pertain to underground miner occupational exposures.
(b) Values pertain to general population environmental exposures.
(c) Lower value is based on ICRP absolute risk projection model, assuming 70 years at risk.
(d) Upper value is based on ICRP relative risk projection model, assuming 70 years at risk.

[c] $f_a$ is the percentage of $218_{Po}$, the first decay product of radon, unattached to carrier aerosols. When expressed as percentages of the radon concentration, the unattached percentages are 1.3 and 5.2%.

[d] Equilibrium factor (F) is the ratio of the nonequilibrium concentration of short-lived radon decay products in air to the equilibrium equivalent concentration.

that the small increase in risk with disequilibrium is of little practical importance in domestic (F $\cong$ 0.4) and mine (F $\cong$ 0.1) exposures. The increase in risk with unattached fraction, predicted in dosimetry modeling of these exposures, is troublesome from the standpoint of increased potential for respiratory tract tumors in the neck and head airways, data noted in the animal experiments. Unattached fractions in domestic environments can easily exceed 10% $f_a$, in contrast to the low values generally found in dusty mines. Therefore, two different household environments with identical radon decay product exposure levels but different unattached fractions probably pose different risks.

Concomitant exposures to radon progeny and uranium ore dust, diesel-engine exhaust, or cigarette smoke increase the incidence of precancerous lesions but, with the exception of cigarette smoke, do not seem to change the incidence of radon-induced lung tumors. Experimental data indicate that added cigarette-smoke exposures, however, may increase (generally when following radon exposures), decrease, or not affect (generally when preceding radon-exposures) radon carcinogenesis.[26] The data are too tentative at this stage to extrapolate to the complexities of the human data from combined exposures to radon and cigarette smoke. New studies are underway at PNL to clarify the respective roles of radon and cigarette-smoke exposures in radon carcinogenesis.

## RADON RISK MODELING WITH REFERENCE TO MAN

### *Statistical Modeling*

Data from experimental animals can contribute most effectively to understanding risks in man if they are analyzed using statistical methods that take account of time-related factors and competing risks and that are comparable to those used in analyzing epidemiological data in humans. Recent analyses of the underground miner data have modeled age-specific lung-cancer death rates as a function of total exposure. They have also evaluated the modifying effects of factors such as exposure rate, age at risk, and time since exposure.

Specific findings of the human analyses were that a linear exposure-response function fit the data reasonably well, although some analyses suggested that a sublinear function or a function that allowed for cell killing at higher exposures provided a slightly better fit. Risks were inversely related to exposure rate, and relative excess risks seemed to increase with increasing age at first exposure. Absolute risks increased with age at risk but not as rapidly as baseline risks; the relative excess risk declined with time since exposure. Most of these findings were interpreted tentatively because of limitations in sample size, exposure assessments, and information on confounding factors.

Similar methods have been used to analyze data on radon-exposed rats from experiments conducted at COGEMA and PNL.[32] Specifically, the age-specific risk was modeled as a function of cumulative exposure and other factors; a Weibull function was used to model baseline risk and the change in risks as animals age. Modifications of methods used in human studies were necessary because age-specific base-line risks in rats are not known as well as those in humans, and because lung tumors in rats are frequently incidental findings rather than the cause of death.

Specific findings from analyses of experimental animal data were that there was little indication of a decrease in the risk per WLM with increasing total exposure, even with cumulative exposures as great as 10,000 WLM. This was true whether the analysis was based on the assumption that tumors were incidental to the death of the animal or that they were fatal. In general, results indicated that risks for PNL exposures at 50 WLM/wk were higher than at 500 WLM/wk, although, as in the raw tumor-risk data, the effect was not seen at about 300 WLM total exposures. The analyses also indicated that risks increased with age, and that the data were inconsistent with the absolute risk model.

Based on the combined exposure rate data and assuming that tumors were nonfatal, the estimated linear lifetime risk coefficient in adult nonsmoking male rats was about $300 \times 10^{-6}$/WLM. If the tumors were fatal, the risk coefficient would be about half this value. For comparison with values obtained from human studies, it may be more appropriate to exclude nonmalignant adenomas from analysis of the rat data; the risk estimate is then reduced from 300 to about 250. This value may be compared with the value of $350 \times 10^{-6}$/WLM, the overall risk estimate given by the BEIR IV Committee,[10] which is intended to apply to a human population including both males and females and both smokers and nonsmokers. The value of $250 \times 10^{-6}$/WLM for rats may also be compared, perhaps more appropriately, with the BEIR IV value of $140 \times 10^{-6}$/WLM for human male nonsmokers. The value given by the NCRP model was $130 \times 10^{-6}$/WLM.[13] Thus, estimates obtained from rat studies on a per WLM basis are roughly comparable to those obtained from human studies. Estimated lung-tumor-related doses per WLM exposure are about the same in humans and in rats.[78] On a dose basis, therefore, overall lung-tumor risk coefficients in humans are also projected to approximately equal the nonsmoking rat values, whereas dose-related coefficients in nonsmoking humans are expected to be about one-half of the nonsmoking rat values.

As previously mentioned, analyses of epidemiological data have addressed the modifying effects of factors such as age at risk and time since exposure and have also addressed evidence for reduced risks at older ages and with increasing time from exposure. Experimental data from rats can provide only limited information on these effects, because less information is available on the time of occurrence of incidental tumors. Also, because all exposures in the rat experiments were started at about the same age, it is difficult to separate the modifying effects of age at exposure and time since exposure. Depending on exposure rate, some exposures continued through a substantial fraction of the life spans of the animals. However, in general agreement with human data, the largest risks occurred in groups where exposure extended to older ages.[32] It is thus possible that the increased risk seen at lower exposure rates is partly explained by a dependence of risk on time since exposure or on time since cessation of exposure.

To summarize, the attenuation of risks observed in simple analyses of raw experimental tumor data at very high cumulative exposure levels was not found when competing risks were accounted for. Quantitative linear lifetime risk estimates were roughly comparable to those obtained from epidemiological data, although estimates from rats were slightly higher if the comparison was made with a nonsmoking human population. Risks observed at low exposure rates were generally higher than those observed at high exposure rates for the same cumulative exposure. However, the implications of this finding for risks at the very low exposure rates and levels typical of indoor radon are uncertain, because the strongest exposure rate differences were seen at higher cumulative exposure levels, and the exposure rate pattern was attenuated at the lowest exposure level examined (about 300 WLM). Because the lower exposure rate groups required more time to attain a given cumulative exposure, the exposure rate effect may have been partly due to time since exposure.

Additional animal experiments could be designed to address questions that cannot be fully resolved with epidemiological data or current experimental data in male laboratory animals. These would include exposures of female rats and younger animals as well as older animals at high exposure rates to clarify whether risks observed at low exposure rates resulted from an exposure rate effect *per se* or because of differences in time since exposure. Overall, the rat seems to provide a reasonable model for human lung-cancer risk resulting from exposure to radon.

### *Carcinogenesis Modeling*

An alternative approach to statistical analyses of exposure-response data is to use a model based on biological considerations. This approach has several advantages: (1) the parameters of the model are directly interpretable in biological mechanistic terms, (2) both cumulative exposure and pattern of exposure can be easily incorporated into the analysis (which is difficult with statistical analyses), and (3) the model makes predictions which are testable in future experiments.

An analysis has recently been completed of PNL experimental data on radon-induced lung tumors in rats within the framework of the two-mutation recessive oncogenesis model.[33] This biological model incorporates two features: (a) transition of target stem cells into cancer cells via an intermediate stage in two rate-limiting and hereditary (at the level of the cell) steps, and (b) growth and differentiation of normal target and intermediate cells. The two rate-limiting steps may be identified with mutations which lead to homozygous loss of antioncogene function, a mechanism substantiated in several human tumors.[34,35,36]

The biological model describes the radon-induced lung-tumor data quite well. The results indicate that fractionation of radon exposure increased the lifetime probability of tumors. This can be explained by the relative effects of radon decay products on the mutation rates and on the growth kinetics of initiated or intermediate cells. The first mutation rate and the net growth rate of intermediate cells is strongly dependent on the rate of exposure to radon decay products. The second mutation rate is much less dependent on exposure rate, suggesting that the nature of the two mutational events is different. The data are also consistent with the view that radon decay products have no direct effect on the second mutation rate. Therefore, the increase in risk is likely due to the increased proliferation of the intermediate cells working in conjunction with the spontaneous mutational events in the animal.

The analysis poses an interesting question regarding applicability of animal data to human exposures. If spontaneous mutational events are important in the development of radon-induced tumors, is the rat model appropriate for human exposures? The PNL data on Syrian golden hamsters exposed to radon environments showed that animal model to be inappropriate, because the hamster was highly refractory to carcinoma induction in inhalation exposures to radon. Are the spontaneous events in hamsters less frequent than in rats or humans? Or, alternatively, are the growth kinetics of intermediate hamster cells different from those in rats and humans? These possibilities need investigation.

For the Wistar rats used in the PNL experiments, the baseline lifetime risk of lung tumors is estimated to be about one or two per thousand. As previously mentioned, the baseline lifetime lung cancer risk for the U.S. population is 0.9% for nonsmokers. Some data suggest even lower values, but it appears that the frequency of spontaneous mutational events in PNL rats is comparable to, if not lower than, that in humans. Thus, even if radon is not a complete carcinogen, i.e., if spontaneous mutational events are necessary to complete the carcinogenesis process, the rat model still appears to be an appropriate model for human exposures.

## DISCUSSION/CONCLUSION

While human epidemiological data provide the most important basis for assessing risks of human exposures to radon, additional insight into the nature of exposure-response relationships is provided by animal experimentation and dosimetric determinations. Because underground miners' exposure levels and those of other radon-exposed populations are uncertain, surrogate animals are a necessary reality in studying components of the radon-exposure problem.

Lung cancer is the risk of greatest concern in radon exposures. The variables influencing the carcinogenic potential of radon exposures are, with the exception of associated cigarette-smoke exposures, well known in animal studies. However, the implications to human exposures are not fully known. The overall lung-tumor-incidence data in rats are similar to current estimated lung-tumor-incidence data in man. This is true on both an exposure (WLM) and a radiation dose (rad) basis.

Carcinogenesis modeling has provided mechanistic understanding of radon carcinogenesis and insight on the appropriateness of animal models for human exposures. Additional insight on the carcinogenecity of radon exposures is provided by molecular- and cellular-level studies, a subject not covered in this chapter.

Observations in animals that have not been reported in human exposures to radon are: (1) the increase in tumor production with increase in radon-progeny unattached fraction and disequilibrium; (2) the importance of the temporal sequence of exposures to cigarette smoke and radon progeny; and (3) the occurrence of extrapulmonary and extrathoracic (head and neck) carcinoma. The absence of these findings, however, does not constitute proof that similar data might not be obtainable in human exposures but may, rather, reflect the paucity of this type of human data. The lack of extrathoracic tumors, in particular, may reflect the low radon-progeny-unattached-fraction levels in underground mines.

At this time, the similarities in the human and animal data outweigh the differences between them. The animal experiments continue to yield valuable information that helps to explain the effects on man of inhaled radon and progeny in domestic and mine environments.

Finally, while cumulative exposures typical of indoor radon have produced lung carcinoma in rats, there are no definitive data in either rats or humans to prove that typical indoor exposure rates produce excess carcinoma. Until such data are obtained in residential epidemiology studies, or inferred from molecular and cellular level studies, it is, of course prudent to assume that risks at low-level exposures are somewhat independent of exposure rate.

## REFERENCES

1. Agricola, G. *De Re Metallica.* Dover Publications, New York, NY. Translated by H.C. Hoover and L.H. Hoover (1950) (in English).
2. Lorenz, E. 1944. Radioactivity and Lung Cancer; A Critical Review of Lung Cancer in the Miners of Schneeberg and Joachimstal. *J. Natl. Cancer Inst.* 5:1-15.

3. Bale, W.F. Hazards associated with radon and thoron, Memo dated March 14, 1951, Division of Biology and Medicine, Atomic Energy Commission, Washington, D.C. (Also found in *Health Phys.* 38:1062-1066, 1951).
4. Hultquist, B. 1956. Studies on naturally occurring ionizing radiations with special reference to radiation dose in Swedish houses of various types. *Kungl. Svenska vetenskapsakademius handligar,* Bd 6/3.
5. National Council on Radiation Protection and Measurements (NCRP). 1987. *Ionizing Radiation Exposure of the Population of the United States,* NCRP Report No. 93, National Council on Radiation Protection and Measurements, Bethesda, MD.
6. U.S. Environmental Protection Agency (U.S. EPA). 1986a. *A Citizens Guide to Radon: What It Is and What to Do About It,* OPA-86-004, U.S. Environmental Protection Agency, Washington, D.C.
7. U.S. Environmental Protection Agency (U.S. EPA). 1986b. *Radon Reduction Methods: A Homeowner's Guide,* EPA-86-005, U.S. Environmental Protection Agency, Washington, D.C.
8. United Nations Scientific Committee on the Effects of Atomic Radiation (UNSCEAR). 1977. *Sources and Effects of Ionizing Radiation,* United Nations, New York, NY.
9. National Research Council (NRC). 1980. *The Effects on Populations of Exposure to Low Levels of Ionizing Radiation* (BEIR-III Report), National Academy Press, Washington, D.C.
10. National Research Council (NRC). 1988. *Health Risks of Radon and Other Internally Deposited Alpha-Emitters* (BEIR-IV Report), National Academy Press, Washington, D.C.
11. International Commission on Radiological Protection (ICRP). 1981. *Limits for Inhalation of Radon Daughters by Workers,* ICRP Publication 32, Pergamon Press, New York, NY.
12. International Commission on Radiological Protection (ICRP). 1987. *Lung Cancer Risk from Indoor Exposure to Radon Daughters,* ICRP Publication 50, Pergamon Press, New York, NY.
13. National Council on Radiation Protection and Measurements (NCRP). 1984a. *Evaluation of Occupational and Environmental Exposures to Radon and Radon Daughters in the United States,* NCRP Report No. 78, National Council on Radiation Protection and Measurements, Bethesda, MD.
14. Samet, J.M. February 1989. Lung Cancer Epidemiology in New Mexico Uranium Miners (work in progress), p. 64, *FY-1988 Radon Research program,* DOE/ER-0405. U.S. Department of Energy, Office of Energy Research, Office of Health and Environmental Research, Washington, D.C.
15. Marks, S. 1988. *Radon Epidemiology: A Guide to the Literature,* U.S. DOE Report No. DOE/ER-0399. U.S. Department of Energy, Washington, D.C.
16. Borak, T.B. and J.A. Johnson. 1988. *Estimating the Risk of Lung Cancer from Inhalation of Radon Daughters Indoors: Review and Evaluation.* U.S.

EPA Report No. EPA 600/6-88/008. U.S. Environmental Protection Agency, Las Vegas, NV.

17. Steinhausler, F. 1988. Epidemiological Evidence of Radon-Induced Health Risks, pp. 311-371. *In:* W.W. Nazaroff and A.V. Nero Jr. (Eds.). *Radon and Its Decay Products in Indoor Air.* John Wiley & Sons, Inc., New York, NY.
18. The Cancer Letter. 1988. NCI conducting four epidemiology studies on residential radon threat. *The Cancer Letter* 14: No. 27, 5-7, July 1988.
19. U.S. Department of Energy (U.S. DOE). 1989. *FY-1988 Radon Research Program,* DOE/ER-0405. U.S. Department of Energy, Office of Energy Research, Office of Health and Environmental Research, Washington, D.C.
20. Puskin, J.S. and Y. Yang. 1988. A Retrospective look at Rn-induced lung cancer mortality from the viewpoint of a relative risk model. *Health Phys.* 54:635-643.
21. National Council on Radiation Protection and Measurements (NCRP). 1984b. *Exposures from the Uranium Series with Emphasis on Radon and its Daughters,* NCRP Report No. 77, National Council on Radiation Protection and Measurements, Bethesda, MD.
22. Alter, H.W. and R.A. Oswald. 1987. Nationwide distribution of indoor radon measurements: A preliminary data base. *J. Air Pollution Control Assoc.* 37:227-231.
23. Jansen, H. and P. Schultzer. 1926. Experimental Investigations into the Internal Radium Emanation Therapy. I. Emanatorium Experiments with Rats. *Acta Radiol.* 6:631.
24. Read, J. and J.C. Mottram. 1939. The"tolerance concentration' of radon in the atmosphere. *Br. J. Radiol.* 12:54.
25. Jackson, M.L. 1940. *The Biological Effects of Inhaled Radon,* Master's Thesis. Massachusetts Institute of Technology, Cambridge, MA.
26. Cross, F.T. 1988. *Radon Inhalation Studies in Animals.* DOE/ER-0396. Report prepared for the U.S. Department of Energy, Office of Health and Environmental Research, Washington, D.C. NTIS, Springfield, VA.
27. Cross, F.T., R.L. Buschbom, G.E. Dagle, K.M. Gideon, R.A. Gies, et al. In press. Inhalation Hazards to Uranium Miners. *In: Pacific Northwest Laboratory Annual Report for 1988 to the DOE Office of Energy Research,* Part I, Biomedical Sciences. NTIS, Springfield, VA.
28. National Council on Radiation Protection and Measurements (NCRF). 1989; Evidence of lung cancer risk from animal studies. *In: Proceedings of Twenty-Fourth Annual Meeting of the National Council on Radiation Protection and Measurements,* Bethesda, M.D.
29. Chameaud, J., R. Masse, M. Morin and J. Lafuma. 1984. Lung cancer induction by radon daughters in rats (present state of the data on low-dose exposures), pp. 350-353. *In:* H. Stocker (Ed.). *Proceedings of the International Conference on Occupational Radiation Safety In Mining.* Canadian Nuclear Association, Ontario.

30. Cliff, K.D., B.L. Davis and J.A. Riessland. 1979. Little danger from radon. *Nature* 279:12.
31. Evans, R.D., J.H. Harley, W. Jacobi, A.S. McLean, W.A. Mills and C.G. Stewart. 1981. Estimate of risk from environmental exposure to radon-222 and its decay products. *Nature* 290:98.
32. National Council on Radiation Protection and Measurements (NCRP). 1989. Lung cancer risk models from experimental animals. *In: Proceedings of Twenty-Fourth Annual Meeting of the National Council on Radiation Protection and Measurements,* Bethesda, M.D.
33. Moolgavkar, S.H., F.T. Cross, G. Luebeck, and G.E. Dagle. 1990. A two-mutation model for radon-induced lung tumors in rats. *Radiat. Res. 121:28-37.*
34. Cavanee, W.K., T.P. Dryja, R.A. Phillips, W.F. Benedict, R. Godbout, et al. 1983. Expression of recessive alleles by chromosomal mechanisms in retinoblastoma. *Nature* 305: 779-784.
35. Solomon, E., R. Voss, V. Hall, W.F. Bodmer, J.R. Jass, et al. 1987. Chromosome 5 allele loss in human colorectal carcinomas. *Nature* 328:616-619.
36. Birrer, M.J. and N.D. Minna. 1988. Molecular genetics of lung cancer. *Semin. Oncol.* 15:226-235.

*Environmental Radon: Occurrence, Control and Health Hazards.* Edited by S. K. Majumdar, R. F. Schmalz and E. Willard Miller © 1990, The Pennsylvania Academy of Science.

# *Chapter Nineteen*
# RADON AND HEALTH

**DAVID R. BRILL**
Chief, Nuclear Medicine
Geisinger Medical Center
Danville, PA 17822

Since the fortuitous discovery of exceptionally high levels of radon gas and radon daughters in a home in Coalbrookdale Township, Montgomery County, Pennsylvania, in late 1984, environmental radiation due to naturally occurring radon has become a focus of concern for the public, the news media, and for officials at virtually every level of government. But for its perceived effect on human health, radon would have no importance whatsoever. The purpose of this chapter is to examine this very topic and place it in a reasonable perspective.

Even as early as the 1500's, founders of modern medicine such as Agricola and Paracelsus noted the unusually high occurrence of lung diseases, which were variously described as "Schneebergkrankheit" or "miner's consumption," among miners in central Europe. These diseases frequently had a fatal outcome. Judging from their description, they probably represented bronchogenic carcinoma (lung cancer). In 1879, Hartung and Hesse identified a much higher than average incidence of pulmonary cancer among underground metal miners in the same region. This phenomenon has been noted frequently since then, in many areas of the world. The relationship is unequivocal, but its fine details have been poorly understood until recently. Careful systemic study is beginning to shed some light upon the applicability of occupational radon exposure data to the general environment.

Although covered in more detail elsewhere in this volume, it is perhaps wise to briefly review the physical properties of Radon 222. Radon 222 is the decay product of Radium 226, part of the decay scheme of naturally occurring Uranium 238. Radon precursors are found to an extent in almost all rocks and soils, but are more prevalent in some than others. Radon 222 is a noble gas which has virtually no physical or chemical reactions. It thus has few barriers to diffusion other than its 3.8 day physical half-life. It decays by alpha emission to Polonium 218, a short-life alpha emitter, and the first of four products generated within one hour after breakdown of Radon 222. While radon gas in the lungs, in equilibrium with the environment, can reasonably be considered a source of radiation exposure, most of the adsorbed radiation is due to these four alpha and beta particle emitting daughters. Although gamma rays are also generated, they contribute a small fraction to the total exposure.

The four daughter products are highly reactive. They may be inhaled directly but are also capable of adhering to particles of dust or aerosols in the air. These particulates, when inhaled, can adhere to the mucosa of the bronchial tree and allow their radioactive passengers to deposit their energy locally. The alpha emissions have a range of 0.022 mm. in tissue, about the depth of the basal cell layer which is the most radiosensitive portion of the lung. This is the principal vector of irradiation of the lung by radon.

The reader should clearly understand that just as the mere *presence* of radon in an environment does not necessarily equate with *exposure* to radiation, exposure does not equate with *adsorption* of the radiation. The reaction of radiation with living tissue is complex. Biologic damage from radiation may be caused by direct interaction with DNA, the chemical backbone of genes, or indirectly, mediated by activated free radicals of water. In either case, mutations can be induced by a change in the chemical structure of the DNA or by a break in the double helix.

The algorithm presented in Figure 1 summarizes the possible events. Ionizing radiation will either interact or not interact with DNA (a "hit" or a "miss"). In the case of X and gamma irradiation, misses are perfectly plausible. The mechanism of exposure of a chest radiograph, for example, is the differential passage of unimpeded X rays through bone, lung, and the heart and great vessels. In the case of beta and alpha particles, the likelihood of interaction is far greater, but since the range of these particles is so short, a significant percentage of emanations may occur in a direction where they are absorbed by the air in the bronchi, or the dust particle to which the daughter has adhered.

Assuming an interaction with a cell, damage may be done to the DNA ("a hit") or to the cytoplasm ("a miss"). Although much of the effect on DNA is due to high energy ions of water, not every interaction involves genetic material. Only those which involve DNA are genetically significant and could relate to the development of cancer.

Among those "hits" that involve DNA, there are three possible consequences. Some interactions are lethal to the cell. The damage is so severe that the cell line cannot reproduce itself and quietly dies at the next mitosis (cell division), being replaced by its neighbors. These interactions produce no clinical effect. At the other end of the spectrum, some damage to DNA can be repaired. If the damage is to one strand of the double helix, the other strand can act as a splint. Since the purine and pyrimidine moieties can only pair in one way, complete repair without sequel is possible. Only nonlethal mutations are of consequence in the induction of cancer, since genetic damage is propagated.

Of course, not all mutations are malignant or premalignant. Those which are not are essentially irrelevant to radiogenic cancer induction. Of the malignant changes, only those escaping recognition and destruction by the immune system are capable of producing clinical cancer.

The foregoing schematic necessarily oversimplifies a highly complex subject, but it is an essentially correct first order approximation of what happens when living tissue is exposed to radiation.

The radiation dose to the lungs associated with radon and its daughters is estimated at about 14 rems for every working level month (WLM)[1] (see glossary) of exposure. Relating exposure to generation of clinical cancer is far more difficult for a variety of reasons. Mere adsorption of radiation is only one of many factors that may result in carcinogenesis; numerous factors interrelate in ways that may enhance or inhibit this effect. In comparatively small doses as are characteristic of most radon exposure, one cannot see an effect on any individual, but must study large homogeneous cohorts to ascertain whether a response has occurred. Finally, cancer is a common disease, one which typically becomes more prevalent in an aging population. Thus, there is no direct evidence in human beings that radon in levels typically found in the environment causes cancer. The basis for concern comes from three sources—data from occupationally exposed metal miners, data from animal experiments, and calculations based upon accepted risk rates used by radiation biologists. These three pillars, epidemiologic, controlled investigation, and theoretical, provide a framework for all estimates of risk for environmental radon.

The greatest body of data is that of long-term studies of underground miners. Until recently, the many existing studies were poorly controlled and retrospective. One could not accurately determine levels or rates of exposure, the existence of cofactors, or compare the results between various groups. In the last decade,

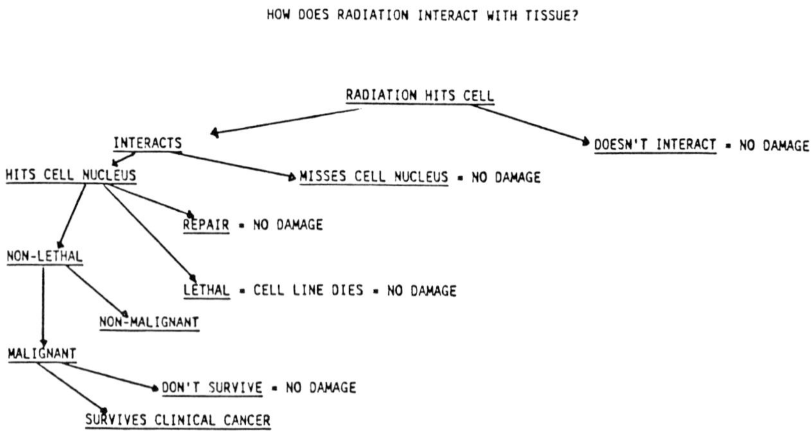


FIGURE 1. Algorith demonstrating possible outcomes of exposure of living tissue to ionizing radiation.

however, five well regarded studies have been published reviewing data from Ontario,[2,3] Sweden,[4] Saskatchewan,[5] Czechoslovakia,[6] and the Colorado Plateau.[7] Each of these studies indicated an increased occurrence rate of lung cancer in rough proportion to radon exposure above levels of about 300 WLM. The Biological Effects of Ionizing Radiation Committee (BEIR-IV), convened by the National Research Council, has obtained and reviewed the original data from four of these primary studies (they are currently attempting to get the Czech data).[8] This work represents the most up-to-date thinking on the subject.

Using exhaustive analyses, BEIR-IV attempted to create a theoretical model that best fit the available data while taking into account the many uncertainties inherent in any such retrospective studies. They concluded that the excess risk of lung cancer correlates directly with cumulative exposure but is slightly greater for young persons than for elderly. Risk appears to diminish with time since exposure in the occupational setting. Females appear to be at a slightly lower risk than males. The greatest single risk factor for radon-induced lung cancer is tobacco abuse. These two risks interact multiplicatively or submultiplicatively.

The authors of BEIR-IV caution against a too literal interpretation of their conclusions. As with any statistical analysis, some uncertainties exist. The "best fit" theories must not be regarded as a "perfect fit". The apparently increased risk among the younger could also relate to the fact that lung cancer is much rarer among young people than among the elderly so that a few cases could profoundly influence the rate of occurrence. The risk differential between male and female may also reflect a differential in the baseline incidence of lung cancer between the two sexes. The effect of tobacco abuse is unequivocal and dominant, probably reflecting the 10:1 lung cancer risk of smokers over nonsmokers far more than the effect of radon.

One must be careful in extrapolating too liberally from occupational studies to the environment. Major differences in the populations exist; miners are almost exclusively adult males who receive their exposures at a younger age and more rapidly than is typical in the domestic environment. Patterns of breathing are different; due to greater exertion, miners breathe more rapidly and deeply which affects the distribution of dust and aerosols in the bronchial tree. Mines are far more dusty than homes, resulting in a more efficient delivery system for radon daughters. Organic fumes and aerosols, which may be cocarcinogenic, or even carcinogenic in their own right, are also present. Finally, miners, as a group, tend to smoke more heavily than the general population, a very important cofactor. Levels of radon exposure among miners overlap the upper level of exposure in the general population. Interpolation of cancer risks to this segment of population may be more valid than extrapolation to the vast majority of people exposed to much lower levels.

The second approach to assessing risk from radon comes from experiments carried out in small animals. Numerous such experiments[9,10,11] have been performed and suggest a risk of cancer occurring in rough proportion to exposure up to levels of 1,000 WLM, where the occurrence rate falls. This phenomenon is related to the somatic effects related to very high levels of radiation.

The data gathered by these investigations is subject to some criticism, not for their validity in the experimental setting but for their applicability to human beings in an environmental setting. In the first place, it is necessary to use nonhuman models such as mice, rats, and beagle dogs. Interspecies differences exist between all three and may well apply to humans as well. In the second place, financial and logistic restrictions require the administration of radiation in quantities and at rates far greater than exist in the environment. Thus, while an unequivocal effect exists at high levels and rates of radon exposure, there is no proof that one exists at a proportionately lower level at exposures more typical of the environment.

A third approach in quantitating risk in the general population is the use of radiobiologic theory. Using data derived from exposed populations such as Hiroshima and Nagasaki, the Marshall Islands, early radiation workers, and others, one can estimate a risk level based upon the number of rems of exposure. That level is one extra case of cancer for every 10,000 man-rems.[1] This approach is based upon the assumption that radiation effects occurring at exposures where no individual response is discernable can be predicted by extrapolating in a straight line between the minimum observable response and zero. This is called the linear no threshold theory and is the most conservative model that can be used. Since, by its very nature, it can neither be proven or disproven, radiation biologists typically use it to predict the maximum possible effect. Unfortunately, public perception has converted theory to fact and maximum possible effect to an absolutely quantifiable fact. Only the uncertainty remains, but has been misconstrued to mean that the effect might be even worse. In fact, many well respected studies strongly suggest that the linear no threshold theory significantly overestimates the risk from low levels of radiation.

Using this approach, various bodies have calculated the risk levels from environmental radon. The Environmental Protection Agency (EPA)[12] assumes about 5,000-20,000 excess lung cancer deaths in a population of 240,000,000 annually. The National Council on Radiation Protection and Measurement (NCRP) calculates 150 extra deaths/million/year for every working level months (WLM).[13] BEIR-IV[8] predicts an excess mortality of 350 per million per WLM per year. There is more similarity than difference between these estimates and all agree that there is an effect.

Probably the most important epidemiologic question about lung cancer relates to tobacco abuse. Cigarette smoking is widely accepted as the most important etiology of bronchogenic carcinoma. The interrelationship between smoking and radon in carcinogenesis is a critical question.

According to the American Cancer Society,[14] bronchogenic carcinoma was a rare disease prior to the widespread use of tobacco. The baseline incidence is about 5 cases per 100,000/year. At present, the occurrence rate among men is 70 per 100,000 and among women, about 30 per 100,000. Lung cancer is the leading cause of cancer deaths among men by a factor of 2.5 and is now tied with breast cancer for the lead in women. The 6 to 14-fold increase has occurred since the early years of this century. Were this increase due to an increase in radon exposure or greater longevity due to conquest of infectious and other diseases, the rates of other cancers might be expected to rise proportionately. They don't. Moreover, there should be no difference between men and women. In fact, the accelerated occurrence among males began in the 1930s, a decade or so following the marked increase of popularity of cigarettes among men. The rapid rise among females began in the 1960s, a generation later, but again followed by a decade or so the marked increase of tobacco use by women.

In 1987, the American Cancer Society[14] attributed 139,000 deaths to lung cancer. Of these, 113,000 were attributed to tobacco, leaving 26,000 attributable to all other causes, including radon. BEIR-IV assumes a tenfold increase of lung cancer risk among smokers, commensurate with ACS statistics. Based upon its analysis of the cohort of Colorado Plateau Miners, the only study for which tobacco use was included, BEIR-IV[8] established a model in which the risks of radon and tobacco were either multiplicative or submultiplicative. Although the exact relationship between tobacco and radon is not clear, it is evident that tobacco use represents a far greater etiologic threat for lung cancer than does radon.

If the observations on tobacco abuse are viewed in the light of the linear no threshold theory, some interesting relationships emerge. The linear no threshold theory assumes that there is no lower limit for biologic damage from radiation. It follows then that the number of cancers in a population of 100,000 exposed to one rem should be the same as in a population of 10,000 exposed to ten rems, or in a population of 1,000,000 exposed to 0.1 rem. If one were to further assume that of the 139,000 lung cancer deaths counted in 1987, the 113, 000 attributed to tobacco were also due to a radon synergy (a rather far-fetched, worst case assumption) and given a tenfold increase of lung cancer risk due to tobacco, then only 11,300 (113,000 ÷ 10) of the 26,000 other lung cancer deaths might have been due to radon. According to data compiled by Nero,[15] the geometric mean of radon levels in homes across the U.S. is only 0.96 pCi/l (picoCuries/liter), and only 8% of the U.S. Homes exceed 4 pCi/l. This observation applies to the presence of, not exposure to, radon. It follows that the vast majority of the population of the U.S. is exposed to very low levels of radon. Assuming that man-rem is a man-rem, most of the radon exposure, and thus risk for lung cancer, occurs in that huge segment of residents whose individual risk is low. Quite logically, anyone whose home has high levels (greater than 200 pCi/l) of radon is at a higher individual risk and should seriously consider remediative measures.

It is perhaps wise to draw a distinction between remediation and testing. The author is skeptical about the benefits of remediating many, many homes which contain low-medium or low levels of radon to very low levels, particularly where the home owner spends little time in that environment. There is no question that high level homes, especially above 200 pCi/l, need prompt attention. There is no way other than testing to determine which homes are at risk. Local geologic anomalies are usually not known and individual vagaries of home construction make prediction unreliable. Widespread testing of homes for radon contamination, particularly in areas such as the Reading Prong which have a higher than average risk, is likely to be cost-effective in identifying those comparatively few homes where rapid remediation is needed.

Massive remediation of all homes to 4 pCi/l or less would be very costly and would likely not produce any meaningful effect either on individual or population risk. If one took seriously the stated goal of Congress' National Indoor Radon Act of 1988,[16] of reducing indoor radon levels to the 0.2-0.7 pCi/l ambient outdoor level, one could possibly reduce the collective risk by about 10,000 cases per year, but at an astronomic cost. Even then, the estimated saving of lives is based upon the linear no threshold hypothesis, whose validity is open to serious question. A far greater life sparing effect could be achieved if tobacco abuse were to be curtailed. The latter claim is not theoretical but based upon epidemiologic observation.

BEIR-IV, as a part of its recommendation, also raises the issue of radiation to other tissues besides the lung from naturally occurring radionuclides and induction of diseases other than lung cancer. The NCRP[13] points out that the risk of gastrointestinal cancer from waterborne radon and more distant daughter products such as Polonium 210 in some foods is several orders of magnitude lower than for lung cancer. Radon precursors such as Uranium 238 and Radium 226 may deposit in bone and kidney, but generally do so in extremely miniscule quantities. Polonium 210 is a component of tobacco smoke but is not considered an important carcinogen, especially in the face of so many other far more serious hazards in tobacco. Even so, BEIR-IV recommends that naturally occurring radionuclides be investigated for their potential in inducing other diseases such as renal and gastrointestinal malignancies and emphysema.

No review of radon and health would be complete without noting that in years past, and even in 1990, radon gas has been marketed as a health aid. For centuries, the world's wealthy have sought the "healing vapors" to be found in spas in, for example, Radium Hot Springs, British Columbia; Bad Gstein and Carlsbad, Austria; Misasa, Japan; and elsewhere. Europeans and North Americans haunt EnGedi, Israel, where they smear themselves with radioactive mud from the phosphate mines near the southern end of the Dead Sea. The majority of the world's population cannot travel to such exotic regions. For them, a variety of notions, potions, and devices have been invented to bring them the "benefits" of radon exposure. One such device was a jug lined with radium ore which was to be filled with water in the evening and allowed to charge up with radon overnight. The next day, the owners were enjoined to drink six healthful glasses of the radon charged water. The author had brief experience with one such device...the radiation levels inside the jug well exceeded the 1 rem/hr. limit of his counter. In a story which may be apocryphal, the Food and Drug Administration is supposed to have prosecuted the manufacturer of a "radium salve" who had deceitfully omitted anything radioactive from his product! The point must be made that radon has not always been, *and indeed still isn't,* universally regarded as a menace.

In summary, radon's effect on public and individual health is not completely known in all its subtleties. A number of conclusions seem warranted, however. First, radon gas and daughters undoubtedly have an effect on human health. This effect, in rough proportion to exposure, is probably negligible for the individual, but may account for as many as 10,000 or 11,000 lung cancer deaths annually in the United States. Second, the vast majority of risk appears to be born by the population whose individual risk is low. This statistical curiosity is due to the extreme preponderance of people whose exposure to radon is minimal and is a consequence of the use of the linear no threshold theory of extrapolating risk of radiogenic cancer from radiation levels for which an effect is apparent to much lower levels where no hard data are available. Third, it is reasonable to remediate homes in which high levels of radon occur. Remediation of all homes to 4 pCi/l maximum, or even to the 0.2-0.7 pCi/l level recommended by Congress, is apt to have a negligible effect on the occurrence of lung cancer. Fourth, widespread testing of homes, especially in high risk areas, is probably a wise commitment of resources, in that it can identify those high risk homes relatively easily. Fifth, although radon has been called the second greatest threat to human health in the United States, the risk for tobacco abuse exceeds that of radon by a factor of at least 10. A far greater effect on the occurrence of lung cancer could be expected from cessation of smoking than from any form of radon abatement. Finally, radon gas and daughters may have an effect other than lung cancer. More work needs to be done to ascertain its effects on the gastrointestinal tract, urinary system, and in the induction of pulmonary pathology such as emphysema.

## GLOSSARY OF RADON RELATED TERMS

*PicoCurie per liter* (pCi/l)—A picoCurie is one trillionth part of a Curie of any radionuclide, which results in 3.7 x $10^{-2}$ radioactive disintegrations per second. It refers to the *presence* of radioactivity, but is only indirectly related to exposure. In the context of radon, it refers only to Radon 222, not the daughter products.

*Working level* (WL)—A "working level" also refers to the presence of radioactivity and is only indirectly related to exposure. This term refers to *daughter products* of Radon 222 and was developed to facilitate discussion of occupational exposure in miners. Depending on the environment, two definitions exist. For underground mines where radon and its daughters exist in complete equilibrium, 1 WL = 100 pCi/l. For the domestic environment where complete equilibrium does not exist, 1 WL = 200 pCi/l. EPA has adopted the latter.

*Working level month* (WLM)—A measure of *exposure.* The "month" is based on 170 hrs., equivalent to a 40 hour week. One simply multiplies the ambient radon daughter level by the time spent in that atmosphere. Estimate of carcinogenesis range from 130-730 excess lung cancer deaths per million per WLM.

*Working level months per year* (WLM/Y)—The *rate* of exposure.

## REFERENCES

1. National Council on Radiation Protection and Measurements. March 1984. Report 77—Exposures from the Uranium Series with Emphasis on Radon and Its Daughters. Bethesda, MD.
2. Muller, J. 1984. Study of Mortality of Ontario Miners, 1955-77. Pt. 1. *In: Proceedings of the International Conference of Occupational Radiation Safety in Mining,* Vol. 1. H. Stecher, (ed.). Toronto.
3. Muller, J. W.C. Wheeler and J.F. Gentleman, et al. 1983. Study of Mortality of Ontario Miners, 1955-77. Pt. 1. Ontario Ministry of Labour, Ontario Workmen's Compensation Board.
4. Radford, E.P. and K.G. St. Clair-Renard. 1984. Lung Cancer in Swedish Iron Miners Exposed to Low Doses of Radon Daughters. N. Engl. J. Med. 310:1485-1494.
5. Howe, G.R., R.C. Nair and H.B. Newcombe, et al. 1986. Lung Cancer Mortality (1950-80) in Relation to Radon Daughter Exposure in a Cohort of Workers at the Eldorado Beaverlodge Uranium Mine. J. Natl. Cancer Inst. 77:357-362.
6. Kunz E., J. Sevc and V. Placek, et al. 1979. Lung Cancer in Man in Relation to Different Time Distribution of Radiation Exposure. Health Phys. 43:65.
7. Hornung, R.W. and T.J. Meinhardt. 1986. Quantitative Risk Assessment of Lung Cancer in U.S. Uranium Mines. National Institute for Occupational Safety and Health. Center for Disease Control, Cincinnati, OH.
8. Committee on the Biological Effects of Ionizing Radiation. 1988. Health Risks of Radon and Other Internally Deposited Alpha Emitters (BEIR-IV). National Academy Press, Washington, D.C.
9. Morken, D.A. and J.K. Scott. 1966. Effects on Mice of Continual Exposure to Radon and Its Decay Products or Dust. University of Rochester Atomic

Energy Project, Report UR-669. National Technical Information Service, Springfield, VA.
10. Chameaud, J., R. Perraud, R. Masse, J.C. Nenot and J. LaFuma. 1976. Lung Cancer Induced in Rats by Radon and Its Daughter Radionuclides at Different Concentrations. *In: Biological and Environmental Effects of Low Level Radiation*, Vol, II. (Pub. STI/PUB 409). International Atomic Energy Agency, Vienna.
11. Cross, F.T., R.F. Palmer and R.E. Filipy, et al. 1978. Study of the Combined Effects of Smoking and Inhalation of Uranium Ore Dust, Radon Daughters, and Diesel Oil Exhaust Fumes in Hamsters and Dogs. Battelle Pacific Northwest Laboratory, Final Report PNL-2744/UC-48. National Technical Information Service, Springfield, VA.
12. U.S. Environmental Protection Agency. 1986. A Citizens' Guide to Radon. OPA 86-004.
13. National Council on Radiation Protection and Measurement. May 1984. Report 78. Evaluation of Occupational and Environmental Exposures to Radon and Radon Daughters in the United States. Bethesda, MD.
14. American Cancer Society. 1988. Cancer Statistics, 1988. CA 38:5-22.
15. Nero, A.V. 1984. Indoor Concentrations of Radon 222 and Its Daughters: Sources, Range, and Environmental Influences (LBL-19346). Proceedings of Seventh ORNL Life Sciences Symposium on Indoor Air and Human Health.
16. National Indoor Radon Act, 1988. Oct. 4, 1988. House of Representatives Report 100-1047.

*Environmental Radon: Occurrence, Control and Health Hazards.* Edited by S. K. Majumdar, R. F. Schmalz and E. Willard Miller © 1990, The Pennsylvania Academy of Science.

# *Chapter Twenty*

# RADON INDUCED LUNG CANCER AND MORTALITY

**DAVID BODANSKY**
Department of Physics
University of Washington
Seattle, WA 98195

## 1. INTRODUCTION

Radon is a major radiation hazard, generally believed to be the cause of a significant fraction of current lung cancer mortality. For example, the U.S. Environmental Protection Agency suggests that inhalation of the daughter products of indoor radon may lead to 5000 to 20,000 lung cancer deaths per year in the United States,[1] roughly 4% to 15% of all lung cancer mortality. However, there are large uncertainties in these estimates. In this Chapter, we will summarize the methods used in obtaining such estimates, the results obtained, and the uncertainties, with reference to three possible approaches:

a. *The dosimetric approach.* The lung cancer risk is here calculated from the radiation dose, using standard relationships between cancer mortality and radiation dose (Section 3).

b. *Extrapolation from miners.* Some groups of miners, especially uranium miners, have been found to have high rates of lung cancer due to radon exposure. The lung cancer rates for miners, as a function of radiation dose, are used to estimate the risk for the general population (Sections 4, 5, and 6).

c. *Population studies.* Correlations are sought between lung cancer mortality and radon exposure among groups in the general population (Section 7).

As will be seen below, the dosimetric approach and the more widely used extrapolation from the experience of miners give results in reasonable overall agreement.* However, there is as yet no substantial supporting evidence from studies of the general population.

*See Note added at end of Chapter for recent indications of an apparent disagreement.

## 2. GENERAL CONSIDERATIONS IN DESCRIBING RISK

### 2.1 Lung cancer incidence and lung cancer mortality

Lung cancer incidence risk and lung cancer mortality risks are almost identical, because lung cancer is usually fatal. In 87% of the cases, lung cancer patients die within five years of diagnosis.[2]

### 2.2 Effective cumulative exposure

The lung cancer risk at age $t$ depends upon the detailed history of exposures at earlier ages, $t_e$. For example, exposures at very recent times are generally excluded from consideration on the assumption of a latent period after exposure, $\tau$, during which no cancer fatalities occur. To allow for this and other adjustments, an *effective cumulative exposure, W(t)*, can be introduced. Although more general forms are possible, $W(t)$ is often taken to be of the form:

$$W(t) = \int_{\tau}^{t} w(T)\, f(T)\, dT \tag{1}$$

where $T$ is the elapsed time since exposure, i.e. $T = t - t_e$, $w(T)$ is the exposure rate, and $f(T)$ is a weighting factor. The magnitude of $\tau$ and the functional form of $f(T)$ are based on empirical studies of cancer incidence. If the simple choice is made that $f(T) = 1$, the exposure given by Eq. 1 is the unweighted total exposure for $T > \tau$.

### 2.3 Excess risk

#### 2.3.1 *Individual age-specific annual risk*

The excess risk in any year of cancer mortality due to radon exposure, $\Delta r$, is the difference between the risk $r(W)$ for exposure $W$ and the background risk, $r_o$, when $W = 0$. The excess risk may depend upon age, $t$, and upon an array of other factors, such as gender and smoking, which we denote collectively by $x$. Thus the age-specific excess risk is

$$\Delta r(W,t,x) = r(W,t,x) = r_o(t,x) \tag{2}$$

where $r_o(t,x) = r(0,t,x)$ and the dependence of $W$ on $t$ is here not explicitly indicated.

#### 2.3.2 *Individual lifetime risk*

The lifetime risk is the sum of the annual age-specific risks:

$$R_x = \sum_{t_i=0}^{t_i=\infty} p(t_i)\, r(W_i, t_i, x_i) \tag{3}$$

where $p(t)$ is the probability of survival until age $t$ and $W$ is a function of $t$. The lifetime risk may depend upon gender or smoking practices, which can be represented by different $x$'s. Eq. 3 may apply alternatively to the background risk ($r_o$ and $R_o$), the total risk ($r$ and $R$), or the excess risk ($\Delta r$ and $\Delta R$). In Eq. 3, terms are suppressed at low ages by taking $r(t) = 0$ for $t$ below some threshold and at high ages by the vanishing of $p(t)$.

### 2.3.3 *Annual fatalities in a population*

For a steady-state population, i.e. one in which the age-distribution never changes, the annual number of excess fatalities, $\Delta F$, is related to the excess individual lifetime risk, $\Delta R$, by:

$$\Delta F = \frac{N\,\Delta R}{L} \tag{4}$$

where $N$ is the size of the population and $L$ is the mean lifetime of the population. It can be shown that Eq. 4 gives a value of $F$ in good agreement with the results of a more detailed calculation for the actual U.S. age distribution, in which the mortality is calculated for each age group and summed.

## 2.4 Linearity assumption

It is commonly assumed, for heavily ionizing radiation such as alpha particles, that the excess risk, $\Delta r$, is linearly proportional to the excess dose, $W$, down to zero excess dose (see, e.g.[10]). While generally accepted for alpha particles, this *linearity assumption* lacks a firm empirical base at very low doses and remains a matter of some uncertainty. We will return to it below.

## 2.5 Absolute and relative risk

Two major variants of risk models are used in studying the effects of radiation: the *absolute risk model* and the *relative risk model*. In the absolute risk model, the age-specific excess risk from radiation is assumed to be a function of radiation dose alone and is the same for all individuals, independent of their risk of cancer from non-radiation causes. Many versions of the absolute risk model incorporate the further assumption of linearity. In this case, the absolute risk is:

$$\Delta r(W) = \alpha W(t) \tag{5}$$

where $\alpha$ is a constant which defines the slope of the dose-response curve.

In the relative risk model, the excess risk from radiation is assumed to be proportional to the background risk. A particularly simple version of the relative risk model is what the BEIR IV study[3] terms the *constant relative risk model:*

$$\Delta r(W,t,x) = r_o(t,x)\, \beta\, W(t) \tag{6}$$

where $\beta$ is a constant. Eq. 6 also represents a linear model.

## 3. DOSIMETRIC APPROACH

### 3.1 Effective dose equivalent, $H_E$

In principle, the probability of fatal cancer can be predicted from the radiation exposure to which an individual or population is subjected. As discussed in Chapter XX, the exposure can be translated into an *effective dose equivalent,* $H_E$. Various estimates of the conversion factor from radon exposure to the lung to the effective dose equivalent, in mSv per WLM, are listed in Table 1. Separate listings are given, when available, for miners and for typical indoor domestic environments. For most of these studies, the authors indicate large uncertainties.

Considering the array of values in Table 1, and the uncertainties inherent in each of them, the conversion factor for the indoor environment is best specified as a range:

$$H_E = 5 \text{ to } 20 \text{ mSv per WLM}$$

Taking into account the fact that the low values in Table 1 derive from a common analysis, it is reasonable to adopt as an approximate central value: $H_E \approx 10$ mSv per WLM. There is at least a factor of two uncertainty in this value.

TABLE 1

*Estimates of conversion factors between radon daughter exposure and effective dose equivalent: for miners and for domestic environment.*

| Group | Year Publ. | $H_E$ (mSv/WLM) Miners | Domestic | Comments |
|---|---|---|---|---|
| NEA-OECD[4] | 1983 | 8.5 | 5.5 | From pp. 54, 65 |
| NCRP[5] | 1984 | 7 | 9.5 | Based on tracheo-bronchial dose, pp. 153, 155 |
| NCRP[6] | 1987 | | ≈ 12 | From Tables 2.1, 2.4 |
| ICRP[7] | 1987 | 8 | 6.3 | Based on pp. 16, 34 |
| UNSCEAR[8] | 1988 | | 6 | From p. 79 |
| James[9] | 1988 | 12 | 15 | Tracheo-bronchial contribution only, p. 296 |

### 3.2 Risk as determined from effective dose equivalent.

To calculate lung cancer mortality from the effective dose equivalent, it is convenient to adopt estimates presented in the 1980 study of the Committee on the Biological Effects of Ionizing Radiation (BEIR III) [10, p. 146] of the excess mortality for a continuous annual whole body exposure to 0.01 Gray per year of lightly ionizing (low LET) radiation. This is equivalent to an effective dose equivalent of 10 mSv per year, or, in the central estimate given above, to an exposure of 1 WLM per year. The corresponding BEIR III estimates of lifetime risk, for a linear dose-response relation, are 0.0112 and 0.0287 for the absolute and relative risk models, respectively.

The annual number of fatalities due to radon follows from Eq.4. If we take the average radon concentration to be about 40 $Bq/m^3$ (1 pCi/liter), the equilibrium factor to be 0.5, and an effective 40 "working-months" per year, the average exposure is 0.2 WLM/yr. Then, for a U.S. population of N = 240,000,000 and a mean lifetime, $\Delta L$ = 73 years, the calculated annual number of U.S. cancer fatalities from indoor radon is approximately:*

$$\Delta F \approx 7{,}500 \quad \textit{BEIR III absolute risk parameters} \tag{7}$$

$$\Delta F \approx 20{,}000 \quad \textit{BEIR III relative risk parameters} \tag{8}$$

There is at least a factor of two uncertainty in these numbers, although a figure of 40,000 lung cancer deaths from radon would be excluded unless many of these could be attributed jointly to smoking and radon (see Section 6.1).

## 4. THE EXPERIENCE OF MINERS

### 4.1 Data base

We list in Table 2 major groups of miners for which detailed studies have been reported on the relationship between radon exposure and lung cancer mortality. In addition to the cohort-specific papers cited in Table 2, there have been reviews by the U.S. National Council on Radiation Protection and Measurements (NCRP),[5] the BEIR IV Committee,[3] and the International Commission on Radiation Protection and Measurements (ICRP).[7] The BEIR IV study included a re-analysis of the data obtained in the individual studies. In the discussion below, we will emphasize results of these reviews.

It should be noted that the data base represented by the available studies

*See Note added at end of Chapter for higher BEIR V results.

is limited. The excess number of lung cancer cases is less than 800. Dividing the miners into subgroups by magnitude of exposure, age at exposure, age at time of cancer death, and smoking practices, leads to small numbers within the subgroups. In addition to the statistical difficulties which this implies, there appear to be systematic differences between studies. For example, the Colorado plateau studies give markedly lower risk coefficients than do the other studies and studies (by the same group) of the Eldorado Beaverlodge and Eldorado Port Radium miners give very different risk coefficients.

### 4.2 Risk parameters from miner studies

#### 4.2.1 *NCRP Report No. 78 analysis for miners*

The main risk projections in NCRP Report No. 78[5] are based on a modified absolute risk model, in which it is assumed that: (a) no lung cancers occur before age 40; (b) repair and other processes give a recovery half-life of 20 years, lessening the effect of past exposure; and (c) the latent period is $\tau = 5$ years. The age-specific absolute risk for $t \geq 40$ is then given by Eq. 5, where $\alpha = 10 \times 10^{-6}$ per person-year of observation and per WLM of exposure (per PY-WLM), and the cumulative exposure is

$$W(t) = \int_5^t w(T) e^{-T \ln 2/20}\, dT \quad \textit{modified absolute risk} \tag{9}$$

With these units, $w(T)$ is the annual exposure in WLM per year. For $t < 40$, $\Delta r = 0$. Results for the risk coefficients from the NCRP and other reviews are listed in Table 3.

#### 4.2.2 *BEIR IV analysis for miners*

In BEIR IV,[3] only relative risk models were reported, both a constant relative

TABLE 2

*Major groups of miners studies for lung cancer mortality.*

| | Mine Region | Recent Type | Reference | LC Fatalities[a] Observed | LC Fatalities[a] Expected |
|---|---|---|---|---|---|
| A. | Colorado Plateau, U.S.A. | Uranium | Hornung[11] | 256 | 59 |
| B. | Malmberget, Sweden | Iron | Radford[12] | 51 | 15 |
| C. | Eldorado Beaverlodge, Can. | Uranium | Howe[13] | 65 | 29 |
| D. | Ontario, Can. | Uranium | Muller[14] | 87 | 57 |
| E. | Czechoslovakia | Uranium | Sevc[15] | 484 | 98 |
| F. | Newfoundland, Can. | Fluorspar | Morrison[16] | 97 | 17 |
| G. | Eldorado Port Radium, Can. | Uranium | Howe[17] | 57 | 25 |

[a]Groups A-D: from BEIR IV re-analysis;[3] groups E-G: from cited papers.

risk model, as in Eq. 6, and a variant termed the *time since exposure* model (TSE). The final analyses were based on the TSE model, which gave the better match to the data. In this model, the age-specific risk is (3, p. 38):

$$\Delta r(W,t,x) = r_o(t,x)\beta\gamma(t)[W_1(t) + 0.5\ W_2(t)] \quad TSE\ model \qquad (10)$$

where $\gamma(t)$ is a factor which takes into account the possibility of a different relative risk of cancer incidence at different ages and where $W_1(t)$ and $W_2(t)$ are the unweighted total exposures for $T$ = 5–15 and $T > 15$, respectively (in years). Based on the miner's data, the following parameters were adopted: $\beta = 0.025$ $WLM^{-1}$, and $\gamma(t) = 1.2$ for $t < 55$, $\gamma(t) = 1.0$ for $t$ = 55–64, and $\gamma(t) = 0.4$ for $t \geq 65$.

The breakup of $W(t)$ into two terms in Eq. 10 acts in the same direction as the exponential recovery factor of Eq. 9 in the NCRP model. Similarly, the parameterization of $\gamma(t)$ moves the TSE model to a position between a pure relative risk and pure absolute risk model, in terms of variation of cancer risk with age. (Although $r_o$(t) increases with age, $\gamma(t)$ decreases the calculated risk at higher $t$.)

Many other studies are couched in terms of a constant relative risk model, and therefore comparisons between studies are most easily made with this parameterization (see Table 3). In the BEIR IV constant relative risk model (Eq. 6), the average risk coefficient from the four miners groups was found to be $\beta = 0.013$ W LM$^{-1}$, where $W$ was taken as the unweighted cumulative exposure (in WLM) and the latent period, $\tau$, is 5 years. The parameter $\beta$ is less in the constant relative risk model than in the TSE model (here 0.013 vs 0.025), because in the TSE model the term $\gamma(t)$ serves to reduce the calculated risk by a large factor for older miners and because the exposure incurred for $T > 15$ years is discounted by a factor of two.

### 4.2.3 *ICRP Report 50 analysis for miners*

TABLE 3

*Risk coefficients for lung cancer in miners due to radon daughters*

| | Miners groups Included (see Table 2) | Absolute Risk Coefficient: $a$ (per PY-WLM)$^a$ | Relative Risk Coefficient: $\beta$ (per WLM) |
|---|---|---|---|
| NCRP 78[5] | A, B, D, E, F, | $10 \times 10^{-6}$ | |
| BEIR IV[3] | A, B, C, D | | |
| average | | | 0.013 |
| range | | | 0.008-0.023 |
| ICRP 50[7] | A, D, E | | |
| average | | $10 \times 10^{-6}$ | 0.010 |
| range | | $(5\text{-}15) \times 10^{-6}$ | 0.005-0.015 |

$^a$Specified per person-year of observation and per WLM of exposure.

In ICRP Report 50,[7] both absolute and relative risk models were considered. As in the NCRP treatment, $\Delta r(t)$ is taken to be zero for $t < 40$, but the adjusted cumulative dose is the unweighted dose, i.e., there is no exponential recovery factor. The ICRP study gives the same value as the NCRP study for the absolute risk coefficient, $\alpha$ and a value in reasonably good agreement with the BEIR IV estimate for the relative risk coefficient, $\beta$. However, a wide range of possible values is indicated (see Table 3).

### 4.3 Cancer induction at low exposures

When intensive studies of miners were undertaken after the 1950s, the initial miners population had received very high radon exposures, typically many hundreds, or even thousands, of WLM. A considerable extrapolation is required to estimate effects for the general population, typically receiving doses of several tens of WLM, spread over a lifetime.

In NCRP 78, it was concluded that while there was no "satisfactory" data showing an excess of lung cancer for cumulative exposures under 60 WLM, it was advisable to assume a linear extrapolation to lower doses. Subsequent studies have suggested excess lung cancer below 60 WLM, e.g. the Eldorado Beaverlodge, Malmberget, and Ontario studies [3, p. 107], although there is no excess in the Colorado plateau studies below 120 WLM. In all these cases, the numbers are very small. In aggregate, they seem to suggest a statistically significant positive effect, for cumulative exposures as low as the 20-50 WLM interval.

However, when the cumulative exposure is low, either due to low radon levels or short duration of employment, the data are particularly vulnerable to confounding factors, i.e. agents other than radon which cause lung cancer. For example, some or all of the excess cancer ascribed to radon exposure might be due to other carcinogens in the mine or to prior occupational exposures.

## 5. PROJECTION FROM MINERS TO THE GENERAL POPULATION

### 5.1 Differences between the mine and the domestic environments

Apart from the differences in typical cumulative exposures and annual exposure rates, there are differences between the mine and home environments which can lead to differences in the conversion factor from exposure (in WLM) to effective dose equivalent (in mSv), as exhibited in Table 1. These include:[7,9]

- Breathing rates are greater for miners than for the average person indoors. More air is inhaled, but the average residence time in the lungs is less. Overall, this leads to a lower dose per WLM in homes than in mines.

- There is usually less dust in homes than in mines and therefore the unattached fraction of radon daughters is greater. This means a higher dose per WLM in homes than in mines.
- The mean size of the dust particles is less in the domestic environment than in mines, implying a higher dose per WLM in homes than in mines.
- In mines, inhalation of dust with uranium and other long-lived radionuclides adds to the total lung exposure. Thus, part of the effect attributed to radon and its short-lived daughters perhaps should be ascribed to other radionuclides. The latter are not present in the home environment.

There is no consensus as to the net effect of the above-cited differences in the environment. As seen in Table 1, analyses by the NCRP and James conclude that the dose per WLM is greater in homes than in mines, while the NEA group of experts and the ICRP reached the reverse conclusion. Considering explicitly the several contributing factors and the uncertainties in their application, the BEIR IV analysis took the middle ground and made the assumption that equal exposures (in WLM) "have equivalent potency in causing lung cancer" [3, p. 52].

## 5.2 Risk parameters for the general population

### 5.2.1 *NCRP Report No. 78*

It is concluded in this report that the risk of lung cancer per WLM of exposure is approximately 40% greater in the domestic environment than in mines. The primary factor which produces this difference is the larger unattached fraction of radon daughters in homes. The factor for the general population is then: $\alpha = 14 \times 10^{-6}$ per PY-WLM. The effective cumulative exposure is still given by Eq. 9.

### 5.2.2 *BEIR IV*

In BEIR IV, the same parameters ($\beta$ and $\gamma$ of Eq. 10) are taken to apply to the general population as to the miners. In its analysis, BEIR IV divided the population into four sub-groups: males and females and, for each sex, non-smokers and smokers. The background lung cancer rates, $r_o(t,x)$, for males and females are well-established and are different. The extraction from these rates of separate rates for smokers and non-smokers is more hypothetical. BEIR IV takes the ratio of risk for smokers to non-smokers as about 12 for males and about 10 for females. Together with an estimate of the numbers of smokers, it is then possible to extract separate relative background risks for these subgroups.

### 5.2.3 *ICRP Report 50*

In contrast to the NCRP and BEIR IV studies, the ICRP report concludes that the dose per unit exposure in the domestic environment is only 64% that of the mine environment due to differences in breathing rates (a factor of 0.8) and in the dust which carries long-lived radionuclides (a factor of 0.8). This reduces the risk coefficients (see Table 3) to $\alpha = 7 \times 10^{-6}$ per PY-WLM and $\beta = 0.0064$ per WLM.

Although in its consideration of miners an unweighted effective cumulative exposure is used, for the general population a distinction is made on the basis of age at exposure. Based on studies of atomic bomb survivors, it is assumed that the relative risk is three times greater for exposures received at $t_e < 20$ than at greater ages. This can be represented by tripling $\beta$ for $t_e < 20$ or by giving triple weight to exposures received before age 20 in calculating the effective cumulative dose. (A similar adjustment is not made in calculating the absolute risk.)

### 5.3 Lifetime risk for constant lifetime exposure

The individual lifetime risk is determined by summing the age-specific risks over the lifetime, as in Eq. 3. For simplicity, a constant rate of exposure is usually assumed, giving the lifetime risk for constant lifetime exposure. The lifetime risks as calculated in the various analyses are presented in Table 4, for a constant annual exposure of 1 WLM per year.

The calculated excess lifetime risk to the individual, averaged over gender and smoking habit, has a broad range in these analyses:

$$\Delta R = 0.009\text{—}0.024 \quad \textit{for constant exposure: 1 WLM/yr} \qquad (11)$$

Smokers have far higher background lung cancer risks than non-smokers, and in the relative risk models they also have higher excess risk from radon (see Table 4).

The relatively low risk estimated in the NCRP study stems in part from the discounting of past exposures. For example, for a 60-year old person, the effective cumulative exposure from NCRP Report No. 78 is

$$W = \int_5^{60} e^{-T \ln 2/20} dT = 21 \text{ WLM}$$

while for BEIR IV it is $W = 0.5 \times 45 + 10 = 32.5$ WLM, and for the ICRP study (with $\tau = 10$) it can be viewed as being $W = 3 \times 20 + 30 = 90$ WLM.

From some standpoints, there is greater disagreement than suggested by the final results for lifetime risk. The NCRP and ICRP studies obtained the same value of $\alpha$ for miners, and almost the same lifetime risk. However, as seen above, the NCRP value of $\alpha$ for the general population is twice the ICRP value. This is outweighed by the fact that in the NCRP case the effective dose is discounted with a 20-year half-life.

## 5.4 Annual fatalities due to radon

Ignoring the small differences between the actual U.S. population and a steady-state population, the annual number of cancers attributable to radon inhalation is given by Eq. 4. For a typical exposure of about 0.2 WLM per year, a U.S. population of 240,000,000, and an average lifetime of 73 years, it follows for $\Delta R$ between 0.009 and 0.024 that $\Delta F$ lies in the range:

$$\Delta F \approx 6000\text{–}16{,}000 \quad \textit{for linear extrapolation} \tag{12}$$

Allowing for the uncertainties in the analysis above, the EPA's estimate of 5000 to 20,000 lung cancer fatalities per year in the U.S. is understandable.

TABLE 4

*Lifetime risk to the general population for lifetime annual exposure of 1 WLM.*

| Study | Model | Group | $R_o$ | $R/R_o$ | $\Delta R$ |
|---|---|---|---|---|---|
| BEIR IV[3] | TSE | Non-smokers[a] | | | |
| | | Male | 0.0112 | 1.54 | 0.0060 |
| | | Female | 0.0060 | 1.57 | 0.0034 |
| | | Smokers[a] | | | |
| | | Male | 0.123 | 1.48 | 0.059 |
| | | Female | 0.0582 | 1.54 | 0.031 |
| | | Total[b] | | | |
| | | Male | 0.067 | 1.512 | 0.034 |
| | | Female | 0.025 | 1.560 | 0.014 |
| | | Both sexes | | | 0.024 |
| ICRP 50[7] | RR | Non-smokers[c] | | | |
| | | Both sexes | 0.0061 | 1.64 | 0.0039 |
| | | Total[c] | | | |
| | | Male | 0.042 | 1.64 | 0.027 |
| | | Female | 0.009 | 1.64 | 0.0058 |
| | | Both sexes | | | 0.016 |
| | AR | All[d] | | | 0.011 |
| NCRP No. 78[5] | MAR | All[e] | | | 0.0091 |
| | RR | Total (all)[f] | | | 0.009 |

[a]From Table 2-4, p. 56 (revised 9/1/88).
[b]From p. 76 and Tables 2-5, 2-8, pp. 63, 69.
[c]*Ro* based on Table 11, p. 44; R/*Ro* from Table 10, p. 42.
[d]From p. 37.
[e]From Table 10.2, p. 156.
[f]Relative risk estimate based on work of Whittemore and McMillan[18]

## 6. RISKS FOR SUBGROUPS OF POPULATION

### 6.1 Smokers and non-smokers

The question of a possible synergism between radon exposure and smoking is an old one,[19] but remains unresolved. Of the studies listed in Table 2, the Colorado plateau study (A) supports a multiplicative model for the relationship between radon exposure and smoking (i.e. a strong synergism) as does a study of New Mexico miners.[20] However, the authors of the Malmberget miners (B) and the Czech miners (E) studies find a nearly additive interaction (i.e., little or no synergism) and the Newfoundland study (F), with caveats on statistical significance, also found no dependence of excess radon risk on smoking. The remaining studies in Table 2 lacked data on smoking status. Other studies, with smaller numbers, do not point clearly in any direction.

Nonetheless, most analysts believe that there is a strong synergism. This is reflected in the much greater excess risks for smokers than for non-smokers found in the BEIR IV and ICRP 50 studies (see Table 4). If these numbers are correct, the "radon problem" becomes a subcategory of the smoking problem. With smoking eliminated, the lifetime risk for lifetime exposure to 1 WLM per year would be only about 0.004 and the annual number of U.S. radon induced lung cancer fatalities would be under 3000. This may be compared to an estimated total of 136,000 lung cancer fatalities in the U.S. in 1987, of which over 110,000 are attributed to smoking.[2]

However, despite adopting a multiplicative model for analytic purposes, BEIR IV points out that "a submultiplicative model...may provide a more accurate description" and concludes that "the interaction between smoking and exposure to radon progeny remain(s) incompletely resolved" [3, pp. 559 and 10]. ICRP 50 concludes only that the data "suggest" a synergism and notes that the "number of lung cancer cases among the non-smoking miners is too low to quantify the magnitude of the synergistic factor with sufficient accuracy" [7, p. 25]. In short, the case for a multiplicative effect remains a tentative one.

Government standards for indoor radon levels, such as the EPA recommendations, do not differentiate between the homes of smokers and non-smokers. No authoritative group has been willing to assure people that, if they refrain from smoking, it is safe to live in houses with high radon levels. This reticence appears to be advisable, in view of the inconclusive nature of the evidence on the interaction of radon and smoking.

### 6.2 Effects of gender

Data on lung cancer in miners pertains to exposure of adult males. Conclusions as to radon effects for women are dependent on the model chosen. In the U.S., the background rates for men are about twice as great as those for women, if one separately considers non-smokers and smokers (see Table 4). Thus, in a relative risk model, as used by BEIR IV, the dangers of radon are twice as great for men as for women, corrected for the same smoking practices.

ICRP 50, on the other hand, raises the possibility that the U.S. difference between the sexes in background lung cancer rates "might be caused by occupational factors" and may not apply to other countries [7, pp. 40]. The ICRP report assumes the same relative risk parameter, $\beta$, for males and females, and appears to use the same background risk. For the population as a whole, with men smoking more, the quoted excess risk is greater for men than women.

NCRP 78 estimates a difference of about 10% between the radon dose to males and females at the same exposure levels, based on size and breathing differences, but essentially ignores this small difference for purposes of further analysis. Background rates are irrelevant in the NCRP absolute risk model, and the excess lung cancer risks for males and females are roughly equal.

In aggregate, there are suggestions that women may be at less risk than men from radon, but there is neither strong evidence nor a consensus opinion that this is the case.

## 6.3 Comparative hazards of childhood exposures

There is at least anecdotal evidence that people are more willing to undertake radon reduction measures if they think that their children are at risk, and the EPA in its brochure for the general public says that "children may be more at risk than adults"[1] This additional risk is not a risk of lung cancer in childhood, but rather of subsequent cancer due to exposures during childhood. Indications that childhood exposures involve greater risk stem from studies of atomic bomb survivors. This had led ICRP 50 to assign a triple risk (or a triple weight to exposures) for exposure before age 20. However, the BEIR IV study gives the same weight to childhood and adult exposures, and the NCRP study discounts childhood exposures through the exponential factor in Eq. 9.

The differences are dramatic in the risk of cancer after, say, age 40. For equal exposures, the ratio of the risk from exposure at age 10 to the risk from exposure at age 30 is *three* in the ICRP 50 analysis and *one-half* in the NCRP 78 analysis.

## 6.4 Comparative risks at older ages

The background lung cancer rate is higher at age 80, say, than at age 60. Therefore, in a constant relative risk model the risk of cancer due to past exposures radon will be greater. However, in the TSE model of BEIR IV, the relative risk factor is a taken to be a factor of 2.5 smaller at age 80, almost cancelling out the difference in background rates, and making the excess risk about the same at the two ages.

## 7. DIRECT STUDIES OF THE GENERAL POPULATION

If indoor radon causes lung cancer, this effect should be manifested in elevated lung cancer rates among groups in the general population which have experienced high exposures. As yet, this has not been seen in a convincing fashion and existing studies give disparate results. The inconclusive nature of the data, as of early 1987, has been reviewed previously.[21] It did not change greatly in the following two years.

A positive effect was found in a study of 292 Swedish women who were diagnosed as having lung cancer.[22] These were compared to 584 referents, matched in age. For each, it was determined if the characteristics of their housing suggested relatively high radon exposure. An indication of higher exposure was found for 22 of the lung cancer cases (7.5%) and 21 of the referents (3.6%), suggesting a relation between indoor radon exposure and lung cancer. Aside from the problems of small numbers, the study suffered from incomplete knowledge of the actual radon levels. Another Swedish study, by Axelson et al., also gave inconclusive suggestions of a positive correlation.[23]

An alternative approach is to search for correlations between lung cancer mortality rates and average radon level by geographical region. Such correlations should exist if the linearity assumption is valid and population mobility is low. In a continuing study, Cohen has considered lung cancer rates and average radon concentration for U.S. counties.[24] He has found a negative correlation between lung cancer rates and average indoor radon concentrations, which persists despite attempts to eliminate confounding factors. If not due to undetected artifacts in the data, this would mean a failure of the linearity assumption at low radon exposures.

None of these studies has been accepted as being definitive. They deal with an issue of prime importance. The importance goes beyond radon itself. If the linearity assumption overestimates the rate of cancer production by alpha particles, a densely ionizing radiation, there would be a strong presumption that it overestimates the effects for lightly ionizing radiation, for example from Chernobyl fallout. It has often been suggested that some sort of sub-linear (e.g., linear-quadratic) dose-response curve is in fact appropriate for lightly ionizing radiation, and this would lend support to that belief.

## 8. SUMMARY AND CONCLUSIONS

### 8.1 What is known

Despite the emphasis on uncertainties in the discussion above, a number of important points have been well established:

- Radon in mines, especially uranium mines, has caused high lung cancer mortality among exposed miners.
- For most members of the general population, the effective dose equivalent from indoor radon is greater than from all other radiation sources combined. For the U.S., a typical dose from indoor radon is 2 mSv per year while the average dose from all other sources together, including medical exposures, is about 1.6 mSv per year.[6]
- Indoor radon causes a substantial number of lung cancers among the general population. The calculated rate for the U.S., assuming linearity, is of the order of 10,000 deaths per year, with an uncertainty of a factor of two or three. Even if the linearity assumption is abandoned, there is a substantial risk in homes where exposures approach those in mines, e.g., the million or so homes in which the exposure exceeds 2 WLM per year.[25]

In short, if ionizing radiation exposures of any sort are a public health problem, indoor radon represents the most serious problem.

## 8.2 What is not well known

Large uncertainties remain, aside from the imprecision of the numerical estimates cited above:

- There is as yet no firm empirical evidence establishing the validity of a linear extrapolation from the miners data to the doses and dose rates of typical homes.
- The relative impact of radon on smokers and non-smokers is not well established. If there is a strong synergism between radon and smoking, the radon risk to non-smokers may be very small.
- Women may be at less risk than men, but this is not clearly established.
- It is not clear how the risk varies with age at exposure (i.e., if childhood exposures are more or less important than adult exposures) or how the risk from past exposures increases (or decreases) with age.

Overall, we lack a satisfactory understanding of how radon-induced lung cancer may be distributed among sub-groups of the population. Even more importantly, we lack certain knowledge that the low doses received by most of the population create an appreciable risk of lung cancer.

## 8.3 Potential for further information

The bulk of information linking radon exposure to lung cancer mortality has come from studies of miners. Obtaining more information from this source will be difficult because present mining practices give much lower exposures than were characteristic of the 1940s and 1950s. The cohorts of miners from that period can continue to be studied to obtain a picture of the effects of radon over long periods, but already much of the pertinent information has been extracted. It remains to be seen whether substantial further information will come from data from other countries (e.g., France and China). Studies of miners exposed at lower concentrations can be informative, but here confounding factors may play a larger role and statistically significant information is harder to obtain.

On the other hand, the potential for studies of the general population has hardly been tapped. Until large-scale epidemiological studies of the general population are undertaken and completed, or until well-founded models are developed for cancer induction, it will not be possible to place great confidence in estimates of lung cancer mortality from indoor radon. Existing estimates are important in highlighting a possible problem, but they do not establish either the existence or the magnitude of the problem with certainty. Nonetheless, pending more conclusive information, prevailing scientific opinion, as well as social prudence, suggests that we adopt policies based on the risk estimates discussed above, i.e. corresponding to some 5,000 to 20,000 US. lung cancer deaths per year from indoor radon.*

*Note added. Subsequent to the completion of this manuscript, new cancer risk estimates have been published in BEIR V.[a] These differ substantially from those of BEIR III.[10] BEIR V uses a linear, relative risk model for lung cancer and for most other cancers. For a constant lifetime effective dose equivalent of 10 mSv per year, the lifetime mortality risk (averaged over sexes) is 0.056. This is approximately twice the BEIR III relative risk model result. If the BEIR V estimate is used, with the approach of Section 3.2, the calculated number of U.S. cancer fatalities from indoor radon is approximately 40,000 per year.

Good agreement was found earlier between results of the dosimetric approach using BEIR III (Section 3.2) and the extrapolation from the experience of miners of BEIR IV[3] and other studies (Section 5.3). The new BEIR V results appear to negate this agreement.

The comparison with BEIR IV depends upon a conversion from exposure in WLM (the key measure in BEIR IV) to dose equivalents in sieverts (the key measure in BEIR V). The disagreement may stem from errors in the standard parameters used in this conversion. Alternatively, it may be related to the difference in the exposure conditions for the Hiroshima and Nagasaki victims (the source of most of the BEIR V data) and for miners. Until the situation is clarified, it would be well to view the range of 5000 to 20,000 U.S. lung cancer deaths per year due to indoor radon as understating, not overstating, the full range of uncertainty (in both directions).

---

[a] *Health Effects of Exposures to Low Levels of Ionizing Radiation, BEIR V.* 1990. Report of the Committee on the Biological Effects of Ionizing Radiations, National Academy Press, Washington, D.C.

## REFERENCES

1. *A Citizen's Guide to Radon, What It Is and What To Do About It.* 1986. Report OPA-86-004. Environmental Protection Agency, Washington, D.C.
2. *Cancer Facts and Figures—1987.* 1987. American Cancer Society, New York, NY.
3. *Health Risk of Radon and Other Internally Deposited Alpha-emitters* (BEIR IV). 1988. Report of the Committee on the Biological Effects of Ionizing Radiations. National Academy Press, Washington, D.C.
4. *Dosimetry Aspects of Exposure to Radon and Thoron Daughter Products.* 1983. NEA Experts Report. OECD, Paris, France.
5. *Evaluation of Occupational and Environmental Exposures to Radon and Radon Daughters in the United States.* 1984. NCRP Report No. 78. National Council on Radiation Protection and Measurements, Washington, D.C.
6. *Ionizing Radiation Exposure of the Population of the United States.* 1987. NCRP Report No. 93. National Council on Radiation Protection and Measurements, Washington, D.C.
7. Lung Cancer Risk from Indoor Exposures to Radon Daughters. 1987. ICRP Publication 50, *Annals of the ICRP* 17, No. 1.
8. *Sources, Effects, and Risks of Ionizing Radiation.* 1988. Report of the United Nations Scientific Committee on the Effects of Ionizing Radiation. United Nations, New York, NY.
9. James, A.C. 1988. Lung Dosimetry, pp. 259-309. *In:* W.W. Nazaroff and A.V. Nero, Jr. (Eds.). *Radon and Its Decay Products in Indoor Air.* Wiley Interscience, New York, N.Y. pp. 518.
10. *The Effects on Populations of Exposure to Low Levels of Ionizing Radiation: 1980* (BEIR III). 1980. Report of the Committee on the Biological Effects of Ionizing Radiations. National Academy Press, Washington, D.C.
11. Hornung, R.W. and T.J. Meinhardt. 1987. Quantitative risk assessment of lung cancer in U.S. uranium miners. Health Phys. 52:417-430.
12. Radford, E.P. and K.G. St. Claire Renard. 1984. Lung Cancer in Swedish Iron Miners Exposed to Low Doses of Radon Daughters. New Eng. J. Med. 310:1485-94.
13. Howe, G.R., R.C. Nair, H.B. Newcombe, A.B. Miller and J.D. Abbatt. 1986. Lung Cancer Mortality (1950-1980) in Relation to Radon Daughter Exposure in a Cohort of Workers at the Eldorado Beaverbrook Uranium Mine. J. Nat. Canc. Inst. 77:357-362.
14. Muller, J., W.C. Wheeler, J.F. Gentleman, G. Suranyi and R. Kusiak. 1985. Study of Mortality of Ontario Miners, pp. 335-343. *In:* H. Stocker (Ed.). *Occupational Radiation Safety in Mining.* Proceedings of the International Conference, Canadian Nuclear Association, Toronto, Ont.
15. Ševc, J., E. Kunz, L. Tomásek, V. Placek and J. Horácek. 1988. Cancer

in Man After Exposure to Rn Daughters. Health Phys. 54:27-46.

16. Morrison, J. et al. 1985. Lung Cancer Mortality and Radiation Exposure Among the Newfoundland Fluorspar Miners, pp. 365-368. *In:* H. Stocker (Ed.). *Occupational Radiation Safety in Mining.* Proceedings of the International Conference, Canadian Nuclear Association, Toronto, Ont.
17. Howe, G.R., R.C. Nair, H.B. Newcombe, A.B. Miller, J.D. Burch and J.D. Abbatt. 1987. Lung Cancer Mortality (1950-80) in Relation to Radon Daughter Exposure in a Cohort of Workers at the Eldorado Port Radium Mine: Possible Modification of Risk by Exposure Rate. J. Nat. Canc. Inst. 79:1255-60.
18. Whittemore, A.S. and A. McMillan. 1983. Lung Cancer Mortality Among U.S. Uranium Miners: A Reappraisal. J. Nat. Cancer Inst. 71:489-499.
19. Robkin, M.A. and D. Bodansky. 1988. Indoor Radon. Adv. in Nucl. Sci. and Tech. 20:155-223.
20. Samet, J.M., D.R. Pathak, M.V. Morgan, M.C. Marbury, C.R. Key and A.A. Valdivia. 1989. Radon Progeny Exposure and Lung Cancer Risk in New Mexico U Miners: A Case-Control Study. Health Phys. 56:415-421.
21. Jackson, K.L., J.P. Geraci and D. Bodansky. 1987. Observations of Lung Cancer Mortality: Evidence Relating Lung Cancer Mortality to Radon Exposure, pp. 91-111. *In:* D. Bodansky, M.A. Robkin and D.R. Stadler (Eds.). *Indoor Radon and Its Hazards.* University of Washington Press, Seattle, WA. pp. 147
22. Svensson, C., G. Eklund and G. Pershagen. 1987. Indoor exposure to radon from the ground and bronchial cancer in women. Int. Arch. Occup. Environ. Health 59:123-131.
23. Axelson, Olav et al. 1988. Indoor radon exposure and active and passive smoking in relation to lung cancer. Scand. J. Work Environ. Health 14:286-292.
24. Cohen, B.L. 1990. An Experimental Test of the Linear-No Threshold Theory of Radiation Carcinogenesis. Chapter 24, 319–330 (this book).
25. Nero, A.V., Jr. 1988. Radon and Its Decay Products in Indoor Air: An Overview. pp. 1-53. *In:* W.W. Nazaroff and A.V. Nero, Jr. *Radon and Its Decay Products in Indoor Air.* Wiley-Interscience, New York, NY. pp. 518.

*Environmental Radon: Occurrence, Control and Health Hazards.* Edited by S. K. Majumdar, R. F. Schmalz and E. Willard Miller 

# *Chapter Twenty-One*

# DOSIMETRY OF RADON DECAY PRODUCTS IN THE HUMAN LUNG

**KENNETH R. KASE**

Department of Radiation Oncology
University of Massachusetts Medical Center
55 Lake Avenue North
Worcester, MA 01655

## INTRODUCTION

The radiation dose to the lung resulting from the inhalation of radon gas is delivered, not by the radon decay, but by the decay of its particulate progeny which lodge in the mucous lining of the bronchial airways. There are three isotopes of radon that occur naturally in the decay of uranium and thorium, $^{219}Rn$, $^{220}Rn$ and $^{222}Rn$. Because of the short half-lives of $^{219}Rn$ (3.9 s) and $^{220}Rn$ (56 s) compared with the half-life of $^{222}Rn$ (3.8 d), only the latter isotope contributes significantly to the radiation dose to the lung.

Since the four immediate decay products of $^{222}Rn$ have half-lives much shorter than the radon, a certain level of equilibrium in the air is usually established between the radon gas and its decay products. The level of equilibrium depends upon the radon source and the removal of the decay products from the air. The relationship between the amount of decay products in the air, and consequently the radiation dose to the lung, and the radon gas concentration will depend on the extent of this equilibrium.

By convention, the concentration of radon decay products in air is measured in units of working levels (WL), and cumulative exposures over time are measured in working level months (WLM). The WL is a specified quantity of alpha particle energy released from the decay products in 1 liter of air and can be related to the concentration of radon gas in the air if the extent of the equilibrium between the radon and its decay products is known. If the equilibrium is complete, 1 WL is equal to $3.7 \times 10^3$ Bq/m$^3$ (100 pCi/L) of radon. A working month is defined as 170 h. Thus, continuous exposure to radon will result in an exposure for about 51.6 working months per year.

The average individual in the U.S. is exposed to an average radon concentration of about 37 $Bq/m^3$ (1 pCi/L). Over a year, this would amount to about 0.52 WLM if the equilibrium between radon and its decay products is complete. As we will see, the equilibrium is seldom complete in outdoor and home environments, and so the average cumulated exposure to radon decay products is generally less than 0.52 WLM.

The majority (>85%) of the radiation dose delivered to the bronchial tissue from the decay of radon progeny arises from the alpha particles emitted by $^{218}Po$ and $^{214}$ Po. Alpha particle tracks are nearly straight lines because the particles are scattered very little by the material through which they pass. The result is that the actual rates of energy loss are close to the average rate, known as the stopping power. It also means that alpha particles of a given energy have a relatively precisely defined range, which is the average distance travelled before coming to rest in a material. For alpha particle dosimetry in bronchial tissue scattering can be ignored and it is sufficient to consider only the stopping powers and ranges of the alpha particles. In some cases it is satisfactory to calculate the average dose (D) to a tissue volume, which is the product of the mean number of alpha particles emitted per unit mass of tissue (C) and their energy E,

$$(D) = (C)E. \tag{1}$$

Equal doses of different radiations do not always result in the same biological effect. The relative biological effectiveness (RBE) has been introduced to compare the doses of different radiations required to produce the same effect. This RBE depends not only upon the radiation, but also upon the dose and the effect being measured. For radiation protection purposes, a single quantity called the quality factor Q, related to the RBE, was introduced so that the risks of various types of radiations could be compared. The dose equivalent H is the quantity that measures the risk of radiation exposure, where

$$(H) = (D)Q. \tag{2}$$

The value of Q for alpha particle radiation compared with gamma radiation as a reference is taken to be 20.[1]

Estimation of the lung dose depends on physical factors related to radon, its particulate decay products and the particles present in the atmosphere, as well as biological factors related to the bronchial airways. Some of the physical factors that are important are:

1) the relative radiological equilibrium between radon and its decay products;
2) the fraction of the particulate decay products, especially $^{218}Po$, that are attached to aerosol particles;
3) the particle size;

4) the model chosen to describe particle deposition in the airways; and
5) the dose calculation model.

The important biological factors related to the lung model are:

1) the breathing pattern;
2) the airway size and structure;
3) the thickness of the mucus layer;
4) the clearance rate of deposited particles; and
5) the location of the target cells that are critical to the induction of cancer.

These factors will be discussed in the following sections which detail the structure of the respiratory tract, the deposition and clearance of particles, and the calculation of radiation dose.

The discussion here is not intended to be exhaustive, nor will it cover the historical evolution of the dosimetry of radon and its decay products. It will summarize in some detail the current models and dose estimations, and give an evaluation of the uncertainties that yet remain in the dose estimates.

## STRUCTURE OF THE RESPIRATORY TRACT

The respiratory tract can be divided into three major regions on the basis of anatomy and function, the nasopharyngeal (N-P), the tracheobronchial (T-B), and the alveolar-interstitial (A-I) as shown in Figure 1. Beginning at the anterior nares, the N-P region extends to the larynx. The T-B region begins at the larynx and includes the trachea, the ciliated bronchial airways and the terminal broncholes. The alveolar ducts, terminal sacs and associated capillaries and lymphatic ducts comprise the A-I region. The T-B region is the portion of the respiratory tract that is of interest in determining the radiation dose related to the risk of lung cancer.

Absorbed dose to the tissues of the respiratory tract will depend upon the deposition of inhaled radon progeny on the airway surfaces. To evaluate this requires a simplified geometrical model of airway size and branching. The most widely used models have been the Weibel "A" model[3] which assumes that branching is symmetrical and that the airways at each branching level are identical, and the Yeh-Schum model[4] which is based upon extensive measurements of airway size in a cast of a human lung. Their model of the first few airway branchings is shown in Figure 2. Yeh and Schum recognized that their measurements overestimated the airway sizes because they were made in a fully inflated lung and recommended that they be scaled down to correspond with a normally inflated lung. More recently Phalen et al.[6] have reported measurements of airway sizes made in casts of 20 human lungs taken from subjects ranging in age from 11 days to 21 years. These measurements allow an estimate of appropriate scaling of airway sizes for age-dependent dose estimates.

It is thought that the dimensions reported by Phalen et al. are more representative of the actual airway dimensions of the lung under normal respiration than are the dimensions given by the Weibel A or Yeh-Schum models. However, it is not known which model is the most realistic overall.[2]

In addition to the dimensions of the respiratory tract, the thicknesses of the bronchial epithelium and mucus layer, and the location of the target cells for cancer induction are critical to the estimation of dose because alpha particles have a very limited range in tissue. This is illustrated in Figure 3 which shows the depth dose in soft tissue for alpha particles of various energies. The 6.0 MeV $^{218}$Po alpha particle has a range of less than 50 $\mu$m, and the 7.7 MeV $^{214}$Po alpha particle has a range of about 70 $\mu$m. Gastineau et al.[8] reported that there is little variation in the thickness of the mucus layer (7 $\mu$m) and cilia length (7.5 $\mu$m) throughout the bronchial airway from the mainstem to the bronchioles in the normal lung. Their measurements of the epithelial thicknesses in the T-B region are given in Table 1.

There is considerable uncertainty about the critical target for cancer induction. Harley and Pasternack[5] assumed that the target cells were likely to be the proliferating basal cells of the bronchial epithelium. This led them to choose

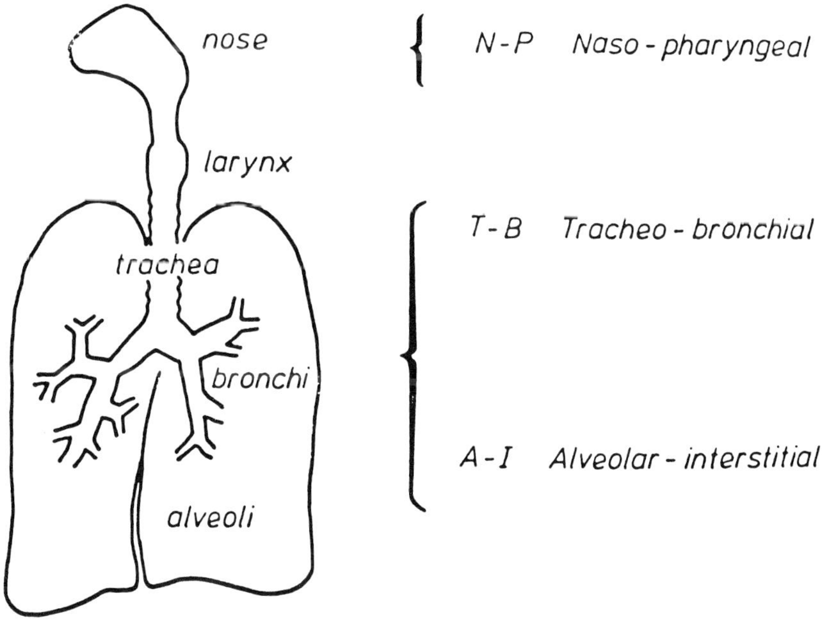


FIGURE 1. Regions of the respiratory tract. From Ref. 2

a depth of 22 $\mu$m below the epithelial surface for their dose calculations. However, Trump et al.[9] indicate that because the induction of cancer is probably a multistage process, direct conversion of epidermal cells can lead to epidermoid carcinoma. Also, Masse[10] has concluded that the cytodynamics of bronchial

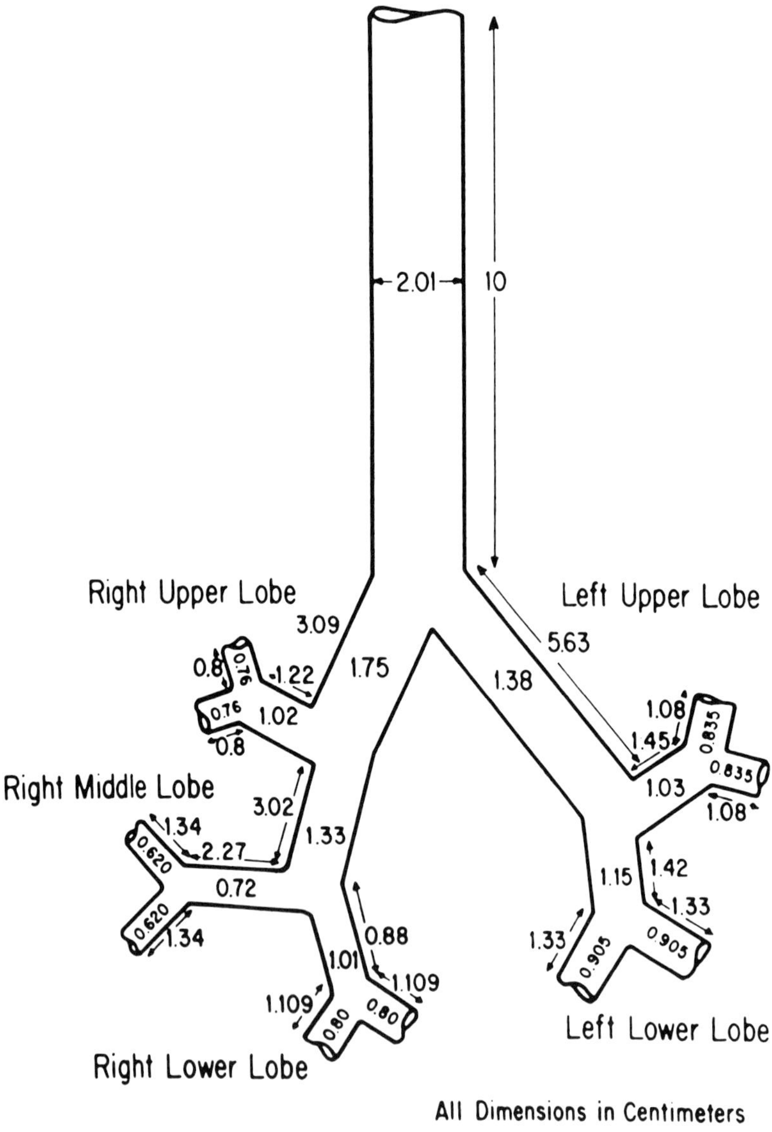


FIGURE 2. Average bronchial morphometric characteristics[4] for the human male lung in the first few airways. From Ref. 5.

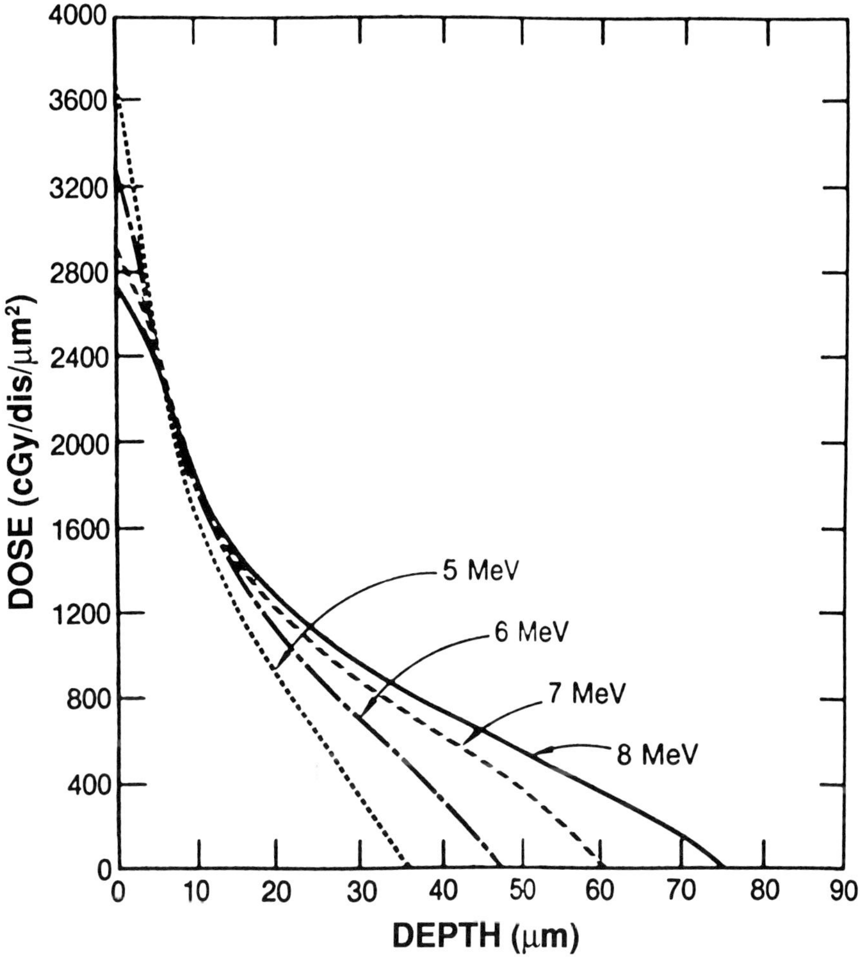


FIGURE 3. Depth dose curves for alpha particles emitted on the surface of cylinders with walls composed of tissue. From Ref. 7.

TABLE 1

*Distribution of Epithelial Thickness*

| Bronchial Airway (approx. generation) | | Thickness (μm) Minimum | Mean | Maximum |
|---|---|---|---|---|
| Main | (1) | 70 | 80 | 85 |
| Lobar | (2) | 25 | 50 | 70 |
| Segmental | (3-5) | 15 | 50 | 100 |
| Transitional | (9,10) | 12.5 | 20 | 30 |

cells and the histogenesis of lung tumors do not allow the exclusion from risk of cells directly below the mucus layer. Consequently, there is a wide range of depth that can be assigned to the cells at risk, and James[2] suggests using the average dose to the epithelium rather than the dose at a specific depth.

## DEPOSITION AND CLEARANCE OF PARTICLES

The radiation dose to the bronchial airways ultimately depends on the deposition and retention of the particulate decay products from radon. Breathing patterns, airway size, particle size and mucociliary clearance will all affect the particle deposition and retention. Estimating the deposition of particles is a complex problem that must account for breathing rate, tidal volume, airway and particle size. Thus, deposition would be expected to be dependent upon the level of physical activity as well as age.

Tidal volume and breathing rate determine the flow rate of air in the trachea. In the upper bronchi an increased breathing rate for a given tidal volume will increase the rate of intake of radon progeny; however, the flow rate also increases causing a decrease in particle deposition. Consequently, the product of fractional deposition, tidal volume and breathing rate is relatively constant. The flow rate, and thus the air velocity, in the smaller airways is less affected by breathing rates and is usually low enough that the airflow can be assumed to be laminar. This means that particle diffusion is the predominant mechanism for particle deposition in the lung, and that the deposition in each airway generation can be calculated using formulations derived by Ingham.[11] However, Cohen[12] has pointed out, based on her measurements, that the diffusion deposition should be corrected for turbulent diffusion in airway generations 0 to 6.

Recently a number of measurements have been published on respiratory tract deposition of particles in the size range from 0.005 $\mu$m to 5 $\mu$m. Comparing these measurements with predictions of models shows generally good agreement as seen in Figure 4. It should be noted that the fraction of the total deposition that occurs in the bronchi increases as the particle size decreases. At particle sizes below 0.005$\mu$m, the size of the unattached radon progeny, almost all the deposition is in the bronchi. The overall particle deposition in the three regions of the respiratory tract is shown in Figure 5 for an adult male working and at rest. The mean particle size for attached radon progeny in ordinary room air is approximately 0.1 $\mu$m.

Radiation dose is dependent, not only on the deposition of particles, but also on their clearance. If a radioactive particle is cleared from the lung before it decays, it does not contribute to the radiation dose. There are very few data on which to base estimates of the clearance of particles from the lung. The variations in transport velocity resulting from various models are compared in Figure 6, together with some limited human data. However, large differences in transit time result in only small variations in the deposited radioactivity and are likely to introduce only a small uncertainty in the dose estimate.

## RADIATION DOSE

The estimation of radiation dose to the lung in humans will depend upon the selection of various physical and biological parameters to describe the characteristics of the aerosol breathed and the lung function. Several such models have been described and as long as the assumptions about the atmosphere are the same, the results are in reasonable agreement. The fraction of unattached $^{218}Po$ is usually assumed to be between 0.05 and 0.07, based on some limited measurements, and the equilibrium factor is taken to be about 0.5. James[2] has pointed out that in estimating the dose to basal cells in the bronchi and segmented bronchi the unattached fraction of the total alpha particle energy determines

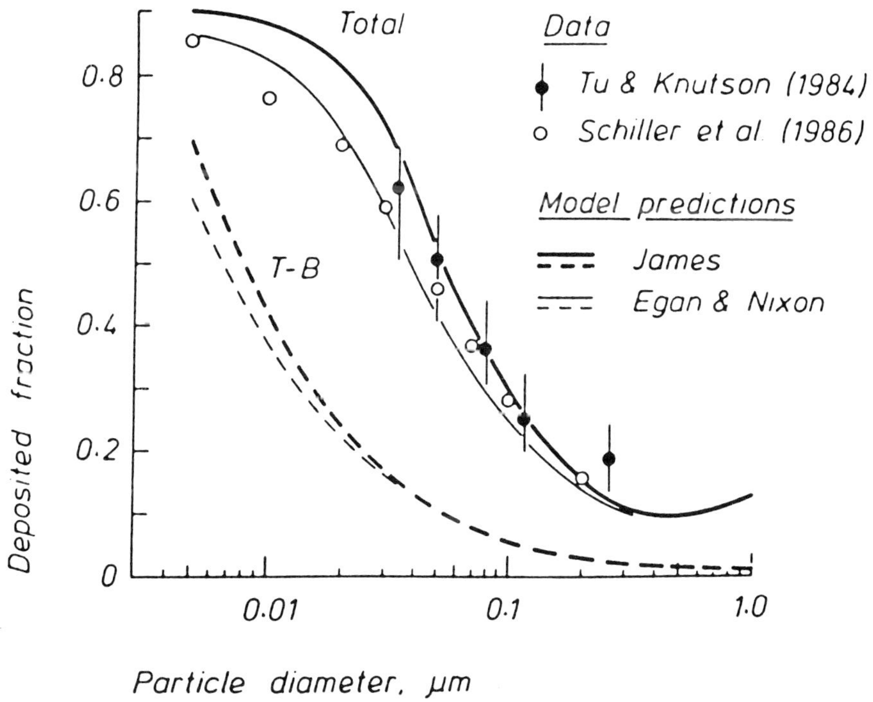


FIGURE 4. Comparison of calculated deposition in the respiratory tract with experimental data (Tu and Knutson[13] and Schiller et al.[14]) for aerosols with particles sizes less than one micrometer. From Ref. 2.

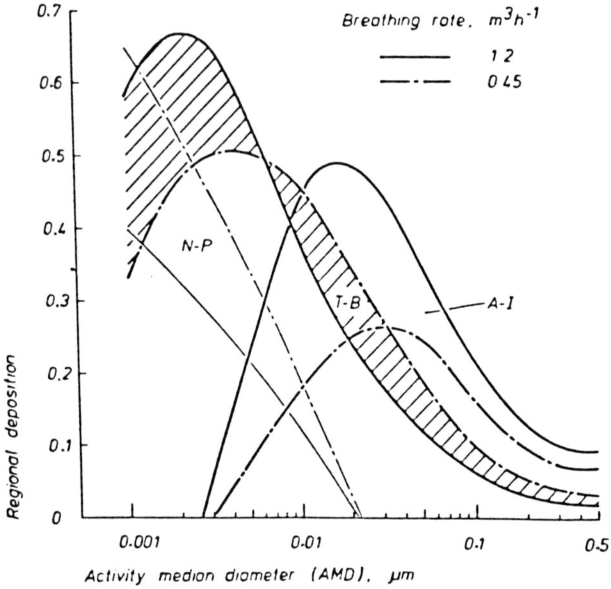


FIGURE 5. Fractional deposition of radon progeny in regions of the adult respiratory tract, calculated as a function of aerosol size and for two breathing rates. From Ref. 2.

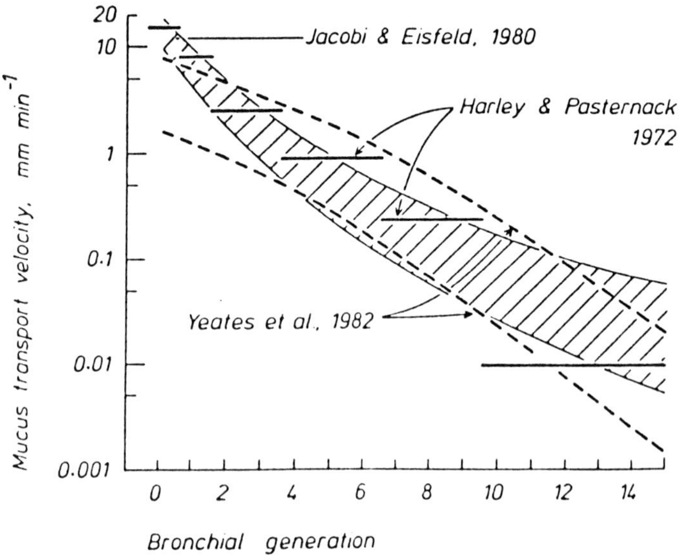


FIGURE 6. Ranges of mucous velocities taken by various authors to represent clearance from each bronchial generation. Yeates et al.[15] derived the range shown from different clearance times observed in human subjects.[16,17] From Ref. 2.

the dose conversion factor (dose per unit exposure to radon progeny) independently of the equilibrium factor.

Unattached radon decay products range in size from 0.001 μm to about 0.003 μm. Aerosol particles will have a distribution in size depending upon the source of the aerosol and existing atmospheric conditions. The mean particle size is usually in the range from 0.01 μm to 0.5 μm.

Breathing rate and bronchial morphometry affect the inhalation and deposition of particles, and so, the dose will depend upon physical activity and age. The estimates of dose will depend upon the lung model used in the calculation. The models used are usually those published by Weibel[3] and Yeh and Schum[4] for the adult male and scaled appropriately for dose estimates for females and children. The estimates of radiation dose to the adult male lung from various sources are compared in Table 2 in units of mGy/WLM.

It can be seen that, for equivalent atmospheres, the dose conversion factors are not too different. The Weibel model of lung morphometry typically results in dose estimates about 30 percent greater than the Yeh-Schum model, which is thought to be a more accurate description of the bronchial airway dimensions and branching.[21]

It should be noted that the Harley and Cohen conversion factor is based on updated modelling of the particle deposition resulting from recent measurements, and assuming a mean particle size of 0.12 μm. The value given in Table 2 is the factor calculated in generation 4 of the bronchial tree and averaged over all lobes, to compare with the other determinations. Averaged

TABLE 2

*Dose Conversion Factors for Radon Progeny*

| Source | $f_a$ | $f_p$ | F | Lung Model | Target | Dose Conversion Factor (mGy/WLM) |
|---|---|---|---|---|---|---|
| Hoffman[18] | 0.2 | 0.035 | 0.5 | Weibel | basal cells (45 μm) | 5.5 |
| Wise[19] | 0.1 | 0.058 | 0.1 | Weibel | basal cells (45 μm) | 21.4 |
| Harley & Pasternack[5] | 0.07 | 0.01 | 0.7 | Yeh-Schum | basal cells (22 μm) | 5.7 |
| NCRP Report | 0.07 | 0.01 | 0.7 | Yeh-Schum | basal cells | 7.0 |
| No. 78[20] | 0.07 | 0.01 | 0.7 | Weibel | (22 μm) | 9.5 |
| Harley & Cohen[21] | 0.07 | 0.012 | 0.5 | Yeh-Schum | basal cells (22 μm) | 4.2 |
| James[2] | 0.07 | 0.01 | 0.7 | Yeh-Schum | epithelium (average) | 7.2 |

$f_a$ is the unattached fraction of $^{218}$Po activity.

$f_p$ is the unattached fraction of the potential alpha particle energy.

F is the fraction of complete equilibrium between $^{222}$Rn and its progeny.

over the entire lung, the dose conversion factor is 4.0 mGy/WLM. The estimated dose is higher in single lobes and in some other generations. Tables of individual dose conversion factors are given in Harley and Cohen[21] for each lobe and 16 generations of the bronchial tree.

The conversion factor attributed to James is for the atmosphere given in the table. However, James[2] states that a value for $f_p$ of 0.05 is more likely correct for the average home environment. This value, together with a mean particle size of 0.1μm would yield a dose conversion factor of 13 mGy/WLM.

As mentioned earlier, location of the target cell, particle size, particle attachment, radiation equilibrium, breathing rate, and age will all influence the radiation dose estimate. From Table 2 it can be seen that the location of the target has only a minor influence on the dose, e.g. comparing the NCRP Report No. 78 value with that of James. The unattached fraction is narrowly distributed in particle size and its size will have little affect on the dose. However, the particle size distribution of the attached fraction can greatly influence the dose because the dose decreases approximately inversely with the particle size. James[2] showed that the dose is not significantly affected by the fractional equilibrium if unattached alpha particle energy is specified rather than unattached fraction of $^{218}$Po. He also gave a complex diagram that relates $f_a$, $f_p$ and F approximately in the following way:

$$f_p = C(F)*f_a/F, \tag{3}$$

$$C(F) = 0.1(1 + 0.070/F)^{-1} \tag{4}$$

The *mean* bronchial epithelial dose per unit exposure from unattached alpha particle energy is about 15 times greater than the dose per unit exposure from particles attached to an aerosol with a mean particle size of 0.1 μm. In the *segmental bronchi* (generations 3-5) the dose to the epithelium from unattached alpha particle energy is about a factor of 30 greater than the dose from 0.1μm particles. Figure 7 shows how the range of dose conversion factors calculated for various lung morphometries and clearance rates varies with particle size. These are average doses calculated to the epithelial cells in the segmental bronchi. The graph also shows that the dose calculated to basal cells at the depth of 22 μm is uniformly a factor of two less than the dose averaged over the epithelium for the same conditions. (Also compare, in Table 2, the dose conversion factors of Harley and Cohen and James.)

Doses in the bronchial and segmental bronchi regions increase by about a factor of three as breathing rate increases from complete rest to active exercise. The influence of particle size is negligible.

Breathing rate and lung dimensions are the major factors influencing the age-dependence of the lung dose. Doses will be higher in the newborn, primarily because of the smaller dimensions of the bronchi. In the newborn, the dose to the epithelium of the segmental bronchi is estimated to be a factor of 2 to 2.5 greater than that to an adult.[2,20] The dose to the lungs of a child depends upon the level of activity and the breathing rate as well as the bronchial dimensions. Estimates for a 10 year-old child vary from a factor of 1.25, assuming only changes in bronchial size,[2] to a factor of 2.3[22] greater than the dose to an adult.

Uncertainties associated with the various models and assumptions used in dose calculations have been estimated.[20] These uncertainties are detailed in Table 3. Based upon this analysis, the estimated dose to the bronchial epithelium from

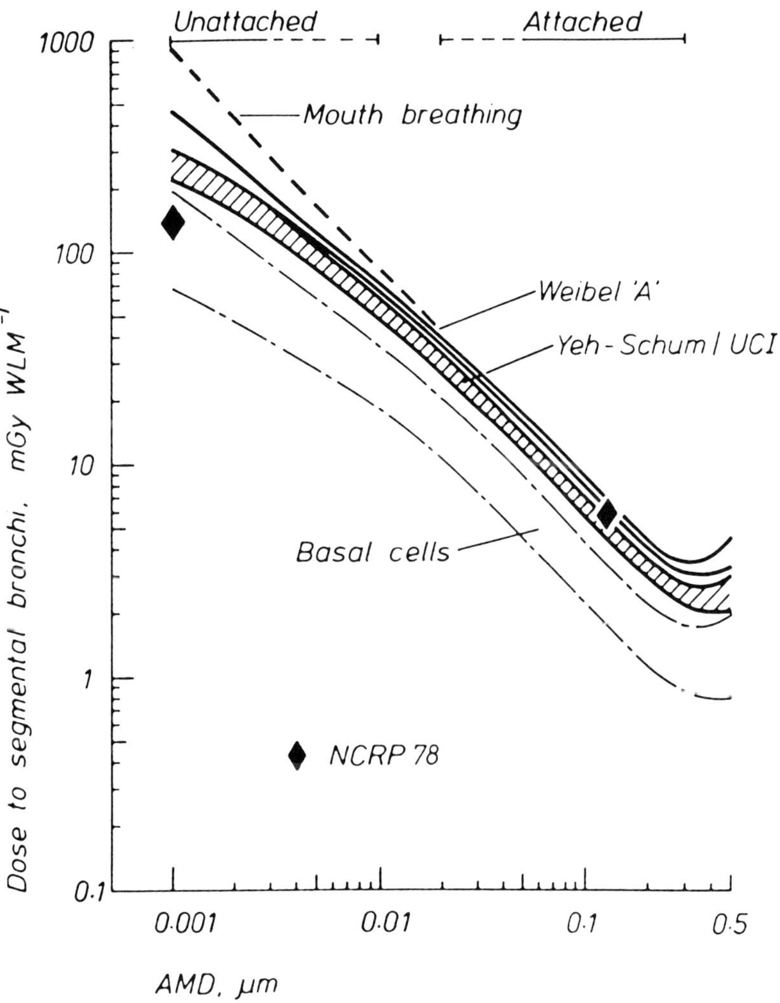


FIGURE 7. Doses averaged over epithelial cells in segmental bronchi per unit exposure to potential alpha energy. Doses derived by the NCRP[20] are shown for comparison. From Ref. 2.

the inhalation of radon decay products is probably accurate within a factor of two.

There is some gamma radiation associated with the decay of radon and its progeny, as well as some transfer of radon gas and its particulate progeny from the lung to other parts of the body. However, the doses to body organs other than the lung from radon and its progeny are about 1 percent of the dose to the bronchial epithelium, and can be neglected when estimating risk.

## SUMMARY

In the previous section, estimates were presented of the dose conversion factor for exposure to radon decay products in air. The final step in determining the annual dose to the lung is to estimate the environmental exposure to radon progeny. From measurements inside homes and out of doors, it is estimated that the average radon level to which people in the U.S. are exposed is about 37 Bq/m$^3$ (1 pCi/L). This is based partly on the assumption that an individual spends approximately 20 h per day indoors. If one assumes an equilibrium factor of 0.7, this radon gas concentration results in a radon progeny concentration of about 0.006 WL, and an annual exposure of 0.29 WLM. The annual lung dose to an adult male will be in the range from 1.20 mGy (4.2 mGy/WLM) to 3.80 mGy (13 mGy/WLM).

TABLE 3

*Variability of bronchial dose from radon progeny as a percentage change from nominal annual dose for males.*

| Factor | Variability | |
|---|---|---|
| 1. Unattached $^{218}$Po (variable from 4% to 20%) | - 10 | (4% unattached) |
| | +30 | (20% unattached) |
| 2. Progeny equilibrium | -20 | |
| 3. Particle deposition models | | a few percent |
| 4. Particle size (change in median particle size from 0.125 μm to 0.05 μm or 0.170 μm) | +100 | (0.05 μm particle) |
| | —20 | (0.170μm particle) |
| 5. Calculated physical dose to a given site in epithelium | | a few percent |
| 6. Breathing pattern | +20 | (entirely active) |
| | —25 | (entirely resting) |
| (mouth vs nose breathing) | (+35) | |
| 7. Bronchial morphometry (child vs adult) | +60 | |
| 8. Mucociliary clearance (normal to complete stasis) | +10 | |
| 9. Mucus thickness (none to twice normal) | +20 | (no mucus) |
| | -30 | (twice normal) |
| 10. Location of target cell (shallow-22μm to average-45 μm) | -80 | |

As has been pointed out, a relationship can be established between aerosol concentration and unattached alpha particle energy $f_p$ that is independent of the equilibrium factor.[2] This allows a single conversion factor from radon concentration to dose. From measurements of average indoor aerosol concentrations, James[2] recommends a dose value of approximately 0.05 mGy/y per Bq/m$^3$ radon concentration. This result is an annual lung dose to an adult male of about 1.9 mGy. A summary of various annual dose estimates to the bronchial epithelium is given in Table IV. The differences in these estimates results from differences in modelling the lung and particle deposition, and differences in the unattached fraction and equilibrium factor.

Because this dose is delivered by alpha particles, a quality factor of 20 has been recommended[1] to arrive at the biological dose equivalent (see Eq. 2). Thus, the annual dose equivalent to the lung for adults will be in the range from 24 mSv to 76 mSv, and for children, the annual dose equivalent will be in the range from 46 mSv to 174 mSv. These values of dose equivalent can be compared with the annual dose equivalent received from other natural sources of radiation given in Table 5.[23]

It is clear that exposure to radon and its decay products contributes a radiation dose to the bronchial epithelium that exceeds the total contribution from all other natural sources by more than a factor of 20. How this radiation dose influences the development of lung cancer is a question that remains to be

TABLE 4

*Annual Dose to the Bronchial Epithelium from Exposure to Radon Gas*

| Source | Dose (mGy/y) | | | |
|---|---|---|---|---|
| | Adult Male | Adult Female | 10 year old | Newborn |
| NCRP[20] | | | | |
| Yeh-Schum model | 2.0 | 2.0 | 4.9 | 3.8 |
| Weibel | 2.8 | 2.5 | 4.4 | 2.5 |
| Harley & Cohen[21] and Harley[22] | 1.2 | 1.6 | 2.8 | 2.3 |
| James[2] | 2.1-3.8 | | 2.6-4.8 | 4.8-8.7 |

TABLE 5

*Estimated Average Annual Dose Equivalents to Various Tissues from Natural Background Radiation (mSv)*

| Source | Bronchial Epithelium | Other Soft Tissues | Bone Marrow |
|---|---|---|---|
| Cosmic Rays | 0.27 | 0.27 | 0.27 |
| Cosmogenic Radionuclides | 0.01 | 0.01 | 0.03 |
| Terrestrial Radionuclides | 0.28 | 0.28 | 0.28 |
| Radionuclides in the Body | 0.36 | 0.36 | 0.50 |
| Total excluding Radon Progeny | 0.92 | 0.92 | 1.08 |

answered. Research should continue to define better the relationships among the various biological and physical factors and the resulting radiation dose, so that measurements of radon in air can be related to radiation dose with more accuracy. This would allow appropriate remedial actions to be taken in situations where the potential dose is high enough to be of concern.

## REFERENCES

1. National Council on Radiation Protection and Measurements. 1987. Recommendations on Limits for Exposure to Ionizing Radiation. NCRP Report No. 91. NCRP Bethesda, MD.
2. James, A.C. 1988. Lung dosimetry. *In:* Radon and Its Decay Products in Indoor Air. W.W. Nazaroff and A.V. Nero, Jr. (Eds.). John Wiley and Sons, New York, NY.
3. Weibel, E.R. 1963. Morphometry of the Human Lung. Academic Press, New York, NY.
4. Yeh, H.C. and G.M. Schum. 1980. Models of human lung airways and their application to inhaled particle deposition. Bull. Math. Biol. 42:461-480.
5. Harley, N.H. and B.S. Pasternack. 1982. Environmental radon daughter alpha dose factors in a five-lobed human lung. Health Physics. 42:789-799.
6. Phalen, R.F.,M.J. Oldham, C.B. Beaucage, T.T. Crocker and J.D. Mortensen. 1985. Postnatal enlargement of human tracheobronchial airways and implications for particle deposition, Anat. Rec. 212:368.
7. Crawford-Brown, D.J. 1987. *In:* Environmental Radon. C.R. Cothern and J.E. Smith, Jr. (Eds.). Plenum Press, New York, N.Y.
8. Gastineau, R.M., P.J. Walsh and N. Underwood. 1972. Thickness of bronchial epithelium with relation to exposure to radon. Health Phys. 23:857-860.
9. Trump, B.F., E.M. McDowell, F.Galvin, L.A. Barrett, P. Becci, W. Schurch, H.C. Kaiser and C.C. Harris. 1978. The respiratory epithelium IV. Histogenesis of epidermoid metaplasia and carcinoma in situ in the human. J. Natl. Cancer Instit. 61:563.
10. Masse, R. 1984. Cells at risk. *In:* Lung Modelling for Inhalation of Radioactive Materials. H. Smith and G. Gerber. (Eds.). Commission of the European Communities, EUR 9384, Brussels.
11. Ingham, D.B. 1975. Diffusion of aerosols from a stream flowing through a cylindrical tube. Aerosol Science. 6:125-132.
12. Cohen, B.S. 1986. Deposition of ultrafine particles in the human tracheobronchial tree: A determinant of dose from radon daughters. Proc. Am. Chem. Soc. Annual Meeting 1986. New York, NY.
13. Tu, K.W. and E.O. Knutson. 1985. Total deposition of ultrafine hydrophobic and hygroscopic aerosols in the human respiratory tract. Aerosol Sci.

Technol. 3:453.
14. Schiller, C., J. Gebhart, J. Heyder, G. Rudolf and Stahlhofen. 1986. Deposition of monodisperse insoluble aerosol particles in the 0.005 to 0.2 $\mu$m size range within the human respiratory tract. (unpublished data)
15. Yeates, D.B., T.R. Gerrity and C.S. Garrard. 1982. Characteristics of tracheobronchial deposition and clearance in man. Ann. Occup. Hyg. 26:245.
16. Harley, N.H. and B.S. Pasternak. 1972. Alpha absorption measurements applied to lung dose from radon daughters. Health Phys. 23:771.
17. Jacobi, W. and K. Eisfeld. 1980. Dose to tissues and effective dose equivalent by inhalation of radon-222, radon-220 and their short-lived daughters. GSF Report S-626. Gesellschaft fur Strahlen-und Umweltforschung, Munich-Neuherberg, FDR.
18. Hoffman, W. 1982. Dose calculations for the respiratory tract from inhaled natural radioactive nuclides as a function of age—II, Health Physics. 43:31-44.
19. Wise, K.N. 1982. Dose conversion factors for radon daughters in underground and open-cut mine atmospheres. Health Physics. 43:53-64.
20. National Council on Radiation Protection and Measurements. 1984. Evaluation of Occupational and Environmental Exposures to Radon and Radon Daughters in the United States. NCRP Report No. 78. NCRP, Bethesda, MD.
21. Harley, N.H. and B.S. Cohen. 1986. Updating radon daughter bronchial dosimetry. *In:* Radon and Its Decay Products. P.K. Hoppke (Ed.). American Chemical Society, New York, NY
22. Harley, N.H. 1984. Comparing radon daughter dose: Environmental versus underground exposure. Radiation Protection Dosimetry. 7:371-375.
23. National Council on Radiation Protection and Measurements. 1987. Ionizing Radiation Exposure of the Population of the United States. NCRP Report No. 93. NCRP, Bethesda, MD.

*Environmental Radon: Occurrence, Control and Health Hazards.* Edited by S. K. Majumdar, R. F. Schmalz and E. Willard Miller © 1990, The Pennsylvania Academy of Science.

# *Chapter Twenty-Two*

# EFFECTS OF INHALATION OF RADON DAUGHTERS IN THE LUNGS

**A.K.M.M. HAQUE**

Department of Physical Sciences & Scientific Computing
South Bank Polytechnic
London, SE1 OAA UK

## INTRODUCTION

There is now world-wide interest in quantifying the hazard due to inhalation of radon and its daughter products. This is mainly due to the discovery in several countries of high exposure in dwellings, change in living conditions, such as, energy saving practice etc. Although the association of radon daughter exposure with the lung cancer of miners has now been established, there are many factors still unknown. Further, the dependence of radon concentration indoors on geology, soil structure, house construction etc., is complicated; in order therefore to have an effective strategy for the mitigation of radon levels and to understand lung cancer incidence research on a broad front has been started, particularly, in the U.S.A. and Europe.

In general terms, on breathing radon and its daughter products, the latter, present as free ions and attached to aerosols, are deposited in various parts of the respiratory tract with varied efficiencies. The alpha emitters among these radioactive atoms are the relevant species. The two alpha particles emitted by Po-218 and Po-214, energy 6 and 7.69 MeV, respectively, with ranges of 47 and 71$\mu$m, irradiate the lung tissue, but the resulting dose cannot be measured. Hence dosimetric modelling, epidemiology and case-control studies have been attempted for risk assessment.

The epidemiological studies of underground miners have not led to a definitive connection between radon daughter exposure and small-cell carcinoma of the lung, although they are more prevalent. A conclusive description of the interaction of cigarette smoking and radon daughters for the induction of lung cancer is also not forthcoming. BEIR IV Committee[1] chose a multiplicative interaction for its risk projections. Other health effects of the exposure such as, chromosomal aberrations increase in lymphocytes, reduction in lung function and adverse reproductive outcomes are inconclusive.

The epidemiological studies and the experimental data on animals have established the carcinogenicity of radon progeny, and although the hazards with mining are well recognized, the estimation of lung cancer risk is a challenging task. The risk estimates have been obtained from diverse studies of underground miners, and may have been influenced by casual factors. There is thus need for the continuation of the epidemiology of miners, together with risk estimates for long-term exposure at relatively low levels to address the effect of radon daughters in homes and workplaces, and for refinements of the estimates in occupational environments.

For the dosimetric modelling, besides the adoption of a better lung model together with realistic physiological parameters and breathing characteristics, accurate characterization of domestic aerosols is needed along with measured values of the unattached fraction, $f_p$, and the equilibrium factor, F, under conditions such as smoking, cooking etc. Measurements of $f_p$ and F under the range of environmental conditions encountered in underground mines are also needed to improve the assessment of dose both in epidemiology and in monitoring of personal exposures by passive devices.

ICRP's recommendation[2] of a mean dose to the basal cells (over generations 2-15 of the lung) contradicting the epidemiological finding of the appearance of tumors in the upper airways together with the recent finding that the secretory cells are just as important as the basal cells, have led to a significant change in the philosophy of dosimetry, i.e., there is need for detailed dose calculation in the sub-regions of the lung.

Recently, case-control studies were performed in Sweden, one in an area of Stockholm with increased probability of radon emanation, Svensson et al.[3] and the other in municipalities with alum shale deposits, Edling et al.[4] and in Port Hope, Canada, where high radon levels were found due to ore-processing residues, Lees et al.[5] These studies were considered inadequate for the purpose of risk estimation, ICRP[6]; BEIR IV.

## REVIEW OF THE METHODS OF LUNG DOSE COMPUTATION

The US National Council on Radiation Protection and Measurements, NCRP,[7] has tabulated data showing the evolution of lung dose calculations due to radon progeny covering the period 1940-1981. Only a few methods will be highlighted in this chapter to draw attention to the importance and complexity of the subject. These are due to: Chamberlain and Dyson,[8] Altshuler, Nelson and Kuschner,[9] Jacobi,[10] Haque and Collinson,[11] Harley and Pasternak[12] and Jacobi and Eisfeld.[13] Further work have since appeared due to Hofmann,[14] James, Jacobi and Steinhausler,[15] Wise,[16] Harley and Pasternak,[17] NEA Experts Report,[18] NCRP Report 78[7] and those due to Haque and co-workers.[19,20,21] Of the more recent work, the calculational models of James and Birchall,[22] Jacobi and Eisfeld,[13] Harley and Pasternak[17] and Haque et al.[20,21] will be discussed in greater detail. (Table 1).

## *Earlier Models*

Chamberlain and Dyson estimated the dose mainly due to unattached Po-218 by observing the deposition of radon daughters in a cast of the trachea and main bronchi. The results were used by ICRP to set the maximum permissible concentration of radon in air.

Altshuler et al. calculated the dose using the Lung Model of Findeisen for a mine atmosphere of 200 pCi $l^{-1}$ of daughters, 4.7% as ions, 55.5% on condensation nuclei and 40% on particles, diameter $>0.1\mu m$. A model of mucus flow, the deposition of the activity by Gormley and Kennedy[23] equations and the results of measurement of epithelial thickness were employed.

The highest dose to the basal cells at 36$\mu$m in the segmental bronchi was taken as the cancer-related dose; the dose due to an individual alpha particle decreased linearly from twice its average value at the surface to zero at the end of range.

Jacobi used the continuous distribution of particle sizes (mean -0.09$\mu$m) in room air and the calculated values of unattached fractions. Employing Findeisen theory, the mean distribution of activity was estimated, ciliary transport and clearance from alveoli being included. The dose distribution was calculated according to the stopping power theory of alpha particles due to Neufeld and Snyder.[24]

The method of Haque and Collinson improved the calculations of Altshuler et al. and Jacobi, and is based on the energy deposited by alpha particles in tissue, 1$\mu$m diameter, and located at various depths. The equilibrium daughter activity on the walls of the lung was calculated from the deposition rates and the mucus movement, the lung model being that of Weibel.[25]

Employing the distribution of radon daughter activity, Mohnen and Stierstadt,[26] on aerosols, of mean radius 0.025$\mu$m, concentration $3\text{x}10^4$ $cm^{-3}$, in conjunction with the work on ion attachment of Lassen et al.,[27] and Croft and Perry,[28] the unattached fractions were calculated. Deposition of the activity was calculated by Gormley and Kennedy expressions with diffusion constants $5.4\text{x}10^{-2}$ and $5.4\text{x}10^{-6}$ $cm^2$ $s^{-1}$, for ions and aerosol particles, respectively. Total daughter activity indoors was calculated by assuming an equilibrium ratio 1: 0.86: 0.76 for RaA: RaB: RaC and realistic ventilation rates.

Finally, the total dose (far wall and near wall contributions) was calculated from energy deposited in tissue, 1$\mu$m dia, and at 15, 20, 25, 30$\mu$m etc., from the top of the mucus layer.

The stopping power data used were those from measured values of Riezler and Schepers[29] in air, normalized to tissue.

Harley and Pasternak employed the Weibel model, Gormley-Kennedy equations and mucus clearance times of Altshuler et al. Activity was considered mixed uniformly with the mucus. Of the total activity, unattached RaA (Po-218) ion content of <5% and the remainder on particles, median size 0.2-0.4μm dia., were taken from the measurements of George et al.[30] in mine atmosphere. Experimental stopping power data in polycarbonate and tissue-equivalent gas were used to calculate the dose in tissue, 1μm dia., at 22μm, for trachea and terminal bronchioles.

### *James-Birchall and Jacobi-Eisfeld Models*

ICRP's recommendation[2] of a mean dose to the basal cells under normal conditions in mines was based on the work of Jacobi et al.[13] and James et al.[22], which formed a report of a Group of Experts, Nuclear Energy Agency, Organization for Economic cooperation & Development.[18] It is therefore appropriate to discuss these together.

For the dose calculations the lung models of Weibel, and Yeh and Schum[31] were employed. A diffusion coefficient of 0.05 $cm^2\ s^{-1}$ for unattached atoms was used, together with 0.08 and 0.02 $cm^2\ s^{-1}$ for comparison. Whereas Jacobi et al. corrected the deposition values from the experimental results of Martin and Jacobi,[32] James et al. used values based on the experimental study of unattached Pb-212 and condensation nuclei in ventilated pig lungs.

Assuming a size distribution of aerosols, range 0.1-0.5μm AMD, Jacobi et al. calculated deposition values with Gormley-Kennedy equations together with corrections, the deposition decreasing with increasing aerosol size (Weibel). James et al.'s calculations for the Weibel, and Yeh-Schum models include the effects on deposition by impaction and sedimentation. For 0.1μm AMD, deposition is predominantly by diffusion; for 0.25μ AMD impaction in the upper airways gives a peaked distribution, while having little effect on total deposition.

The retention of the activity deposited will be controlled by radioactive decay, muco-ciliary transport, diffusion through epithelium to blood and retention in the mucus. Jacobi et al. adopted a model in which the velocity of the mucus decreases with airway diameter. They estimated a range of mucus velocities in each generation of the Weibel model from the reported values of mucus velocity, 1.8 to 1 cm $min^{-1}$, in the trachea, the time range to clear all the activity from the bronchi, 0.25-2 days, and a power law relationship between mucus velocity and bronchial diameter. These values are consistent with the experimental results of Yeates et al.[33] and Foster et al.[34] In order to determine the action of other processes acting concurrently for retention, a compartment model was used. James et al.'s generalized model of retention accommodated human and animal data on the clearance of $Pb(NO_3)_2$ aerosols and unattached Pb-212 ions. The deposited radon daughters was partitioned between a gel phase, for ciliary transport, and an aqueous sol phase of mucus for transport to epithelial tissue. They assumed protracted retention of two types of half-time of about 10 hours for approximately 30% of the activity: rapid transfer with epithelial retention or a slow transfer through mucus sol followed by rapid adsorption into blood.

Histological studies have shown that the lung tumors of uranium miners were predominantly of small-cell type and commonly found in segmental bronchi. The bronchial epithelium and the basal cells were considered to be at risk.

Gastineau et al.[35] made measurements on the surgical specimens of normal bronchi, giving the distribution of epithelial thickness, 25-70$\mu$m (50$\mu$ median) in gen. 2, going down to 5-30$\mu$m (15$\mu$ median) in the bronchioles. James et al. weighted the mean dose to basal cells according to the depth distribution for five classes of bronchi. Jacobi et al. assumed a continuous decrease in the depth of basal cells from about 60-30$\mu$m in gen. 0 to 20-10$\mu$m in gen. 15, and calculated the dose at the mean depth for each generation.

Jacobi et al. obtained the dose to the basal cells using the calculated values of Harley and Pasternak, while James et al. assumed a thinner 7.5$\mu$m thick mucus layer and theoretical range-energy relations. The weighted dose rates to the basal cells in lobar and segmental bronchi from Po-214 were half as much as the mean values of Jacobi et al., whereas the average dose rates to bronchiolar stem cells were about a factor of 2 higher than those of Jacobi et al.

According to these studies the dosimetric quantities related to risk are the mean dose or dose equivalent to the sensitive cells of T-B and P regions of the lung. This mean dose has been derived as the mean dose to the basal cells by averaging with equal weight doses in the lobar bronchi (gen. 2) through to the fifteenth generation.

### *Harley-Pasternak 5-Lobed Model:*

The five-lobed lung model calculation of Harley and Pasternak[17] is basically the same as the earlier version, except for Yeh-Schum morphometry. The reference atmosphere had an equilibrium ratio 1: 0.9: 0.7: 0.7, with an unattached fraction 0.07 of diffusion coefficient 0.005 $cm^3\ s^{-1}$ (range 0.09 - 0.004) and AMD of 0.12$\mu$m for the aerosol.

### *Haque et al. Models:*

The computation of Haque and Geary[19] was carried out along the same lines as in 1967 paper, but with up-to-date data and experimental results on tissue-equivalent solids, liquids and gases. Weibel lung model together with James's results on pig lungs were used. The total dose was calculated for activity mixed with a mucus layer of thickness 12$\mu$m.

The calculations of Haque and Al-Affan[20,21] were carried out with new information on activity deposition and lung parameters; experimental stopping power together with the effect of the physical state and straggling were considered. Dose rates in gen. 0 - 16 were obtained for Weibel, and Yeh and Schum models for a radon concentration of 23 $Bqm^{-3}$ (UK average of dwellings). The characteristics of the dwellings and the information on the natural aerosols, including those from cigarette smoking, cooking and a stove were taken from Vanmarcke et al.[36] and Reineking et al.[37]

For activity deposition in near airways, Gormley-Kennedy equation with Martin-Jacobi correction factors were considered adequate; Yeh-Schum values were taken for pulmonary region.

The results of Jacobi and Eisfield,[13] Yeates et al.[33] and Foster et al.[34] were combined to give a mucus velocity of

$$V_m = \exp\,(0.511 - 0.4265\ G_i)\ \text{cm min}^{-1}$$

where $G_i$ is the generation number, the transit time being calculated from $t$ = airway length/$V_m$.

The effect of the following factors on the dose has been considered:

Diffusion coefficient of unattached daughters, D: On the basis of recent work, Haque and Al-Affan have recommended a value of 0.03 $cm^2\ s^{-1}$ for D for unattached RaA (Po-218) and 0.004 $cm^2\ s^{-1}$ for RaB (Pb-214) and RaC (Bi-214). Dose calculations have been carried out with these values together with values 0.085 $cm^2\ s^{-1}$ and 0.0025 $cm^2\ s^{-1}$. The choice of the maximum value for D instead of the minimum will enhance the dose by as much as a factor of 2 in gen. 1-5 and by a factor of 0.75- 1 in gen. 6-16.

Aerosol size Distribution: The effect of 3 AMD's 0.12, 0.18 and 0.06$\mu$m, characteristic of city, country and when cigarette smoke etc. is present, have been studied, lowest AMD producing a dose twice as high as 0.18$\mu$m.

Room Conditions etc: For low ventilation, the dose rate is higher in gen. 3-8. Introducing cigarette smoke will produce a higher dose rate specially in gen. 7-16. For natural aerosols, Weibel model gives a higher dose for gen. 1-13 than Yeh-Schum.

## INTERACTION OF ALPHA PARTICLES WITH THE LUNG

The two alpha particles of interest having energies 6 and 7.69 MeV, traversing the lung tissue, each producing approximately $2x10^5$ ion-pairs, $5x10^6$ excited molecules and free radicals; the radiolysis products include $0, 0_3$, OH, $H_2O_2$, $CH_2O$ and other highly reactive chemicals.

It is believed that alpha particles from Th-232 in thorotrast patients and from

Ra-226 in watch-dial painters have produced cancers of the liver and bone respectively, May.[38] The epidemiological studies of uranium miners have yielded a clear correlation between radon daughter exposure and excess lung cancer frequency which cannot be explained by other influences. This finding is confirmed by experimental studies on animals exposed to radon daughters, ICRP.[6]

In England and Wales, the proportion of lung cancers in men between the ages of 45 and 64 is about half as much as total fatal cancer; the proportion is about a third in the USA. A recent survey of 2800 cancer deaths in the USA showed that epidermoid tumors and small-cell carcinomas appearing in the larger airways accounted for about 48%, whereas adeno carinomas and large-cell carcinomas appearing in small airways accounted for 34%, James.[39]

Early studies of lung cancer incidence in the uranium miners of the Colorado plateau indicated that about two-thirds of the tumors were of small-cell type, Saccomanno[40]; subsequent follow-up of this population led to the conclusion that there is no unique link between exposure to radon daughters and small-cell carcinoma (BEIR IV).

Cigarette smoking produces in the lung effects which are well-characterized and relevant for the induction of cancer. Some of these are: hypertrophy and hyperplasia of mucus gland, dysplasia and metaplasia of epithelial cells, increased epithelial permeability, impaired muco-ciliary transport; greater deposition of particles etc. (BEIR IV). The consequences of combined exposure to radon daughters and cigarette smoke must take into account the various effects as well as the interaction between the two agents. A definitive conclusion on the net effect of these smoking-related changes on the dosimetry of radon daughters cannot be arrived at present (BEIR IV).

The effect of radon daughters in the presence of smoking has been modelled employing multiplicative and additive interactions. The following treatment is that of BEIR IV committee. Let us consider a simple model of the form

$$r(d,a) = r_o(a)[1+Bd] \tag{1}$$

where r(d,a) is the age- specific risk i.e. the chance of dying of lung cancer in 1 year at age 'a', and $r_o(a)$ is the age-specific background risk, $B$ is the slope of dose-response relation and d is the cumulative dose.

The multiplicative model can be expressed as

$$r(d,a,s) = r_o(a)\ [1+\Theta(s)][1+Bd] \tag{2}$$

to incorporate an effect of smoking, s; and the additive model will take the form

$$r(d,a,s) = r_o(a)\ [1+\Theta(s)+Bd] \tag{3}$$

where is $\Theta$ is zero for non-smokers. In these models, $r_o(a)$ represents the age-specific lung cancer rates among non-smokers not exposed to radon daughters. For persons not exposed, the ratio of disease rates for smokers to non-smokers, r(o,a,s)/r(o,a,o) is $[1+\Theta(s)]$. Simultaneously the relative risk at the radiation dose 'd' among non-smokers is $(1+Bd)$ for either of these models.

The relationship between radon daughter exposure and smoking is complex, and the best description of the association could vary from supramultiplicative to subadditive. A few analyses which have been published are briefly discussed.

Damber and Larsson[41] carried out a case-control study of 69 ion miners from Kiruna and Gullivare areas. The results suggested that radon daughter exposure and smoking combine multiplicatively on a relative scale. The work of Radford and Renard [42] for 50 cases of lung cancer among a cohort of 1415 miners of Malmberget led to an additive model.

Whittmore and McMillan[43] and Saccomanno et al.[44] studied two groups of uranium mines of Colorado plateau, the former looked at 194 cases out of 3362 miners and the latter had 489 cases drawn from 9817 miners. Both studies suggested multiplicative effects.

BEIR IV committee analyzed further three populations: New Mexico uranium miners, Japanese A-bomb survivors and the cohort study of Colorado plateau with follow-up to 1982. Both multiplicative and additive models are consistent with New Mexico and Japanese survivors data, whereas the multiplicative model fit well the Colorado plateau data.

This short review suggests that several areas of study have to be followed to clarify the combined effect of cigarette smoking and exposure to radon daughters.

## CELLULAR RADIOBIOLOGY

The general expression describing cellular radio-biological effects in vivo or in vitro is given by

$$Y = (a + a_1 D + B_1 D^2) \exp(-a_2 D - B_2 D^2) \quad (4)$$

where Y is the response observed in a population, a is the spontaneous incidence, $a_1$ and $B_1$, are the coefficients for the observed effect and D is the dose; the exponential term also known as the survival fraction, S, expresses the loss of observed response due to competing effects, such as, cell killing with $a_2$ and $B_2$ as its coefficients. For low doses, when competing effects are unimportant,

$$Y = (a + a_1 D + B_1 D^2) \quad (5)$$

This equation is known as linear-quadratic, and describes for low to moderate doses a wide variety of cellular radiobiological end-points.

The main health hazard of exposure to radon daughters is the induction of lung cancer; all the other organs and tissues of the body are also subjected to a radiation burden as radon and part of the inhaled daughter products pass from the lungs into the blood stream.

The study of chromosome aberrations in peripheral blood lymphocytes is also actively pursued, as radiation induced transformation may lead to cancer, Pohl-Ruling,[45] Edwards et al.[46] The line of approach of cell culture systems allows us to study dose-related and time-dependent transformation of single cells. Such transformed cells in vitro give rise to tumors when injected into hamsters, whereas untreated cells show no spontaneous transformation. The system has therefore implications with respect to in vivo carcinogenesis (BEIR IV).

The data on in vitro transformation produced by alpha particles are beginning to be available and the following is a brief summary of recent results. Although there are no data on the tumorigenicity of the transformed cells, they are considered relevant for radiation protection concerning exposure to natural and/or artificial alpha-emitters.

Lloyd et al.[47] studied the mouse embryo C3H10T1/2 cells to 5.6 MeV alpha particles. The transformation frequency increased approximately as the cube of the dose to a maximum of about 4 per cent of the surviving cells in the dose range 205-342 rad. Robertson et al.[48] exposed A31-11 mouse BALB/3T3 cells to 5.3 MeV alpha particles, and the biological effects were compared with those of 220 $kV_p$ X-rays. The transformation frequency increased exponentially with dose in the range, 25-250 rad. The main conclusions from the experiments were: that the carcinogenic damage induced by high LET radiation was inefficiently repaired compared with X-rays and that the carcinogenic effect might be cumulative. Hieber et al.[49] studied the transformation frequencies with $\gamma$-rays at 0.5 Gy $min^{-1}$ and with alpha particles at high, 0.2 Gy $min^{-1}$, and low dose rates, 0.83 - 2.5 m Gy $min^{-1}$, in C3H10T1/2 cell system. Alpha particles were substantially more effective than $\gamma$-rays, the RBE for cell inactivation and for neoplastic transformation was of similar magnitude, and ranged from about 3 at an alpha dose of 2 Gy to values about 10 at 0.25 Gy.

Battega et al.[50] compared transformation of C3H10T1/2 cells irradiated with 4.3 MeV alpha with that for low LET protons. The alpha particles were more effective, the RBE decreased from 15 to 4, while the dose ranged from 0.1 to 2 Gy. There was no effect of the dose rate at total doses of 0.05 and 0.1 Gy, with dose rates of 0.11 and 0.005 Gy $min^{-1}$.

Pohl-Ruling et al.[45] and Edwards et al.[46] have reviewed investigations on chromosome aberrations in peripheral blood lymphocytes. The main problems of the study of chromosome aberrations caused by radon and its daughters at their most frequently existing dose levels are (i) to get statistical significance at very low doses, (ii) to study their induction by internal exposure to alpha-emitters only. Extensive collaboration between several laboratories is therefore pursued in this field.

Application of Transformation data to radon problem:

Al-Affan and Haque[51] have calculated, employing the data of Hieber et al., the cummulative fraction of transformed cells for radon concentrations in the range 23-230 Bq $m^{-3}$, which increased exponentially with the concentration. It has been assumed that causes other than radiation increase the rate of death of mature cells, and hence the turn over of stem cells to replace them. For radon concentrations of 50 and 100 Bq $m^{-3}$, the fractions of transformed cells are 0.2 and 6 per cent, respectively, for an exposure over 70 years.

## RISK ESTIMATES

Excess lung cancers have been diagnosed in uranium miners in the USA, Czechoslovakia, France and Canada and in other underground miners exposed to radon daughters, such as, Swedish metal miners, British iron and tin miners etc.

The epidemiological studies have shown an increasing risk as cumulative exposure to radon daughters increases. Data from underground miners do not directly address the risks at homes, offices etc., which are now of increasing concern.

The BEIR IV committee has recently carried out risk estimates of exposures to radon daughters by reanalyzing four principal data sets: two Canadian uranium-miner cohorts, Eldorado and Ontario, Swedish iron miners, Malmberget; Colorado plateau uranium miners. In this section, the basic models for risk analysis will be outlined first, to be followed by brief discussions on the analyses of BEIR IV,[1] ICRP Task Group,[6] NCRP,[7] Samet et al.[52], Pushkin and Yang,[53] and Hornung and Meinhardt.[54]

There has been recent important developments in statistical methods of analysis of data from occupational-cohort studies. The reader's attention is drawn to the excellent treatment and references of BEIR IV Report for details.

### *Basic Models*

These are statistical models which are of minimal regression type necessary for proper analysis of data.

The rate of lung-cancer mortality per person-year at risk (or simply risk), r(d,a,t,e), is given by

$$r(d,a,t,e) = r_o(a)\,[1 + f(d)\,\varrho\,(a,t,e)] \qquad (6)$$

where, $r_o(a)$ is the age-specific background risk, f(d) represents the effect of dose, d, 'a' is age at risk, 't' is time since exposure and 'e' is age at exposure; $\varrho$ (a,t,e) is excess relative risk for unit increase of f(d).

For an individual, r is the age-specific risk, the chance of dying of lung cancer in 1 year at age 'a'. If the risk is expressed as

$$r(d,a,t,e) = r_o(a) + f(d)\,\tau\,(a,t,e) \qquad (7)$$

so that $\varrho(a,t,e) = \tau\,(a,t,e)/r_o(a)$. The equation 6 is a model for the relative risk and the equation 7, for absolute risk. The distinction between the two types of models arise when restrictive forms for the functions $\varrho$ and $\tau$ are taken.

*BEIR IV Model*

As a general description of the risk for the miners the following relative-risk model has been recommended:

$$r(a) = r_o(a)\,[1 + 0.025\,\gamma(a)(W_1 + 0.5\,W_2)] \qquad (8)$$

where r(a) is the age-specific lung-cancer mortality rate from all agents, $r_o(a)$ is the age-specific background lung-cancer mortality rate, $\gamma(a)$ is 1.2 when age 'a' is less than 55 years, 1 when 'a' is 55-64 years and 0.4 when 'a' is 65 years or more, $W_1$ is the cumulative WLM incurred between 5 and 15 years before age 'a' and $W_2$ is the WLM incurred 15 years or more before this age.

Several assumptions are required to transfer risk estimates to the indoor domestic environment. The committee assumed that the epidemiological findings could be extended across the entire life-span and that cigarette smoking and exposure to radon daughters interact multiplicatively.

The committee investigated how the relative risk depended on the following potential risk factors: cumulative exposure, duration of exposure, age at which risk is evaluated, age at first exposure, time since exposure, and time after cessation of exposure. This was done by first making a cross-classification by cohort, age, year, cumulative exposure and other variables, and then counting the lung cancer deaths and person-years at risk. Excess relative risk in the model is linear in cumulative exposure.

The committee did not find consistently significant effects on relative risk due to age at first exposure or duration of exposure after adjusting for cumulative exposure. The analysis did, however, indicate that the age at risk and the time since cessation of exposure were significant factors modifying the excess relative risk. Therefore, a time-since exposure (TSE) model seemed more plausible, and the model which fitted best the combined data is given by the eqn[8].

The committee made use of life table analyses to estimate the lifetime risk of lung cancer and the expected number of years of life lost. For the baseline case of lifetime exposure to a general population, the model gave results comparable to other risk estimates given in Table 2.

TABLE 1

*Summary of Dose Calculations*

| Authors | Exposure conditions | Target | Unattached fraction $f_p$ | Unattached fraction diameter μm | Attached fraction AMD μm | F | Dose conversation factor mGy/WLM |
|---|---|---|---|---|---|---|---|
| *Harley & Pasternak[17] | Home: Nose | Segmental bronchi 22 μm depth | 1% | 0.003<br>0.001 | 0.12 | 0.7 | 5.2<br>5.7 |
| *NEA Experts Report[18] | Mine: Nose | J-B | 1-5% | | 0.1-0.5 | 0.4 | 5-8 |
| | | J-E | (2.5%) | | (0.2) | | |
| | Home: Nose | J-B | 1.5-6% | | 0.1-0.2 | 0.4 | 4-5 |
| | | J-E | (2.5%) | | (0.17) | | |
| *NCRP Rep. 78[7] | Mine: Nose | Segmental bronchi 22 μm depth | 0.9% | 0.001 | 0.3 | 0.3 | 4 |
| | Home: Nose | | 1% | 0.001 | 0.125 | 0.7 | 7 |
| Haque and Al-Affan[20,21] | Home: Nose | 22 μm depth | 5-12% | | 0.06-0.12 | 0.25-0.39 | 7-12 |

*Taken from James[39]

*ICRP Task Group: Proportional Hazard Model*

This model assumes that a single exposure $E(t_e)$ at age $t_e$ leads, after a constant lag of $\tau$, to an excess proportional to exposure and base-line, $\lambda_o$, i.e. the total rate

$$\lambda(t,t_e) = \lambda_o (t)[1 + r(t_e) E (t_e)] \tag{9}$$

$r(t_e)$, the relative risk coefficient, depends on the age at exposure. For chronic exposure to radon daughters

$$\lambda(t) = \lambda_o (t) [1 + r E(t-\tau)] \tag{10}$$

where r is the age-averaged relative excess risk coefficient and $E(t-\tau)$ is the cumulative exposure up to age $(t-\tau)$. For promoting influence of smoking, the total age-specific lung cancer rate,

$$\lambda_s(t) = \lambda_{o,ns} (t) [1 + S_s(t)] [1 + r E(t-\tau)] \tag{11}$$

where $\lambda_{o,ns}$ is the base-line rate for non-smokers and $S_s$ characterizes enhancement by smoking. The relationship is based on the study of Whittmore et al. on Colorado uranium miners and is claimed to be a good fit to excess lung cancer frequency among Czechoslovak uranium miners.

The report also derives: a). the individual lifetime risk of lung cancer from inhaled radon daughters; b). the radiation-induced annual lung cancer frequency among current populations; c). the loss of life expectancy attributable to lung cancer risk from inhaled radon daughters.

*Main results:*

The relative attributable lifetime risk per unit exposure rate are similar for males, females and non-smokers. At indoor exposure rate below about 3 WLM $yr^{-1}$, the relative risk coefficient is $1 \times 10^{-6}$ per Bq $hm^{-3}$ per unit annual exposure $\equiv$ 0.64 per WLM.

For the estimate of absolute risk, a reference population of males, females and non-smokers with a normal lung cancer frequency 600, 120 and 80 cases per $10^6$ have been assumed. Attributable lifetime for these groups are given as a function of annual indoor exposure.

Chronic exposure of the reference population amounts to a mean loss of life expectancy of $1.4 \times 10^{-4}$ days per Bq $hm^{-3}$ per unit annual exposure. For an equilibrium equivalent radon concentration of 15 Bq $m^{-3}$, this is about 16 days.

## *NCRP Model*

The NCRP Model assumes that a). the tumor rate, following a latent period, decreases exponentially from the time of exposure, the half-life being about 20 years; b). the rate of appearance of lung cancer after a single exposure is uniform with time, and the excess rate increases with age at exposure; c). the lung cancer incidence is rare before the age of 40; d). the median age of appearance of lung cancer is about 60 in non-smokers and over 50 in smokers; and e). the time for tumor growth is about 5 years.

For a single annual exposure at $t_o$, the excess radiation-assisted risk is given by

$$A(t,t_o) = R(P_t/P_o) \exp[-\lambda(t-t_o)] \tag{12}$$

where, $A(t,t_o)$ is the attributable annual tumor rate at age t ($t \geq 40$) (if the exposure occurs after 40, risk commences at $t_o + 5$ years; if before 35, risk commences at 40); R is the risk coefficient, $10 \times 10^{-6}$ per year per WLM; $P_t/P_o$ is the life-table correction to account for death from other causes; and $\lambda$ is the decrease in rate of risk expression, due to repair, cell death etc. ($\lambda = (\ln 2/20)$ $yr^{-1}$). To obtain the lifetime risk, integration is carried out from $t = 40$ to 85 for a single exposure at $t_o$, or over each single exposure.

The model predicts that the excess risk declines with time, which contradicts BEIR IV's work. The other difficulties of the model are: choice of $\lambda$; occurrence of lung cancer limited to 40 years or over.

TABLE 2

*Lung Cancer Morality Due to a Lifetime Exposure to Radon Progeny*

| Study | | Cancer deaths per $10^6$ person WLM |
|---|---|---|
| BEIR IV | 1988 | 350 |
| ICRP | 1987 | 230 |
| NCRP | 1984 | 130 |
| BEIR III | 1980 | 730 |
| UNSCEAR | 1977 | 200-450 |

Samet et al.[52] conducted a case-control study in a cohort of New Mexico uranium miners with 65 cases and 230 controls. For exposures below 1000 WLM, the excess relative risk was 1.1% per WLM, with multiplicative interaction between smoking and exposure.

Pushkin and Yang[53] studied the contribution to lung cancer deaths from indoor radon exposure. Applying EPA coefficients it was estimated that 14% (8-25%) of all lung cancer deaths were attributable to radon daughter exposure.

Hornung and Meinhardt[54] looked at the updated data of US uranium miners.

Relative excess risks (internal referent group) were 0.9-1.4 per 100 WLM. Other findings were; low exposure rate is more harmful per unit cumulative exposure; smoking and radon daughter exposure association is slightly less than multiplicative.

## SUITABILITY OF WLM FOR INDOOR EXPOSURE

It is now realized that the inhalation of radon daughters constitutes the most important contribution to the radiation exposure of the public. Until recently, radon was considered only in the context of mines.

The exposure of the miners was inferred from Equilibrium Equivalent Radon Concentration, $EEC_{Rn}$, which is dependent on the relative concentration of the daughter products. In practice, it is difficult to measure the daughter concentrations, which may vary with time. The average radon concentration, measured over a long period to eliminate short-time variations is convenient to monitor in domestic situation. The US EPA and the UK NRPB have therefore decided to use radon concentration instead of $EEC_{Rn}$, to express recommendations, etc. Exposure to potential alpha energy is estimated by assuming an average equilibrium factor, F, between $EEC_{Rn}$ and the measured radon concentration. It has been suggested that a direct conversion between radon exposure and dose is valid under domestic conditions since $f_p$ tends to be negatively correlated with F. The constancy of such direct conversion is discussed in the light of the work of Vanmarcke et al.[55] and James et al.[56].

Experimental Work: Vanmarcke et al. Concentrations of radon daughters, collected on filter paper, were measured by alpha spectroscopy, together with the ventilation rates by observing the decay of released $N_2O$ with an infra-red spectrometer.

The aerosol size distribution and concentration was monitored with an aerosol spectrometer, aerosol conditions being changed frequently by smoking cigarettes, cooking etc. The unattached fraction in the indoor environment lay between 5 and 15%, decreasing below 5% following smoking, cooking etc. The dose was found independent of the equilibrium factor, F.

James et al.[56] reviewed the results of Stranden et al.,[57] Strong et al.,[58] Porstendorfer et al.,[59] and Vanmarcke et al.,[60] plotting $f_p$ vs. F. An inverse relationship between $f_p$ and F was found, average values of $f_p$ being about 0.1 and weakly correlated with F. The effective dose equivalent for unit exposure to radon daughter potential alpha energy was estimated for various conditions. It was concluded that radon concentration is a better index of dose than $EEC_{Rn}$. However, measured values of $f_p$ and F are sparse, and should be supplemented by further studies.

These measurements, are also recommended for conditions encountered in underground mines for the assessment of dose in epidemiological studies and in monitoring individual exposures by passive devices.

### *Relationships between $EEC_{Rn}$, F, $f_p$ and WLM (NEA, ICRP)*

The potential alpha energy of a radon daughter atom, $\epsilon_p$, in the decay chain is the total alpha energy emitted during the decay of this atom to Pb-210. For Po-218, Pb-214, Bi-214 and Po-214, these are 13.7, 7.69, 7.69 and 7.69 MeV, respectively.

The potential alpha energy concentration, $c_p$, of any mixture of the radon daughters is the sum of the potential alpha energy of all the daughter atoms present per unit volume of air, and is given by

$$c_p = \sum_i (c_{act,i})\, \epsilon_{p,i} / \lambda_{r,i}$$

where, $c_{act,i}$ is the activity concentration of the daughter i, $\epsilon_{p,i}$ its potential alpha energy and $\lambda_{r,i}$ its decay constant.

1 Working Level (WL) corresponds to the potential alpha energy concentration of radon daughters in equilibrium with radon concentration of 3700 Bq $m^{-3}$ and is equal to $2.08 \times 10^{-5}$ $Jm^{-3}$.

The $EEC_{Rn}$ of a non-equilibrium mixture of radon daughters in air is that activity concentration of radon in equilibrium which has the same potential alpha energy concentration as the non-equilibrium mixture.

Equilibrium factor, $F = EEC_{Rn}/c_{act,Rn}$

Radon daughter exposure, $E_p(T)$

$$= \int_0^T c_p(t)\, dt = \text{potential alpha energy exposure}$$

The unit of $E_p$ is $Jhm^{-3}$. The radon daughter exposure of miners is expressed in WLM (Working Level Month), corresponding to an exposure of 1 WL for 170 h. Hence 1 WLM = 170 WLh = $3.5 \times 10^{-3}$ $Jhm^{-3}$.

## DISCUSSION

Surveys of radon levels in houses in several countries are in different stages of completion. One important result is the large variation of radon levels from house to house, from a few to several thousand Bq $m^{-3}$. In a few extreme cases levels as high as 100,000-200,000 Bq $m^{-3}$, or over, have been found, for example, in eastern Pennsylvania and New Jersey, Guimond,[61] or in Cornwall and Devon in the UK, Cliff et al.[62]

It is estimated that 5000 to 20,000 lung cancer deaths per year in the USA and 1500 in the UK, O'Riordan,[63] are related to exposure to radon daughters, which are preventable. The survey of radon results reinforce the difficulty in predicting the concentration in any specific dwelling. However, regional geology and indoor radon level appear to be related, but soil and rock permeability may be a major controlling factor. Consequently, measurements are the only reliable method for the detection of houses with high radon levels.

Now to the question, "what is the lower boundary of radon level below which no actions are necessary?", a simple cost-benefit analysis is not sufficient. Guidelines of national authorities must be in accordance with the demands of public health, bearing in mind the collective detriment to health.

Individual freedom and decision-making should also be considered. There is, therefore, urgent need to inform and advise the public clearly. Further, the public may be unwilling to measure radon levels, because if high levels are detected the value of the property might be affected. To minimize this effect, there has to be urgent development of low cost techniques to mitigate against high radon levels in existing homes and build new homes with low radon levels.

The possible lung cancer risk associated with indoor exposure is an important criterion for the final choice of the action level. Epidemiology in conjunction with dosimetric modelling will lead to better estimates of the risk.

Continued epidemiology of miners, more information on the interaction between cigarette smoking and radon, characterization of mine environment, together with a better lung model will lead to improved dosimetry. Long-term exposure at relatively low levels have to be studied to address the effects of radon daughters at homes. Epidemiology in domestic environment, its extensive characterization, experimental values of F and $f_p$—their dependence on ventilation and the presence of aerosol sources from activity such as, cooking and smoking are in progress. Comparative dosimetry in mine and domestic settings will be important. It is felt that the proposed ICRP lung model will play an extremely important part in future dosimetry.

The animal studies have provided invaluable data on the dosimetry of radon daughters and also on disease incidence as a function of exposure. This type of work must be continued for low exposure and low exposures rates, as the animals are reasonable surrogates for humans.

The other fields of study which need increasing attention are the radiobiology of human lung cells and of increased occurrence of chromosomal aberration in lymphocytes. At present, the transformation work are being carried out in mammalian cells in vitro. Over and above the study of the action of high LET radiation at low dose and low dose rates, work is necessary on human lung cells.

The actual process of lung cancer induction is complex. It is a challenge to understand carcinogenesis, for which we may have to wait a long time.

To quote from Jacobi:[64]

"The future improvement of risk models should take into account that carcinogenesis is a multifactorial multistage process. There is strong evidence that ionizing radiation, particularly at low doses, acts primarily on earlier stages of the process. On the other hand, the effects of aging and the influence of other promoting factors affect the later stages. Thus the latency distribution of radiogenic lung cancer is mainly determined by aging and other factors, like smoking, which stimulate the tumor growth."

## REFERENCES

1. BEIR IV. 1988. Committee on the Biological Effects of Ionizing Radiations, Health risks of radon and other internally deposited alpha-emitters. National Academy Press, Washington, D.C. pp. 602.
2. ICRP 32. 1981. International Commission on Radiological Protection, Limits for inhalation of radon daughters by workers. Pergamon, Oxford, pp. 24.
3. Svensson, C., G. Eklund and G. Pershagen. 1987. Indoor exposure to radon from the ground and bronchial cancer in women. Int. Arch. Occup. Environ. Health. 59:123-131.
4. Elding, C., G. Wingren and O. Axelson. 1986. Quantification of the lung cancer risk from radon daughter exposure in dwellings—an epidemiological approach. Environmental International. 12:55-60.
5. Lees, R.E.M., R. Steel and J.H. Roberts. 1987. A case-control study of lung cancer relative to domestic radon exposure. Int. J. Epidemiology. 16:7-12.
6. ICRP, 50. 1987. International Commission on Radiological Protection, Lung cancer risk from indoor exposures to radon daughters, Pergamon, Oxford, pp. 60.
7. NCRP. 1984. National Council on Radiation Protection and Measurements, Envaluation of Occupational and Environmental Exposure to Radon and Radon daughters in the United States. Report No. 78. Bethesda, MD, pp. 204.
8. Chamberlain, A.C. and E.D. Dyson. 1956. The dose to the trachea and bronchi from the decay products of radon and thoron. Br. J. Radiol. 29:317-325.
9. Altshuler, B., N. Nelson and M. Kushner. 1964. Estimation of lung tissue dose from the inhalation of radon and daughters. Health Phys. 10:1137-1161.
10. Jacobi, W. 1964. The dose to the human respiratory tract by inhalation of short-lived Rn-222 and Rn-220 decay products. Health Phys. 10:1163-1174.
11. Haque, A.K.M.M. and A.J.L. Collinson. 1967. Radiation dose to the respiratory system due to radon and its daughter products. Health Phys. 13:431-443.

12. Harley, N.H. and B.S. Pasternak. 1972. Alpha absorption measurements applied to lung dose from radon daughters, Health. Phys. 23:771-782.
13. Jacobi, W. and K. Eisfeld. 1980. Dose to tissues and effective dose equivalent by inhalation of radon-222, radon-220 and their short-lived daughters. GSF Report S-626. Gesellschaft fur strahlen-und Umweltforschung, Munich. pp. 131.
14. Hofmann, W. 1982. Cellular lung dosimetry for inhaled radon decay products as a base for radiation-induced lung cancer risk assessment I. Calculation of mean cellular doses. Radiat. Environ. Biophys. 20:95-112.
15. James, A.C., W. Jacobi and F. Steinhausler. 1981. Respiratory tract dosimetry of radon and thoron daughters: The state-of-the-art and implications for epidemiology and radiobiology, pp. 42-54. *In:* International Conference. M. Gomez (Ed.). Radiation hazards in Mining: Control, Measurement and Medical Aspects. October 4-9, 1981. Colorado School of Mines, Golden, CO (USA).
16. Wise, K.N. 1982. Dose conversation factors for radon daughters in underground and open-cut mine atmospheres. Health Phys. 43:53-64.
17. Harley, N.H. and B.S. Pasternak. 1982. Environmental radon daughter alpha dose factors in a five-lobed human lung. Health Phys. 42:789-799.
18. NEA. 1983. Nuclear Energy Agency Group of Experts, Dosimetry Aspects of Exposure to Radon and Thoron Daughter products. OECD, Paris, France. pp. 107.
19. Haque, A.K.M.M. and M.J. Geary. 1980. The radiation dose to the respiratory system due to the daughter products of radon, pp. 113-127. *In:* Seventh symposium on Microdosimetry. J. Booz, H.G. Ebert and H.D Hartfield (Eds.). Sept. 8-12, 1980. Oxford, UK.
20. Haque, A.K.M.M. and I.A.M. Al-Affan. 1988. Radiation dose to the lung to inhalation of alpha emitters, pp. 171-176. *In:* Health effects of low dose ionizing radiation—recent advances and their implications. Proceedings of international conference, BNES and ENS, May 11-14, 1987. British Nuclear Energy Society, pp. 208. London.
21. Haque, A.K.M.M. and I.A.M. Al-Affan. 1988. Main factors affecting the calculation of radiation dose to the lung from inhalation of radon daughters. The Science of the Total Environment. 74:279-288.
22. James, A.C. 1984. Dosimetric approaches to risk assessmentfor indoor exposure to radon daughters. Radiat. Prot. Dosim. 7:353-357.
23. Gormley, P.G. and M. Kennedy. 1949. Diffusion from a stream flowing through a cylindrical tube. Proc. Roy. Irish Acad., Sec. A. 52:163-169.
24. Neufeld, J. and W.S. Snyder. 1961. Estimates of energy dissipation by heavy charged particles in tissue, pp. 35-44. *In:* Selected Topics in Radiation Dosimetry. IAEA Publication STI/Pub/25. IAEA, Vienna.
25. Weibel, E.R. 1963. Morphometry of the human lung. Academic Press, New York, NY.

26. Mohnen, V. and K. Stierstadt. 1963. Die Verteilung der naturlichen radioaktivitat auf das Grossenspektrum des naturlichen aerosols. Zeitschrift fur physik. 173:276-293.
27. Lassen, L. and G. Rau. 1960. Die anlagerung radioaktiver atome and aerosole (Schwebstoffe). Z. Physik. 160:504.
28. Croft, J.F. and K.E.G. Perry. 1962. An experimental study of the behaviour and detection of hazardous radon releases in air. AEEW - R149, HMSO.
29. Riezler, W. and H. Schepers. 1961. Ionization and Energieverlust von *a*-Teilchen in gasen. Annalen der Physik. 8:270-277.
30. George, A.C., L. Hinchliffe, R. Epps and T.J. Shepich. 1970. Respiratory tract deposition of radon daughters in humans exposed in uranium mine. Atomic Energy Commission, Health and Safety Report TM 70-7. USAEC, Washington, D.C.
31. Yeh, H.C. and G.M. Schum. 1980. Models of human lung airways and their application to inhaled particle deposition. Bull. Math. Biol. 42:461-480.
32. Martin, D. and W. Jacobi. 1972. Diffusion deposition of small-sized particles in the bronchial tree. Health Phys. 23:23-29.
33. Yeates, D.B., T.R. Gerrity and C.S. Garrad. 1982. Characterisation of tracheabronchial deposition and clearance in man. Ann. Occup. Hyg. 26:245-257.
34. Foster, W.M., E.G. Langenback and E.H. Bergofsky. 1982. Lung Mucociliary function in man. Ann. Occup. Hyg. 26:227-244.
35. Gastineau, R.M., P.J. Walsh and N. Underwood. 1972. Thickness of bronchial epithelium with relation to exposure to radon. Health Phys. 23:857-860.
36. Vanmarcke, H., A. Janssens and F. Raes. 1985. The equilibrium of attached and unattached radon daughters in the domestic environment. The Science of the Total Environment. 45:251-260.
37. Reineking, A., K.H. Becker and J. Porstendorfer. 1985. Measurements of the unattached fractions of radon daughters in houses. The Science of the Total Environment. 45:261-270.
38. Mays, C.W. 1988. Alpha-particle-induced cancer in humans. Health Phys. 55:637-652.
39. James, A.C. 1988. Lung Dosimetry, pp. 259-309. *In:* Radon and its Decay Products in Indoor Air. W.W. Nazaroff and A.V. Nero, Jr. (Eds.). J. Wiley & Sons, pp. 518.
40. Saccomanno, G., V.E. Archer, R.P. Saunders, L.A. James and P.A. Beckler. 1964. Lung cancer of uranium miners on the Colorado plateau. Health Phys. 10:1195-1201.
41. Damber, L. and L.G. Larsson. 1985. Underground mining, smoking and lung cancer, A case-control study in the ion-ore municipalities in northern Sweden. J. Natl. Cancer Inst. 74:1207-1213.
42. Radford, E.P. and K.G. St. Clair Renard. 1984. Lung Cancer in Swedish iron miners exposed to low doses of radon daughters. N. Eng. J. Med.

310:1485-1494.
43. Whittemore, A.S. and A. McMillian. 1983. Lung cancer mortality among U.S. uranium miners: A reappraisal. J. Natl. Cancer Inst. 71:489-499.
44. Saccomanno, G., C. Yale, W. Dixon, O. Auerbach and G.C. Huth. 1986. An epidemiological analysis of the relationship between exposure to Rn progeny, smoking and bronchogenic carcinoma in the U-mining population of the Colorado plateau—1960-1980. Health Phys. 50:605-618.
45. Pohl-Ruling, J., P. Fischer and E. Pohl. 1987. Effect on peripheral blood chromosomes, pp. 487-501. *In:* Radon and its decay products, occurrence, properties and health effects. P.K. Hopke (Ed.). ACS, Washington, D.C. pp. 609.
46. Edwards, A.A., D.C. Lloyd and J.S. Prosser. 1989. Chromosome aberrations in human lymphocytes—A radiobiological review, pp. 423-432 *In:* Proc. 14th L.H. Gray Conference, 11-14 Sept. 1988, Oxford, Low Dose Radiation: Biological Basis of Risk Assessment, K.F. Baverstock and J.W. Stather (Eds.), Taylor and Francis, London, pp. 606.
47. Lloyd, E.L., M.A. Gemmell, C.B. Henning, D.S. Gemmell and B.J. Zabransky. 1979. Transformation of mammalian cells by alpha particles. Int. J. Radiat. Biol. 36:467-478.
48. Roberson, J.B., A. Koehler, J. George and J.B. Little. 1983. Oncogenic Transformation of mouse BALB/3T3 cells by plutonium—238 alpha particles. Rad. Res. 96:261-274.
49. Hieber, L., G. Ponsel, H. Roos, S. Fenn, E. Fromke and A.M. Kellerer. 1987. Absence of a dose-rate effect in the transformation of C3H 10T1/2 cells by $\alpha$-particles. Int. J. Radiat. Biol. 52:859-869.
50. Bettega, D., P. Calzolari, A. Ottolenghi and L. Tallone Lombardi. 1989. Oncogenic transformation induced by high and low LET radiations. *In:* Proc. Tenth Symp. on Microdosimetry, 21-26 May, 1989. Rome (In Press).
51. Al-Affan, I.A.M. and A.K.M.M. Haque. 1989. Transformation of lung cells from inhalation of radon daughters in dwellings: a preliminary study. Int. J. Radiat. Biol. 56:413-422.
52. Samet, J.M., D.R. Pathak, M.V. Morgan, M.C. Marbury, C.R. Key and A.A. Valdivia. 1989. Radon progeny exposure and lung cancer risk in New Mexico U. miners: A case-control study. Health Phys. 56:415-421.
53. Puskin, J.S. and Y. Yang. 1988. A retrospective look at Rn-induced lung cancer mortality from the viewpoint of a relative risk model. Health Phys. 54:635-643.
54. Hornung, R.W. and T.J. Meinhardt. 1987. Quantitative risk assessment of lung cancer in U.S. uranium miners. Health Phys. 52:417-430.
55. Vanmarcke, H., P. Berkvens and A. Poffijn. 1989. Radon versus Rn daughters. Health Phys. 56:229-231.
56. James, A.C., J.C. Strong, K.D. Cliff and E. Stranden. 1988. The significance of equilibrium and attachment in radon daughter dosimetry. Radiat. Prot.

Dosim. 24:451-455.
57. Stranden, E. and T. Strand. 1987. A dosimetric discussion based on measurements of radon daughter equilibrium and unattached fraction in different atmospheres. Radiat. Prot. Dosim. 16:313.318.
58. Strong, J.C. and K.W. Tu. 1987. Design and Calibration of a multichannel diffusion battery for the measurement of radon daughter activity-size distributors, pp. 21-24. *In:* Aerosols, their behaviour and applications. Proc. 1st conf. 31 Mar. - 1 Apr. The Aerosol Society, Loughborough.
59. Porstendorfer, J., A. Reineking and K.H. Becker. 1987. Free fractions, attachment rates and plate-out rates of radon daughters in houses, pp. 285-300. *In:* Radon and its decay products: Occurrence, properties and health effects. P.K. Hopke, (Ed.). American Chemical Society, Washington, D.C.
60. Vanmarcke, H., A. Janssens, F. Raes, A. Poffijn, P. Berkvens and R. Van Dingenen. 1987. The behaviour of radon daughters in the domestic environment: Effect on the Effective dose equivalent, pp. 301-323. *In:* Radon and its decay products: Occurrence, properties and health effects. P.K. Hopke, (Ed.). American Chemical Society, Washington, D.C.
61. Guimond, R.J. 1988.Reducing radon risks: The approach in the United States. Radiat. Prot. Dosim. 24:483-485.
62. Cliff, K.D., J.R. Britten, D.W. Dixon, P.H. Gardner and P.A.T. Richardson. 1988. Radon control and remedies in buildings. Radiat. Prot. Dosim. 24:251-255.
63. O'Riordan, M.C. 1988. Notes on radon risks in homes. Radiological Protection Bulletin, NRPB, No. 89:13-14.
64. Jacobi, W. 1988. Environmental radioactivity and man — The 1988 Sievert Lecture. Health Phys. 55:845-853.

*Environmental Radon: Occurrence, Control and Health Hazards.* Edited by S. K. Majumdar, R. F. Schmalz and E. Willard Miller © 1990, The Pennsylvania Academy of Science.

# *Chapter Twenty-Three*
# RADON LUNG DOSIMETRY

**MAURICE A. ROBKIN**

Department of Environmental Health
University of Washington
Seattle, WA 98195

## INTRODUCTION

### Radon Daughters

The decay chain of U-238 and the history of lung disease associated with uranium mining is discussed elsewhere in this volume. One of the chain members, Ra-226, generates the radon isotope, Rn-222, that escapes the soil and rocks and accumulates in the atmosphere of enclosed spaces.

The damage caused by radon is due to the alpha particles produced by the short-lived daughters that are inhaled and deposited in the lung. The air concentration of these daughters arise directly from radon decaying in the air. Awareness of the health risks from inhaling these short-lived daughters comes from consideration of the incidence of lung disease among uranium miners and the elevated levels of radon in which they worked. These original studies are of occupational exposure, and the unit used for radon concentration reflects this. This unit, called the "working level" (WL), is defined to be a radon air activity concentration of 100 pCi/l (3,700 $Bq/m^3$).

### An Index of Harm

The index of potential for harm is taken to be proportional to the total alpha particle energy per unit volume that will be produced when all of the short-lived daughters present in the air at any one time finally decay. When the short-lived daughters are in equilibrium with radon at 100 pCi/l, the number of atoms per liter of each is shown in Table 1.

Every atom of Po-218 will yield a 6.00 MeV alpha from its own decay, and then a 7.69 MeV alpha from the decay of its daughter, Po-214. Every atom of Pb-214 and Bi-214 will decay to Po-214 and then also yield a 7.69 MeV alpha. Table 1 shows the energy density from alpha particle decays that potentially will be released in air from the short-lived daughters in equilibrium with 100 pCi/l of radon. As is shown, 100 pCi/l of radon in equilibrium with its short-lived daughters will result in those daughters potentially generating about $1.3 \times 10^5$ MeV/l of alpha particle kinetic energy.

TABLE 1

*Potential Alpha Energy Concentration*

*Daughters in Equilibrium with 100 pCi/l of Radon*

| Daughter Nuclide | Half-Life (sec) | $N_i$ (Atoms/l) | $E_i$ (MeV) | $N_iE_i$ ($1.0 \times 10^5$ MeV) |
|---|---|---|---|---|
| Po-218 (RaA) | 187 | 996 | 13.69 | 0.136 |
| Pb-214 (RaB) | 1608 | 8583 | 7.69 | 0.660 |
| Bi-214 (RaC) | 1188 | 6342 | 7.69 | 0.488 |
| Po-214 (RaC′) | 0.0002 | 0.001 | 7.69 | $<<0.001$ |
| Total | | | | 1.284 |

The number of working levels in an atmosphere indicates the actual concentration of potential alpha energy in the air. Because of removal processes such as ventilation and deposition on surfaces, the indoor concentration of short-lived daughters is usually less than the equilibrium values. Indoors, it is common for the actual potential alpha energy concentration to be about half of the equilibrium value, but the ratio depends on the ventilation rate, the concentration and size distribution of particulates in the air, and other adsorbing surfaces. This ratio, called the "equilibrium factor", is an important parameter in indoor radon dosimetry.

The number of working levels present can be calculated from knowledge of the radon concentration [Rn] pCi/l, and the equilibrium factor, F, from the expression

$$\mathrm{WL} = \frac{\mathrm{F \times [Rn]\ pCi/l}}{\mathrm{100\ pCi/l}} \quad (1)$$

The product, F x [Rn], is often referred to as the "equilibrium effective concentration" and is treated as though this were the true radon concentration and all of the short-lived daughters were in equilibrium. The usefulness of this expedient is to provide a ready comparison among different environments with different balances between the short-lived daughters and the parent radon.

If the concentrations of the short-lived daughters are measured instead of the radon concentration, there is an alternate expression for the working level called the potential alpha energy concentration or "PAEC". In units of working levels, the PAEC is

$$\mathrm{PAEC} = \frac{137\,[\mathrm{A}] + 660\,[\mathrm{B}] + 488\,[\mathrm{C}]}{1.3 \times 10^5}\ \mathrm{WL} \quad (2)$$

where [A], [B], and [C] are the instantaneous air concentrations of Po-218 (RaA), Pb-214 (RaB), and Bi-214 (RaC) in pCi/l.

Specifying the effective working level is only one part of defining the potential risk to an exposed person. The other parts are to define the radiation dose rate due to the exposure and the length of time of exposure. The working level was originally developed to indicate the hazard in the working conditions in a mine. Assuming that a miner will work for 40 hours per week, his exposure time will be approximately 170 hours per month.

In an environment in which a person is breathing air at WL working levels, the accumulated exposure is expressed in terms of the "working-level-months" or WLMs, where

$$\text{WLM} = \frac{\text{WL x hours exposed}}{\text{170 hours}} \tag{3}$$

This same equation is now applied to all persons in all exposure environments, including domestic housing. The general population, for the most part, spends most of its time indoors. It is usually assumed that people are indoors 80% of the time, or about 7000 hours per year. This time is the equivalent of about 40 working months for a miner.

## PHYSICAL PARAMETERS

### Atmospheric Radioactivity

The short-lived daughters of radon, Po-218, Pb-214, and Bi-214, that are produced in the air from the decay of radon are heavy metals. When they are formed they are highly ionized and attach very readily to surfaces, particularly to the particles suspended in the air. When these particles are inhaled, they deposit in the lung where their decay results in irradiation of lung cells. Nearly all of the damage is due to the alpha particles emitted by Po-218 and Po-214.

### Atmospheric Aerosol

The distribution of deposition of breathed activity in the lung depends on the sizes of the particles carrying the activity. The atmospheric aerosol in homes typically occurs in two size ranges. The larger size group occurs in a range strongly peaked between 100 and 200 nanometers. The smaller size group occurs in a range between 4 and 15 nanometers (Ge80). These particle distributions are generated by normal human indoor activities (Be84, Jo84, Kn88, Re85, St84).

Raabe (Ra69) suggests that when atoms of the decay daughters are produced in the air they quickly react to form a mixture of free ions, oxides, or other chemical forms, usually as very small (less than nanometer) agglomerates with several water molecules. These ultra fine particles are much more likely to attach to the aerosol particles in the larger size group than to those in the smaller group. (Kn88).

The short-lived daughters that have attached to particles of the atmospheric aerosol comprise the "attached fraction". The remaining unattached decay daughters comprise the "unattached fraction". The unattached daughters are very reactive and attach readily to surfaces. They deposit entirely in the upper respiratory tract. The attached fraction has much smaller deposition and penetrates much more deeply into the lung.

In the humid environment of the respiratory system, small particles would tend to grow due to the accumulation of water molecules. Larger particles penetrate more deeply so that deposition measurements done in dry mock-ups of the respiratory tree may not accurately represent the deposition in the living lung.

Once the radioactivity has been deposited, the energetic alphas emitted in the decay of the short-lived daughters irradiate the cells of the lining of the airways. The depth of penetration of these alpha particles is limited by their short range. The 6.0 MeV alpha from Po-218 has a range in tissue of about 47 microns and the 7.69 MeV alpha of Po-214 a range of about 71 microns (Le47). Thus, cells that lie deeper within the wall of the airways than these ranges are not irradiated by activity on the surface of the mucus.

## Structure of the Respiratory System

### *Gross description of the lung*

The respiratory system is commonly described as composed of three sections. The uppermost section, called the "naso-pharyngeal" or "N-P" region, is made up of the nose, mouth, throat, pharynx and larynx. This is the region in which air is moving at the greatest speed and the air flow is the most likely to be turbulent.

The next part of the respiratory system begins at the trachea and is made up of the larger airways. By and large, except for the initial divisions, each airway divides into two smaller airways. Each size cohort is called a "generation" and given a numerical index indicating how many subdivisions have occurred since the trachea (generation zero). In the human lung, the first generation on the right side gives rise to three second generation airways while on the left only two are formed. This difference ultimately results in the right lung having three lobes while the left lung has two.

This region of subdividing airways, called the "tracheo-bronchial" or "T-B" region, forms the conduit to carry air to the deeper regions of the lung. It runs from the trachea down through 16 subdivisions. In this region, except for the first generation below the trachea, the airways branching tends to be binary, although not necessarily into two equal sized branches. Each of the five lobes of the lung has its own generational pattern of airway lengths and diameters. Below the 16th subdivisions, the airways divide further until finally they bud out into the alveolar sacs where gas exchange occurs. This deepest region of the lung is called the "pulmonary" or "P" region.

### *Airway dimensions*

There are three detailed descriptions of the lung in common use. These are the Weibel-A model (We63), the Yeh-Schum model (Ye80), and the University of California, Irvine (UCI) model (Ph85). These models differ in the way that they represent the divisions of the airway and the airway dimensions.

*a. The Weibel-A model (We63):* In this model, the lung is represented as a single lobe. It is assumed that every time that the airway in this lung divides it does so into two exactly equal parts. Weibel obtained the dimensions for this model from casts of human lungs and from histological specimens.

*b. The Yeh-Schum model (Ye80):* In this model, the lung is treated as five separate lobes, each with its own characteristic set of bifurcation patterns and dimensions. Yeh and Schum obtained their dimensions of the airways from casts of human lungs. Because the bifurcations are not uniform, there are many different combinations of airway dimensions experienced by a unit volume of air as it progresses from the trachea down to the alveoli.

In each lobe, these variations are averaged into a "typical path" representation in which each bifurcation is symmetric and the dimensions of the airways in each generation are the same. In an additional simplification in this model, the five lobes are further averaged into a single lobe with a typical path pattern of bifurcations and dimensions. In this averaged single lobe representation, the airway dimensions are slightly larger than in the Weibel single lobe model.

*c. The UCI model:* This recent model is based on a series of measurements of human lung airway dimensions obtained from a large number of casts of lungs from persons of various ages. The dimensions reported are generally smaller than those of either the Weibel-A or the Yeh-Schum models.

### *The T-B epithelium*

In most of the T-B region, starting from the lumen of the airway and progressing into the epithelium, the layers of the structure are: 1. a viscous mucous layer; 2. a serous (watery) mucous layer; and 3. a cellular epithelial layer mostly made up of a combination of basal cells, ciliated cells and goblet cells. These cells are attached to a basement membrane attached to deeper supporting structures. The layers vary in thickness in each generation throughout the T-B region and become progressively smaller at the deeper generations.

The goblet and ciliated cells extend from the basement membrane to the serous layer. The cilia of the ciliated cells extend through the serous layer with their tips imbedded in the viscous layer of mucus. The basal cells are pyramidal and extend only a short distance upwards from the basement membrane. The lung cancers caused by exposure to radon daughters seem to develop mainly, although not exclusively, in the epithelium of the third, fourth and fifth generations of airways.

Gastineau, et al. (Ga72) made measurements of the various layers of the bronchial epithelium throughout the T-B region. At the level of generations 3-5, their measurements indicate a median length for the cilia of 7.5 microns. It is usually assumed that the serous and viscous mucous layers are approximately equal in thickness and slightly thinner than the length of the cilia. Their reported median value for the thickness of the epithelium in this region is 50 microns. They give a median thickness of 7.5 microns for the basal cell layer. They estimate that only about 20% of the basal cells are exposed to the 6.00 MeV alphas from Po-218 and about 70% to the 7.69 MeV alphas from Po-214 for activity located on the surface of the mucus.

In the past it had been assumed that it was only the dividing basal cells that were at risk for becoming cancerous due to irradiation. This concept was based on the assumption that it was only these basal cells that divided. In the first of a series of papers on the respiratory epithelium, McDowell and her colleagues (McD78) point out that both basal cells and the various forms of mucous secreting cells can divide. In a recent review, Masse (Ma84) concludes:

> "Cells at risk in the lung are mainly made of cell population derived from basal and mucous cells. Cytodynamics of bronchial cells and histogenesis of lung tumors does not allow to exclude that cells at risk are immediately located beneath the mucous blanket where radon and progeny dissolve and deposit—a wide range of depths can be assigned to cells at risk."

Thus, there are cells at risk within the range of the all of the alpha particles emitted from both Po-218 and Po-214.

## Deposition and Clearance in the Lung

In order to calculate the dose delivered to the cells at risk, it is necessary to know the deposition rates of the unattached and attached fractions and the various mechanisms and rates of clearance and redistribution of the radioactivity following deposition.

### *Deposition*

The ultrafine particles that make up the unattached fraction range in size up to a few nanometers. The smallest of these are the free ions of the short-lived daughters and the larger ones represent small agglomerates of the ion with condensable atmospheric components such as water molecules.

In indoor air, the particles of the attached fraction range in size from about 10 to 1000 nanometers (St84) with the sizes most commonly in the 50 to 200 nm range (Ge80, Re85). These particles deposit throughout the respiratory system. The mechanisms for deposition include: diffusion by Brownian motion to the surface of the airways where they attach; settling to the wall of the airways by sedimentation under the influence of gravity; and impaction and adherence to the wall of particles unable to negotiate the bends at the bifurcation of airways as they are carried along by the airstream. The larger the particles, the greater is the influence of sedimentation and impaction on deposition. The deposition of the attached fraction calculated with the Yeh-Schum model is less in the upper generation and larger in the deeper ones below generation 13 compared to the results with the Weibel-A model (NEA83).

Because of the strong dependence of the deposition in the various regions of the lung on the particle diameter, the prediction of deposition requires knowledge of the size distribution of the atmospheric aerosol particles. For example, the study of Reineking, et al. (Re85) of the aerosols in houses in Southern Germany indicated sizes ranging from about 20 nanometers to about 200 nanometers. The larger sizes were observed in rooms without particulate sources. The smaller sizes were observed when sources such as fresh cigarette smoke, candle burning, cooking or an operating electrical motor were present. The small particles generated by cigarette smoke agglomerated with timc to produce particle sizes comparable to those found in the absence of sources. Over all, for typical ventilation rates from 0.3 to 1.0 air changes per hour (ACH), most of the particles were in the 50 to 200 nanometers range as has been found in many other studies (Be84, Jo84, Kn88, St84). These are the particle sizes most associated with the attached fraction of the short-lived daughters of radon.

James (Ja88) summarizes the deposition probabilities of particles of various sizes in the respiratory tract for two breathing rates, 0.45 and 1.2 $m^3/h$, corresponding to a person resting or working, respectively. For the ultrafine sizes of the unattached fraction, up to a few nanometers, about half (range from 35% to 65%) deposits in the N-P region and about half (range from 35% to 65%) deposits in the T-B region. There is a rough complementarity of the deposition in these two regions with the deposition probability for the two summing to about unity. None of the unattached fraction penetrates to the pulmonary region.

For the particles in the size range of the attached fraction, from 50 to 200 nanometers, the deposition in the N-P region is negligible. Table 2 gives estimates of the deposition fractions for some sizes of the atmospheric aerosol for the two breathing rates (Ja88). Although deposition of the ultrafine unattached fraction is essentially complete and about equal in magnitude in the N-P and T-B regions, deposition of the larger aerosol particles to which the short-lived daughters are attached is relatively small in the T-B region and modest in the

P region. It is because of the greater deposition that the unattached fraction has such large dosimetric significance.

TABLE 2

*Percentage Lung Deposition for the Atmospheric Aerosol* for Breathing Rates of 0.45 and 1.2 m³/h*

| | Lung Region | | | | | | | | | | |
|---|---|---|---|---|---|---|---|---|---|---|---|
| | N-P | | T-B | | P | | Total Dep. | | Nom Dep.+ | | |
| | 0.45 m³/h | 1.2 m³/h | 0.45 m³/h | 1.2 m³/h | 0.45 m³/h | 1.2 m³/h | 0.45 m³/h | 1.2 m³/h | N-P | T-B | P |
| Size nm | Percent Deposition | | | | | | | | | | |
| 1 | 65 | 40 | 33 | 58 | 0 | 0 | 98 | 98 | 53 | 46 | 0 |
| 50 | 0 | 0 | 19 | 12 | 25 | 35 | 44 | 47 | 0 | 16 | 30 |
| 100 | 0 | 0 | 10 | 6 | 16 | 22 | 26 | 28 | 0 | 8 | 19 |
| 200 | 0 | 0 | 5 | 3 | 10 | 13 | 15 | 16 | 0 | 4 | 12 |

*Data abstracted from James (Ja88, fig. 7.16)
+Nominal deposition taken as mid-value of 0.45 and 1.2 m³/h values.

### *Clearance*

Once the activity has deposited on the surface of the mucous layer, there are various processes that modify the concentrations of the short-lived daughters. Activity attached to particles deposited in the tracheo-bronchial or pulmonary region can desorb and dissolve in the body fluids. These particles can transfer across the epithelium to the blood which removes them from the lung. Particles trapped in the mucous layer of the T-B region are swept up the respiratory system by the action of the beating cilia. These are carried to the throat where they are swallowed and eliminated via the gut. The deposited unattached activity is assumed to behave similarly to the attached activity that has desorbed from its carrier particle.

Particles that deposit in the pulmonary region of the lung can be engulfed by phagocytes. These specialized cells trap and transport foreign particles up the respiratory system to the mucociliary escalator which then transports them to the throat for elimination. In the pulmonary region, activity can be transferred to the blood or to the lymphatic system. In addition to these physiological processes, simple radioactive decay also removes activity.

## DOSIMETRY MODELS

### Introduction

There have been three main radon dosimetry models in use in recent years.

These are the Harley-Pasternack (H-P) model (Ha82), the Jacobi-Eisfeld (J-E) model (Jac80), and the James-Birchall (J-B) model (Ja80). These models take the dividing basal cells to be the cells at risk. The models differ in the depth in the epithelium that they assume for these cells and in their assumptions regarding deposition and clearance of the radioactive short-lived daughters of radon.

## The Dosimetry Models

### *Harley-Pasternack*

The Harley-Pasternack model uses the Yeh-Schum lung model. Deposition is due only to diffusion with enhancement due to turbulence in the first few generations. The ultrafine particles of the unattached activity are assumed to grow hygroscopically in the respiratory tract with resulting decrease in their diffusion coefficient. In this model, these particles penetrate further into the T-B region than in the other models.

After deposition, no distinction is made between attached and unattached activity. The difference in their dosimetric contribution is based on the difference in their deposition probabilities. As in the other models, the attached activity is carried all the way to the alveolar region with incomplete deposition. The unattached activity is completely deposited in the N-P and T-B regions. Activity is either cleared by the muco-ciliary escalator or remains in the mucus until it decays.

The target cells are the basal cells. These are located at 22 microns below the epithelial surface in the first 10 generations and at 10 microns in the last 7 generations of the T-B region.

### *Jacobi-Eisfeld*

The Jacobi-Eisfeld model uses the Weibel-A lung model. Deposition is assumed to be due only to diffusion with some enhancement due to turbulence in the first few generations. The diffusion coefficient for the unattached fraction in the J-E model is larger than in the other models. As a result, there is more deposition and dose in the earlier generations and less in the deeper T-B generations.

Unattached activity that deposits or activity that desorbs from particles in the mucus can be rapidly transferred to the blood with half-times between 4 and 40 minutes. The desorption rate is slow compared to the mucous transfer rate. The rapid transfer to the blood for unattached activity reduces its contribution to the lung dose.

The basal cells at risk are located at depths that decrease continuously with increasing generation number. These depths are taken as the mean of Gastineau's (Ga72) minimum and maximum epithelial thickness in each generation. In comparison to the H-P model above, a mean depth of 22 microns is not reached until about generation 8.

*James-Birchall*

The James-Birchall model is based on the Yeh-Schum lung model. Deposition is mainly due to diffusion but there is some enhancement for larger particles from impaction and sedimentation. The deposited activity is mixed in a 7 micron viscous mucous layer above a 7 micron serous mucous layer.

Activity is removed from the mucus by the muco-ciliary escalator or by transfer to the epithelium. Of the latter, about 10% is cleared to the blood with a half-time of about 15 minutes and about 30% is held up in the epithelium or slowly cleared to the blood with a half-time of about 10 hours. The target basal cells are located according to the Gastineau (Ga72) distribution of epithelial thicknesses at each generation.

*Current models:*

The current viewpoint on the dosimetry of radon daughters is that cells at risk are distributed throughout the bronchial epithelium (i.e., Ja88). The same issues of deposition and clearance still apply. However, it is the mean epithelial tissue dose that is appropriate rather than the dose at some specific depth. The dosimetry of the pulmonary region is not affected by this new viewpoint since the tissue is very thin and the alpha particle can penetrate through several alveoli. Thus, for the pulmonary region, the mean tissue dose is the appropriate quantity for risk estimates.

The mean epithelial tissue dose requires some elaboration. The alpha particle range in tissue was given as 47 and 71 microns for the Po-218 and the Po-214 activity respectively. For a model in which the activity is contained on the surface of the mucus, only the first 33 or 54 microns of the epithelium, respectively, can be irradiated.

In the segmental bronchi (generations 4-6), the epithelium thickness ranges up to 100 microns (median thickness 50 $\mu$m) beneath a mucous layer up to 20 microns, while in the bronchioles (generation 16), the mucous layer is much thinner and the epithelial thickness ranges only up to 30 microns (median thickness 15 $\mu$m) (Ga72). In this part of the T-B region, the entire thickness of the epithelium is well within the range of the alpha particles. Thus, in the deep lung, the average tissue dose and the average dose to the irradiated cells are probably similar. However, most (but not all) of the radon induced lung cancers appear to arise in the segmental bronchi. In this part of the T-B region, the average tissue dose is much less than the average dose to the irradiated cells.

## REGIONAL LUNG DOSE

### General discussion

Because of the large difference in the deposition probability between the unattached and the attached activities, their contributions to the lung dose are treated separately. The unattached fraction is usually described by its airborne potential alpha energy concentration in terms of the fraction of the total PAEC contributed by the unattached RaA, RaB, and RaC. Denoting this fraction by $f_p$ and the dose conversion factor (mGy/WLM) for unattached and attached activity as $D_u$ and $D_a$ respectively, the dose to a lung region can be expressed as

$$D = (1\text{-}f_p)D_a + f_p D_u \text{ mGy/WLM} \tag{4}$$

In the size range above 50 nm, James (Ja87, Ja88) has shown that the bronchial dose conversion factor is inversely proportional to the size of the carrier aerosol (Ja87, Ja88). He notes that there is evidence that the inhaled aerosol is hygroscopic and grows rapidly in the moist environment of the respiratory tract.

Since no unattached activity penetrates to the pulmonary region, the dose there is due only to the attached fraction. Here also, the dose conversion factor is a strong function of the particle size. In the size range between 50 and 200 nm, the mean epithelial dose conversion factor for the alveolar region decreases exponentially with size (Ja88). The values are less than 10% of the values for the bronchial epithelium.

In his review of the dosimetry, James considers the three lung models discussed above and proposes a mean bronchial epithelium dose conversion factor for both the unaltered aerosol and one that has doubled in size from hydration. He also gives values for the alveolar region. His values are shown in Table 3. As can be seen, the alveolar contribution is less than 10% of the total dose, a contribution much smaller than the uncertainties in the calculations leading to these values.

TABLE 3

*Mean Dose Conversion Factors for the Bronchial Epithelium**

| | mGy/WLM | | |
|---|---|---|---|
| Aerosol size | Bronchial Dose for Aerosol | | Pulmonary Dose |
| nm | size in air | twice size in air | size in air |
| unattached | ≃150 | ≃100 | |
| 50 | 20 | 10 | 0.8 |
| 100 | 10 | 5 | 0.5 |
| 150 | 7 | 4 | 0.4 |
| 200 | 5 | ≃3 | 0.3 |

*Data from James (Ja88)

## Regional Lung Dose

Fitting the James prescription for the inverse dependence of the average bronchial epithelium dose on particle size in the range from 50 to 200 nm, the T-B regional dose can be expressed as

$$D = 1000/d \text{ mGy/WLM} \tag{5}$$

where d is the aerosol particle diameter in nanometers. Most of the measurements of the indoor aerosol reported in the literature for a broad range of domestic activities cluster in this size range. Literature values for the unattached fraction have ranged from 0.02 to 0.15 (NEA83, Ja88, Va85, Re85) with most values in the neighborhood of 0.05.

Taking 100 nm for the aerosol particle size in the atmosphere and 0.05 for the unattached fraction as representative values, and assuming that the effective particle size for deposition falls midway between the atmospheric value and twice that due to hydration in the respiratory tract, the dose conversion factors can be obtained from Table 3. These are $D_u$ = 125 mGy/WLM and $D_a$ = 7.5 mGy/WLM. The resulting dose conversion factor from equation (4) is

$$D = 0.05 \times 125 + 0.95 \times 7.5 = 13 \text{ mGy/WLM.}$$

The dose equivalent conversion factor is obtained by taking the currently accepted value of 20 for the quality factor for alpha irradiation of the lung epithelium. Thus, DE = 260 mSv/WLM.

## Effective Dose Equivalent

It is usual for the lung dose to be converted to an effective whole body dose based on equal overall risk of cancer. That is, a weighting factor is applied to the lung dose equal to the ratio of the risk of developing lung cancer anywhere in the lung following lung irradiation to the risk of developing any cancer following uniform whole body irradiation.

The ICRP weighting factor (ICRP77) does not specify the spatial distribution of the irradiation of the organ. The weighting factor should reflect the number of cells at risk as well as the relative radiation sensitivity of the cells. For non-uniform irradiation of the lung, the weighting factor to convert the lung dose to an effective whole-body dose should be reexamined.

The ICRP (ICRP77) recommends a weighting factor for the whole lung of 0.12. To put the regional lung doses on a common health hazard basis with other radiation exposure, the ICRP weighting factor of 0.06 (ICRP81) is used for both the T-B and the P regions to convert the regional doses to an effective dose equivalent, $H_E$. For the T-B region, the $H_E$ = 0.06x260 mSv/WLM = 16mSv/WLM. The addition of the pulmonary dose does not change this value.

James (Ja88) argues that 16 mSv/WLM is substantially above a commonly used value of 5 mSv/WLM and that a single "rounded" value of 10 mSv/WLM is appropriate for both houses and mines. Given the uncertainties involved, the value is reasonable.

## Average Population Dose

Surveys (Ne86, Co86) of homes in the United States indicate an arithmetic average radon concentration of about 55 $Bq/m^3$ (1.5 pCi/l). With an equilibrium factor of 0.5, the average radon level corresponds to 0.0075 working levels. For 40 equivalent working months of domestic exposure per year, the annual exposure is about 0.3 WLM. Using 10 mSv/WLM for the dose conversion factor, 55 $Bq/m^3$ for the radon concentration and 0.5 for the equilibrium factor, the annual average population dose from the radon daughters accumulated in indoor air is 3 mSv (300 mrem).

The average population effective dose equivalent from natural radioactivity and medical and dental procedures is about 1 mSv/y. Thus, the average effective dose equivalent due to the short-lived daughters of radon to which the population of the United States is exposed is three times that from all other sources. This ratio is not significantly different anywhere else in the world for averages over populations who live in housing similar to that in the United States.

## REFERENCES

Becker, K.H., A. Reineking, H.G. Scheibel and J. Porstendorfer. 1984. Radon daughter activity size distributions. Rad. Prot. Dos. 7:147-150.

Donaldson, A.W. 1969. The epidemiology of lung cancer among uranium miners. Health Physics. 16:563-569.

Gastineau, R.M., P.J. Walsh and N. Underwood. 1972. Thickness of bronchial epithelium with relation to exposure to radon. Health Physics. 23:857-860.

George, A.C. and A.J. Breslin. 1980. The distribution of ambient radon daughters in residential buildings in the New Jersey-New York area. *In:* T.F. Gesell and W.M. Lowder (eds.). *The Natural Radiation Environment, III, Vol. 2.* CONF-780422, pp. 1272. Oak Ridge Technical Information Center, Oak Ridge, TN.

Holaday, D.A. 1969. History of the exposure of miners to radon. Health Physics. 16:547-552.

Anonymous. 1977. *Recommendations of the International Commission on Radiological Protection: ICRP Publication 26.* Pergamon Press, Oxford.

Anonymous. 1981. *Limits for Inhalation of Radon Daughters by Workers: ICRP Publication 32.* Pergamon Press, Oxford.

Jacobi, W. and K. Eisfeld. 1980. Dose to Tissues and Effective Dose Equivalent by Inhalation of Radon-222, Radon-220 and their Short-Lived Daughters. GSF-Report S-626. Gesellshaft für Strahlen-und Umweltforschung MBH, Munich.

James, A.C., J.R. Greenhalgh and A. Birchall. 1980. A dosimetric model for tissues of the human respiratory tract at risk from inhaled radon and thoron daughters, *in: Radiation Protection, A Systematic Approach to Safety.* Vol. 2, pp. 1045-1048.

James, A.C. 1987. A reconsideration of cells at risk and other key factors in radon daughter dosimetry. *In:* P.K. Hopke (ed.). *Radon and its Decay Products: Occurrence, Properties, and Health Effects.* ACS Symposium Series 331, pp. 400-4181. American Chemical Society, Washington, D.C.

James, A.C. 1988. Lung Dosimetry. pp. 259-309. *In:* W.W. Nazaroff and A.V. Nero, Jr. (eds.). *Radon and Its Decay Products in Indoor Air.* Wiley, New York, NY.

Johansson, G.I., C.Samuelsson and H. Petterson. 1984. Characterization of the aerosol and the activity size distribution of radon daughters in indoor air. Rad. Prot. Dos. 7:133-137.

Knutson, E.O. 1989. Modeling indoor concentrations of radon's decay products. pp. 161-202. *In:* W.W. Nazaroff and A.V. Nero, Jr. (eds.). *Radon and Its Decay Products in Indoor Air.* John Wiley & Sons, New York, NY. p. 171.

Lea, D.E. 1947. *Action of Radiation on Living Cells.* Cambridge University Press, Cambridge, MA.

Masse, R. 1984. Cells at risk. pp. 227-245. *In:* H. Smith and G. Gerber, (eds.). *Lung modeling for inhalation of radioactive materials.* Proceedings of a meeting jointly organized by the Commission of the European Communities and the National Radiological Protection Board. Oxford, 26-28 March, 1984, EUR-9384.

McDowell, E.M., L.A. Barrett, F. Glavin, C.C. Harris and B.F. Trump. 1978. The respiratory epithelium. I. Human bronchus. J. Natl. Cancer. Inst. 61:539-549.

Phalen, R.F., M.J. Oldham, C.B. Beaucage, T.T. Crocker and J.D. Mortensen. 1985. Postnatal enlargement of human tracheobronchial airways and implications for particle deposition. Anat. Rec. 212:368-380.

Reineking, A., K.H. Becker and J. Porstendorfer. 1985. Measurements of the unattached fractions of radon daughters in houses. Sci. Tot. Env. 45:261-270.

Stoute, J.R.D., G.C.H. Groen and T.J.H. deGroot. 1984. Characterization of indoor atmospheres. Rad. Prot. Dos. 7:159-163.

Vanmarcke, A., F. Janssens F. Raes. 1985. The equilibrium of attached and unattached radon daughters in the domestic environment. Sci. Tot. Env. 45:251-260.

Weibel, E.R. 1963. *Morphometry of the Human Lung.* Springer-Verlag, Berlin.

Yeh, H.C. and G.M. Schum. 1980. Models of human lung airways and their application to inhaled particle deposition. Bull. Math. Biol. 42:461-480.

*Environmental Radon: Occurrence, Control and Health Hazards.* Edited by S. K. Majumdar, R. F. Schmalz and E. Willard Miller © 1990, The Pennsylvania Academy of Science.

# *Chapter Twenty-Four*

# EXPERIMENTAL TESTS OF THE LINEAR - NO THRESHOLD THEORY OF RADIATION CARCINOGENESIS

**BERNARD L. COHEN**

Department of Physics
University of Pittsburgh
Pittsburgh, PA 15260

In a 1979 paper[1], I analyzed various potential methods for testing the linear-no threshold theory and concluded that by far the most powerful approach was through radon in homes. There is a great deal of data on this at high dose from studies of miners exposed to high doses, and there is a tremendous amount of potential data at low dose from radon in homes which was (and still is) estimated to cause about 7% of all lung cancers. In that paper I analyzed data from the early 20th century, with due consideration for variations in diagnostic efficiency (a topic on which there is a substantial literature) and concluded that the predictions of the theory used at that time (absolute risk model) grossly over-predicted the observed effects if histological data were taken into account. Many discounted these arguments because of lack of confidence in the information on variations of diagnostic efficiency with time and on histological assignments. Another problem was that opinion in the scientific community was shifting from support of an absolute risk model to a relative risk model.

In a paper written in 1985,[2] I pointed out that all geographic areas that had been found to have high radon levels had relatively low lung cancer rates. This included Sweden and Finland (vs the rest of Europe where radon levels were measured), Skaraborg County in Sweden (vs the rest of Sweden), Southeastern Finland (vs the rest of Finland), Cornwall vs the rest of England, two provinces in China, the Reading Prong Area of Eastern Pennsylvania, and Cumberland County in Central Pennsylvania. That paper included analyses showing that this situation was highly improbable if the theory was correct.

At that point I started on a vigorous program of measuring radon levels in homes, accompanying each measurement with a questionnaire to provide additional information. That program has been the principal source of data for the preliminary study described here.

The finding of this study is that there is a very strong tendency for counties with high average radon levels to have low lung cancer mortality rates, and vice versa. This does not necessarily mean that people with high radon exposure have low lung cancer risk. That conclusion would be what epidemiologists call "the ecological fallacy." For example, suppose that only radon levels above 50 pCi/L could cause lung cancer. Then the lung cancer rate would depend only on the fraction of the population that is exposed to levels above 50 pCi/L. This is not necessarily correlated with the average radon level in a county which is always below 8 pCi/L. This points out the principal problem with ecological studies for determining cause-effect relationships: the people whose radon exposure we measure (whole population) are not the people at risk from radon (in our example, those exposed above 50 pCi/L).

*However,* if the linear-no threshold theory is valid, the lung cancer rate is determined by the *average* radon exposure for the whole county, and that is exactly what we measure. Thus, while our study cannot determine a cause-effect relationship which is the usual purpose of an epidemiological study, it *can* test the linear-no threshold theory. The "ecological fallacy" does not apply.

The data consist of:

1. About 39,000 purchased measurements of radon levels in the living areas of homes (livingroom, diningroom, kitchen, bedroom, family room, or hall) in which there had been no previous radon measurement and no house within 5 miles was known to have a high radon level. These gave 411 counties with 10 or more such measurements.
2. About 29,000 purchased measurements as in #1 except in basements, giving 253 counties with 10 or more measurements (15% of them not included in #1).
3. Random selection—no charge measurements in 40 counties with the highest and lowest lung cancer rates in U.S. for an intermediate population range (no big city, no rural counties).
4. State-sponsored studies in NJ, NY, FL, and SC. In addition we have very preliminary data on a case-control study, presented at the end.

Since people do not spend their whole lives and accumulate all of their radon exposure in their county of residence at death, a correction for migration is applied to the theory estimates. This is done by assuming that average radon exposure is as though people spend *50%* of their lifetimes in their county of residence at death, and the remainder in areas of U.S. average radon exposure. The "50%" is a guesstimate, but an experimental determination of this percentage by a random-digit-dialing survey is in progress.

The linear-no threshold theory predicts that the average lung cancer mortality rate, $\overline{m}$, in counties with mean radon level r, should be given by

$$\overline{m} = a + br \qquad (1)$$

Values of the slope b can be obtained from various published analyses, as given in Table 1.

TABLE 1

*Estimates of b*

| Absolute risk model | | Relative risk model | Male | Female |
|---|---|---|---|---|
| UNSCEAR - 1977 | 1.8 | EPA-1986 | 2.6-10.4 | 0.4-1.6 |
| ICRP - 1981 | 1.8 | BEIR-IV-1988 | | |
| NCRP - 1984 | 1.6 | Simple RR | 3.6 | 0.6 |
| BEIR-III - 1980 | 5.0 | TSE Model | 1.9 | 0.3 |

We use b = 1.9 for males and b = 0.3 for females, which are essentially the lowest values consistent with any published estimate. (Absolute risk model requires similar values of b for males and females).

Purchased measurements have many biases, especially on socioeconomic bases, although there is little a priori reason to expect them to correlate with radon levels (this question will be faced later). On the other hand, there is a great advantage in using purchased measurements because they provide large quantities of data at very low costs.

In order to check the relationship between purchased and random selection-no charge (RS-NC) measurements, we determined correlations between mean radon levels in counties where determinations by both methods were available. There were 35 counties for which we had at least 10 purchased and 10 RS-NC measurements and the coefficient of correlation between the means of the two types was 85%. In addition there were 17 counties in which we had at least 5 measurements of each type, but not more than 9 of both; the correlation was 84%.

The correlation between mean radon levels in counties obtained from state-sponsored RS-NC studies and our purchased measurements where at least 10 were available was 95% for 17 counties in Florida, 89% for 16 counties in New Jersey, and 90% for 21 counties in a New York State study in basements. For counties in which we had only 5 to 9 purchased measurements, these correlations were respectively 84%, 59%, and 63%. There was also a New York State study of first floors but it was stated that areas sampled were chosen for geological reasons and are not representative of the whole county; for these, the correlation with our purchased measurements was 65% for the 28 counties where we had at least 10 measurements, but only 5% for the 15 counties where we had 5-9 measurements.

Based on these results, we felt justified in using purchased measurements in counties for which we had at least 10 measurements. There were 411 of these counties.

Figures 1 and 2 show, for males and females respectively, a plot of age-adjusted

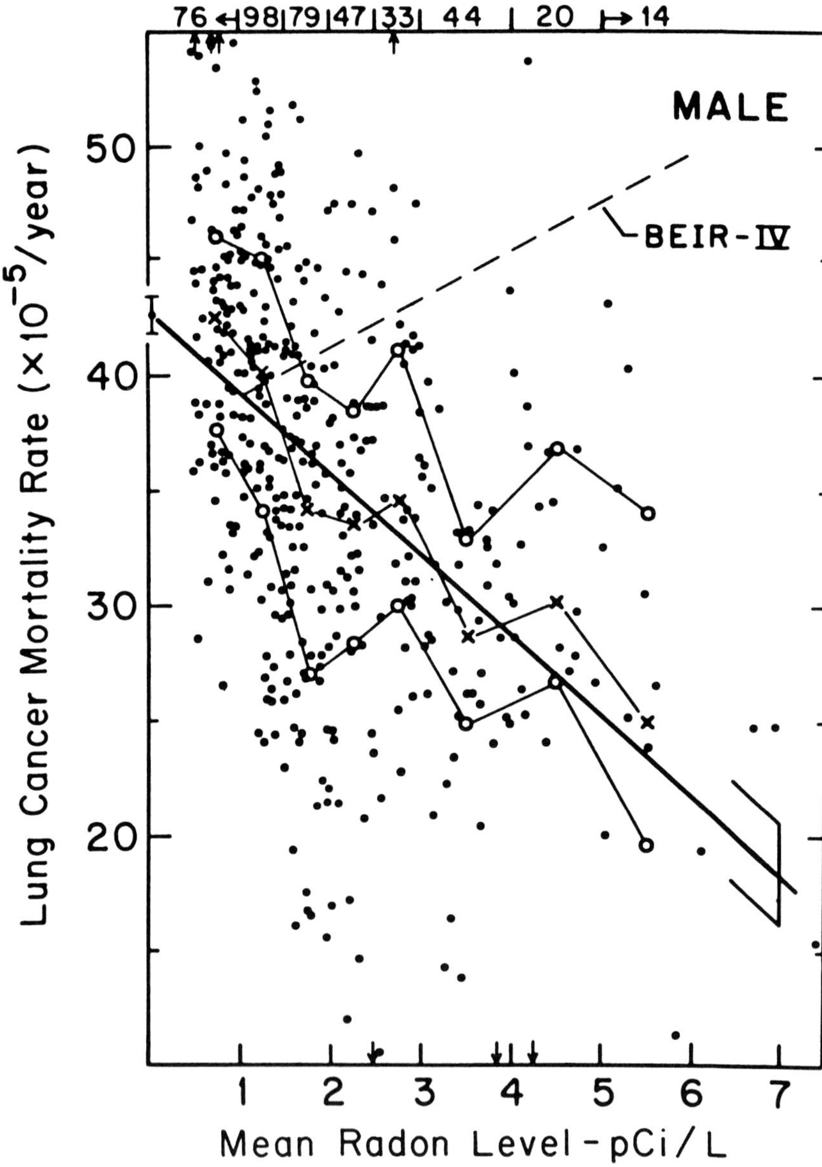


FIGURE 1. Age-adjusted lung cancer mortality rates vs. mean radon level in males in living areas of homes.

lung cancer mortality rate vs mean radon level in living areas of homes for each of our 411 counties. To help see the trends, the abscissa is divided into small divisions containing the number of counties shown at the top of the figure, and for each division the median ordinate is shown by a cross and the first and third quartiles are marked by open circles. It is apparent that the trend for these agrees well with the thick solid line which represents the best least squares fit to the data. The standard deviation of this slope is shown near the right end of the line. Clearly the slopes of these lines, b in (1), are *negative* by many standard deviations. Also shown in Figure 1 and 2 as dashed lines are the predictions of linear-no threshold theory, which have a relatively strong *positive* slope, b in (1). The remainder of this paper represents attempts to resolve this discrepancy between the strong negative slopes found in the experiments and the strong positive slope predicted by the theory.

In a limited geographic area, a negative slope could occur by chance if a section of that area that has heavy smoking—and hence a high lung cancer rate—happens to have a geology that yields low radon levels, but where a large variety of geographical areas is involved this would be highly improbable. To test how commonly a negative slope occurs, the data were divided into 23 state groupings. The distribution of slopes (b) is shown in Figure 3. Note that they are predominantly negative, with a reasonably symmetric distribution about $b_o$, the slopes from Figure 1 and 2. Also shown in that part of Figure 3 are the values of b from our basement studies, b', suitably corrected for the fact that basement radon levels are typically double those of the main living areas of houses. The values of b expected from the BEIR-IV version of linear-no threshold theory are also shown there, labelled TSE (time-since-exposure) and RR (relative risk model).

Our data were also subdivided into 7 national regions. The results, shown in the upper part of Figure 3, indicate that b is negative for 6 of the seven regions, with the only exception, New England, having a positive value of b substantially smaller than the prediction of the theory.

Also included in the upper part of Figure 3 are completely independent sets of data from 5 state-sponsored studies in New York, (2) New Jersey, South Carolina, and Florida. Four of the five studies give a negative value for b, quite consistent with our findings. The exception with a positive value of b, is Florida which also has a positive value of b in our data.

Figure 4 shows a completely separate set of data from our random selection, no charge studies in all states in a given population range (selected to exclude large cities and rural areas) with the highest and lowest lung cancer rates. The least squares fit to the data, shown by the solid line, has a strong negative slope, whereas the theory line, like that in Figure 2, has a positive slope.

From Figure 1-4, it is clear that negative values of b are far more common than positive values. The question at this point is whether they can be explained by confounding factors. For example, our studies have shown that average radon levels in rural homes are substantially higher than in urban areas (perhaps due to differences in house construction practices; not due to apartment living in cities as very few of our measurements are from above the second floor). It is well known that urban areas have higher lung cancer rates than rural areas, perhaps due to more smoking and/or air pollution in the former. These two unrelated facts conspire to make the slope b negative.

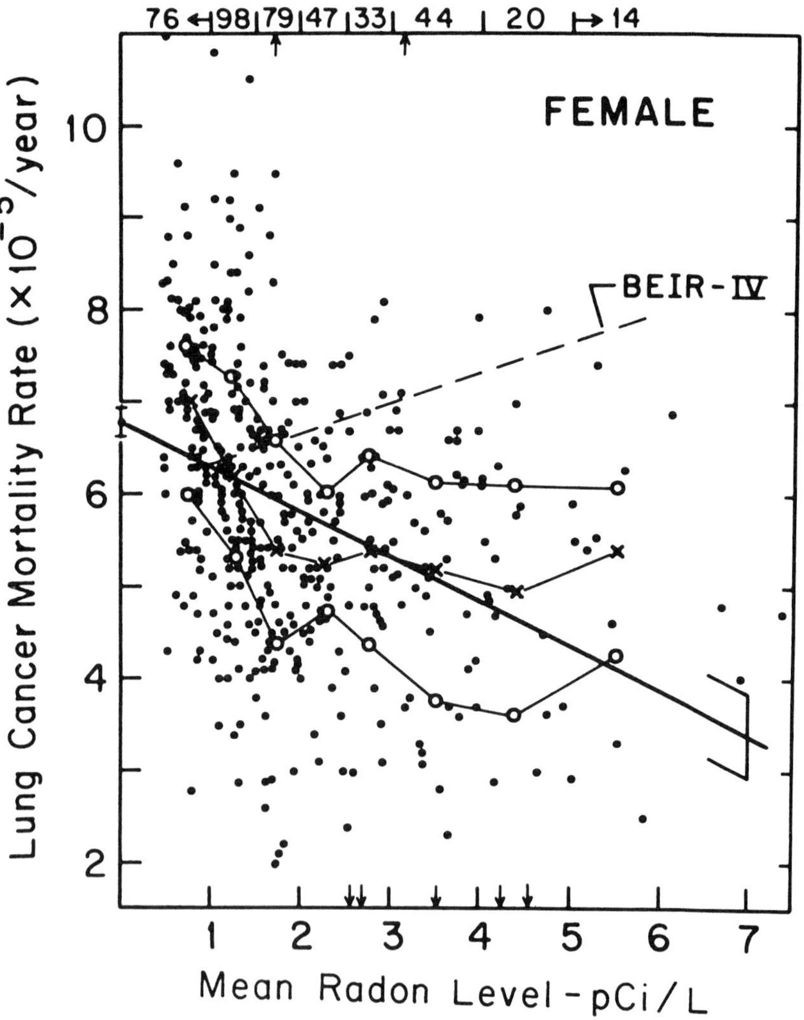


FIGURE 2. Age-adjusted lung cancer mortality rates vs. mean radon level in females in living areas of homes.

In order to compensate for this and other factors that may correlate both with radon levels and with lung cancer rates, the following list of potential confounding factors was investigated:

-P, total population, which indicates whether the county contains a large city
-I, median family annual income, an economic indicator
-E, percent of population of age over 25 that has graduated from high school, an indicator of educational status
-H, percent of housing that was owner occupied in 1960, an indication of the housing situation
-U, percent of population living in urban areas
-A, median age of the population
-M, number of physicians per 100,000 population, an index of medical care
-H′ percent of housing that was owner occupied in 1980, to check on changes in this factor from H
-J, percent of labor force that is unemployed, an indicator of the job situation

We then did least square fits of the data to

$$\overline{m} = a + br + c_1P + c_2I + c_3E + c_4H + c_5U + c_6A + c_7M + c_8H' + c_9J \quad (2)$$

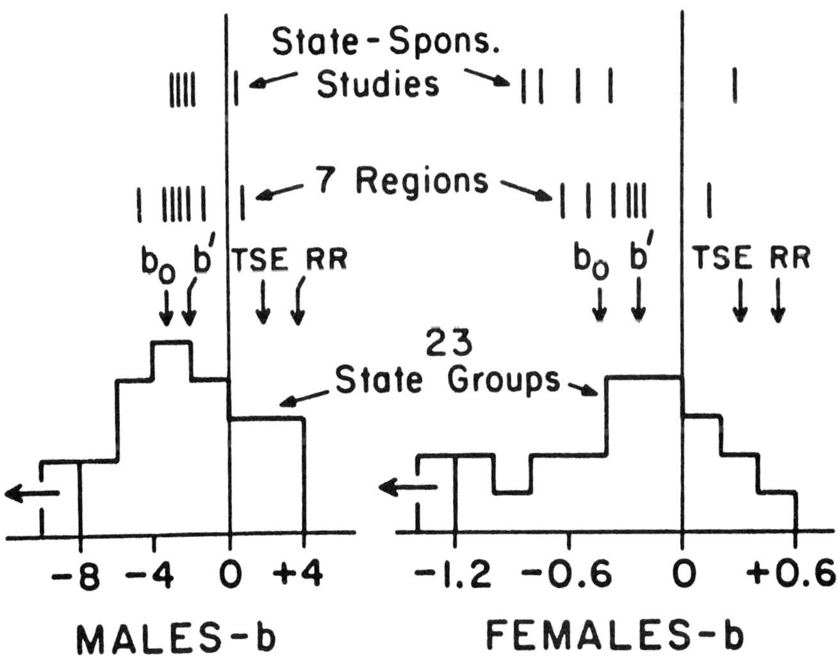


FIGURE 3. Cancer rate in a variety of geographical areas.

to determine the constants a, b, $c_1$, $c_2$, - - - , $c_9$. This is a process called "multiple regression". Note that inclusion of the factors P and U should take care of the urban-rural confounding relation described above.

The results of this analysis on our 411 counties are shown in Table 2. The symbol t is the number of standard deviations (SD) by which b differs from zero, and $R^2$ is the percentage of the variations in the data that is explained by the factors considered. For males, b is negative by 10 SD and discrepant with theory by 16 SD if only radon is used in the regression, and including our 9 confounding factors it is still negative by 6 SD and deviates from the theory by 12 SD. For females, the deviation from theory is by 12 SD with radon alone,

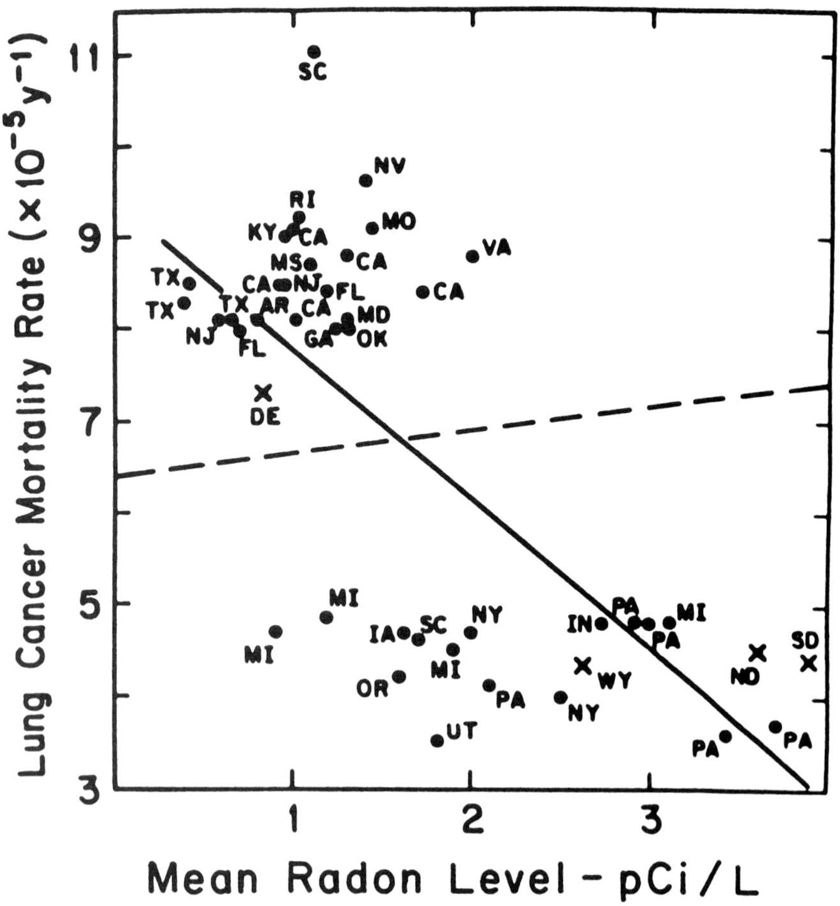


FIGURE 4. Lung cancer mortality rate vs. mean radon level.

and by 8.5 SD when the 9 confounders are included. Considering the confounding factors thus reduces the huge discrepancy with theory only by about 25-30%. Since these factors were chosen to be broadly representative of all socioeconomic factors, it is highly doubtful if inclusion of more such factors would do much to remove the discrepancy with theory.

TABLE 2

*Results of regression analyses on 411 counties*

| Factors included in the regression | males b | males t | males $R^2$ | females b | females t | females $R^2$ |
|---|---|---|---|---|---|---|
| Radon only - Eq. (1) | −3.4 | −10.2 | 21 | −.45 | −7.2 | 11 |
| All 10 factors - Eq. (2) | −1.8 | −5.9 | 46 | −.24 | −3.8 | 25 |
| BEIR-IV - TSE | +1.9 | | | +.30 | | |

Another approach to removing confounding or extraneous factors is to delete portions of the data, or stratify it on various bases. One obvious problem is that many people retire, move to Florida, Arizona, or California, and eventually, die there, contributing to the lung cancer statistics of those states although their radon exposure was received elsewhere. The obvious solution to that problem is to delete data from those states. The results of multiple regression on all 10 variables (Eq. (2)) are listed in Table 3. We see that deleting FL, AZ, and CA counties has so significant effect on the results.

TABLE 3

*Effects of deleting or stratifying*

| Counties included: Limitations | Number | Males b | Males t | Males $R^2$ | Females b | Females t | Females $R^2$ |
|---|---|---|---|---|---|---|---|
| None | 411 | −1.8 | −5.9 | 46 | −.24 | −3.8 | 25 |
| Delete FL, AZ, GA | 382 | −1.8 | −6.0 | 48 | −.22 | −3.4 | 24 |
| Pop<45,000 | 104 | −1.2 | −2.1 | 40 | −.17 | −1.1 | 19 |
| Pop 45-100,000 | 102 | −1.4 | −2.0 | 26 | −.17 | −1.4 | 20 |
| Pop 100-262,000 | 104 | −2.1 | −4.1 | 29 | −.23 | −2.9 | 34 |
| Pop>263,000 | 101 | −0.9 | −1.5 | 47 | −.16 | −1.4 | 35 |
| Delete sea coast | 338 | −1.4 | −4.3 | 44 | −.09 | −1.4 | 27 |
| + Pop>300,000 | 283 | −1.3 | −3.8 | 40 | −.08 | −1.2 | 22 |
| $\overline{m}$ for 1970-79 | | | | | −.83 | −5.3 | 27 |
| Theory | | +1.9 | | | +.30 | | |

Since population is known to be an important confounding factor, its effects were further studied by stratifying the data by population into four separate files of equal size. The results, shown in Table 3, indicate that b is strongly negative for each of the four files separately (t-values are reduced because the reduced number of data points increases the SD).

It was noticed that sea coast counties tend to have high lung cancer rates (salt spray in the air ??) and low radon levels (sand is low in uranium ??). To assess the importance of this problem, all counties washed by sea water (including Potomac River to Washington, DC, San Francisco Bay, etc.) were deleted. As seen in Table 3, this reduces the discrepancy with theory by 11% for males and by 28% for females, which is a long way from removing it.

The combined influence of sea coast location and population was studied by deleting counties with population over 300,000 from the non-sea coast file. We see from Table 3 that this does very little to reduce the discrepancy with theory.

Since our use of 1950-1969 mortality data has been criticized, the effect of using more recent (1970-1979) data was studied for females where the difference is by far the largest. As indicated in Table 3, this increases the discrepancy with theory.

Since smoking is the major cause of lung cancer, the principal effect of our confounding factors is to represent variations in smoking. However, it would be highly desirable to introduce smoking frequency directly into our multiple regression analysis. I know of no data on smoking frequency by Counties, but for states there are data on cigarette sales from tax collections. In order to utilize it, our analysis was repeated for states rather than for counties. The results are shown in Table 4.

TABLE 4

*Regression analyses of data on States*

| Factors included | Males | | | | Females | | | |
|---|---|---|---|---|---|---|---|---|
| in regression | b | t(r) | t(s) | $R^2$ | b | t(r) | t(s) | $R^2$ |
| Radon only | −6.8 | −6.0 | — | 44 | −.69 | −4.0 | -- | 26 |
| Smoking only | — | — | +3.9 | 26 | — | — | +4.8 | 34 |
| Radon and smoking | −6.1 | −6.1 | +4.1 | 60 | −.56 | −3.8 | +4.7 | 51 |
| All 10 factors | −4.4 | −4.5 | +2.7 | 76 | −.38 | −2.6 | +3.0 | 69 |
| Counties - Table 2,3 | −1.8 | | | | −.24 | | | |
| Theory | +1.9 | | | | +.30 | | | |

When radon only is included in the regression (Eq (1), the slopes are more negative than for the previous analysis by county (Figure 1 and 2, Table 2) by factors of 2.0 for males and 1.5 for females (the t-values are smaller because the much smaller number of data points leads to large SD). When smoking only is used in the regression vs. lung cancer rates, the slope is positive as expected—more smoking leads to more lung cancer—but it is shocking to see that the negative slope vs radon is at least as strong in terms of number of SD from zero slope as is the positive slope vs smoking, and the $R^2$ (fraction of the variation explained) is at least as large for radon as for smoking.

When both radon and smoking are included in a double regression, the slopes vs both radon and smoking are almost the same as in the single regressions, and the $R^2$ values are close to the sum of those found in the single regressions. It seems that radon and smoking have separate effects on lung cancer rates and there is almost no interaction between the two.

Even when all of the other potential socioeconomic confounding factors are included, the discrepancy with theory is not substantially reduced; in fact it is reduced by less than was the case for counties, where smoking was not included in the regression.

Up to this point we have been considering confounding factors on the level of counties or states. It is also relevant to consider confounding factors on the level of individual houses. This is suggested by the fact that our studies have found that average radon levels in houses of smokers are about 10% lower than in houses of non-smokers. The correction for this can be understood from Figure 1 and 2: If we assume that high lung cancer counties have 70% smokers while low lung cancer counties have only 20% smokers, the radon level in the former is depressed by (.70-.20) x 10% = 5% by that fact. This artificial depression can be corrected for by shifting high lung cancer points in Figure 1 and 2 to the right by 5%. It is readily seen that this does little to change the slope of the line through the data. It reduces t from −10.2 to −9.5 for males, and from −7.2 to −6.6 for females. If the average radon level were lower for smokers than for non-smokers by 20%, 30%, or 50%, t for males would be reduced only to −8.9, −8.4, and −7.4 respectively.

We have studied the correlation of radon levels with a variety of house characteristics, socioeconomic factors, and environmental characteristics (e.g., air pollution), and in all cases these correlations are rather weak, affecting radon levels by less than 50%. From the above analysis of the smoking correlation we therefore conclude that confounding factors on the level of individuals cannot have an important effect on our results.

## SUMMARY

Our data indicate that lung cancer rates in Counties tend strongly to *decrease* as average radon levels increase, and vice versa. The effect seems to be statistically unquestionable.

This is in sharp contrast to the prediction of the linear-no threshold theory, that lung cancer rates should *increase* substantially as average radon levels increase.

The only reasonable explanation, other than failure of the theory, is that there are *confounding factors* that correlate strongly with both radon levels and lung cancer rates.

Nine socioeconomic factors were selected as most likely confounders, but when they are included in the regression analysis, the discrepancy with theory is reduced by only 25-30%.

Stratification and deletion of parts of the data was explored, but in no case was the discrepancy with theory dramatically reduced.

A similar study of data on states gave even larger discrepancies with theory, and when cigarette sales were introduced into the regression, the discrepancy with theory was not reduced.

The results are easily explained without abandoning linear-no threshold theory, if there is an added hormesis effect at low dose rate.

## CASE-CONTROL STUDY

Beginning in January 1988, questionnaires filled out before measurements are made have included questions about cancer mortality in the house. Houses are included in our study only if the radon level has not been measured previously and if the householder knows of no house within 5 miles that has a high radon level.

Linear-no threshold theory predicts that the average radon level should be at least 10% higher in houses where an occupant has died of lung cancer than in houses where an occupant has died of some other type of cancer. Only a small fraction of the data have been collected, but our latest analysis gives the following mean radon levels (number of measurements in parentheses):

For living areas of houses (not basements, etc.),
lung cancer – 1.42 pCi/L (215), other cancer – 1.41 pCi/L (359)
For basements,
lung cancer – 2.42 pCi/L (275), other cancer – 2.68 pCi/L (492).

There is clearly no indication from these results of the difference predicted by the theory, but the statistics are not yet very robust.

## REFERENCES

1. Cohen, A.F. and B.L. Cohen. 1980. Test of the Linearity Assumption in the Dose-Effect Relationship for Radiation-Induced Cancer. Health Phys. 38:53-59.
2. Cohen, B.L. 1987. Test of the Linear, no threshold Dose-Response Relationship for High-LET Radiation. Health Phys. 52:629-636.

*Environmental Radon: Occurrence, Control and Health Hazards.* Edited by S. K. Majumdar, R. F. Schmalz and E. Willard Miller © 1990, The Pennsylvania Academy of Science.

# *Chapter Twenty-Five*

# THE ENVIRONMENTAL PROTECTION AGENCY'S RADON ACTION PROGRAM

**MARGO OGE[1] and MARK DICKSON[2]**

[1]Director, Radon Division
[2]Policy Analyst
ANR-464
U.S. EPA
Office of Radiation Programs
Washington, D.C. 20460

## INTRODUCTION

In December 1984, very high radon levels were discovered in a house in the Reading Prong, a geologic region extending from eastern Pennsylvania into New Jersey and New York. Although scientists in Lawrence Berkley Laboratory had expressed concerns about indoor radon prior to 1984, it was not until the Reading Prong incident that the federal and state government officials focused their attention to this problem. Measurements taken in the house found radon concentrations as high as 130 times the Federal occupational exposure standard for underground uranium mines. Subsequently, thousands of other houses in the area were also found to be contaminated by naturally-occurring radon. High levels have now been found in nearly every State.

After the discovery of the radon problem in the Reading Prong, EPA quickly realized the national implications of the radon issue. In 1985, EPA established the Radon Action Program to provide technical assistance to States and to coordinate Federal radon activities. As the Agency began to develop its strategy, it requested the advice of scientists and State officials regarding basic scientific and public information needs, and how EPA could best address this important public health problem. Together, they identified seven key needs:

1. An assessment of national and regional radon problems, including methods for predicting potential high-risk areas;
2. Standardized radon measurement methods;

3. A system for evaluating the competence of firms and laboratories offering measurement services;
4. Cost-effective techniques for reducing radon levels in existing and new houses;
5. Programs for transferring technical assistance capabilities to States and the private sector;
6. Action level guidelines for reducing radon in houses; and
7. Effective communication tools to present health concerns and recommended solutions to the public.

EPA developed its program strategy around these seven needs. The goal of the program is to reduce significantly the health risks of radon. EPA realized that the traditional regulatory approach was inappropriate in the case of a naturally-occurring contaminant such as radon.

To accomplish its goal, EPA decided to work in partnership with the States to structure the Radon Action Program as a non-regulatory technical assistance and public information program. The program has four key elements:

- *Problem Assessment,* to determine the magnitude and distribution of the radon problem;
- *Mitigation and Prevention,* to develop cost-effective mitigation and prevention technologies;
- *Capability Development,* to transfer technologies to State and local governments and the private sector; and
- *Public Information,* to communicate radon information to the public.

## PROBLEM ASSESSMENT

EPA began its Radon Action Program by gathering information about the extent of the radon problem. These activities have included identifying homes with high radon levels, identifying geographic areas with high radon exposure potential, and determining human health risks associated with radon exposure.

MEASUREMENT PROTOCOLS. To help both the private sector and homeowners test for radon, EPA first wanted to develop accurate, reliable, and comparable methods of measuring radon. In its "Indoor Radon and Radon Decay Product Measurement Protocols," EPA has evaluated numerous measuring devices and has prepared standardized measurement procedures for eleven different devices considered acceptable for making radon measurements. The most common device, the charcoal canister, provides a 2 to 7 day screening measurement for homeowners. The protocols offer guidance for other short-term devices such as grab sampling and working level monitors. If these measurements yield high levels, residents should follow up with one of several long-term devices. These protocols, revised most recently in March 1989, are continually updated.

STATE/EPA SURVEYS. The Agency has helped twenty-five States and several Indian Lands design and complete statistical surveys of radon measurements in homes to determine the radon potential in those States. In these surveys, homeowners placed charcoal canisters in their homes for two days to obtain a screening measurement (such measurements may maximize radon annual levels) of their homes' radon potential. In ten States, including Pennsylvania, over 25 percent of the sampled homes had screening levels that exceeded EPA's action level of 4 picocuries per liter (pCi/l), and in another four States, 20 percent of the homes had similarly elevated levels. Only four States surveyed to date—Alabama, Alaska, Georgia, and Arizona—had fewer than 10 percent of their homes over 4 pCi/l. Figure 1 summarizes these findings.

As a result of these surveys, EPA held a national press conference in October, 1989, to convey the following key messages:

- One in four homes has elevated screening levels in the 25 States tested.
- EPA confirms the severity of radon problems.
- Homes should be tested as soon as possible to protect families.
- Radon is a readily fixable problem.
- State and local government action is vital in reducing radon risks.

At an earlier press conference in September 1988, EPA and the Office of the Surgeon General had issued the following National Radon Health Advisory: Indoor radon gas is a national health problem. Radon causes thousands of deaths each year. Millions of homes have elevated radon levels. Most homes should be tested for radon. When elevated levels are confirmed, the problem should be corrected. This year, nine more States are conducting surveys, and those results should be available this Fall.

NATIONAL SURVEY. While the State surveys have provided valuable information about screening measurements in selected States, EPA's National Survey will determine the distribution of annual average radon concentrations throughout the country and investigate the relationship between housing construction and resulting radon levels. Last summer, the Agency placed alpha-track detectors in over 7,000 residences to measure radon for one year. Results from the National Survey will be available in the Spring, 1991.

RADON IN SCHOOLS. EPA completed a study to measure radon in 130 schools across sixteen States in March, 1989. Nearly 20% of the rooms tested had levels over 4 pCi/l, and over one half of the schools had at least one room

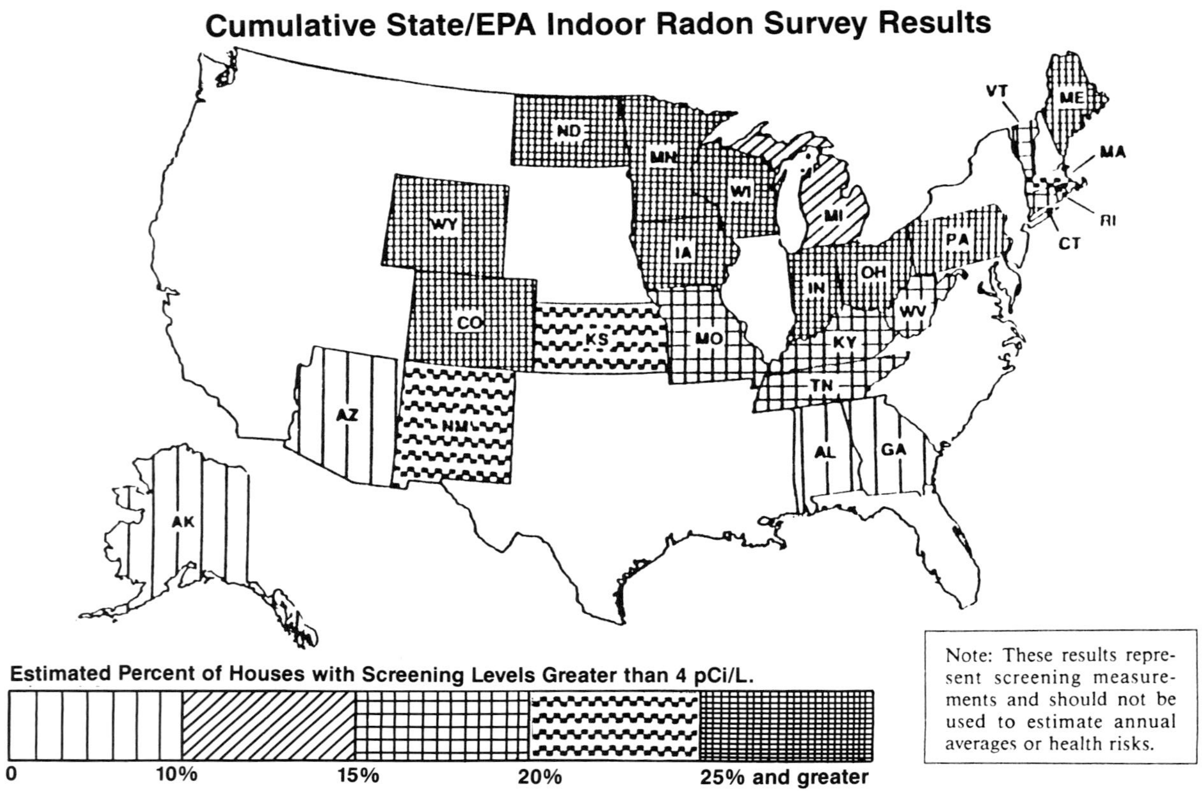


FIGURE 1. Cumulative State/EPA Indoor Radon Survey Results

**SCHOOL PROTOCOL DEVELOPMENT STUDY SCREENING RESULTS**

| | |
|---|---|
| No. States | 16 |
| No. Schools | 130 |
| No. Rooms<br>Rooms > 4 pCi/L<br>Rooms > 20 pCi/L | 3000<br>19 percent<br>3 percent |
| Schools with at least 1 Room > 4 pCi/L | 54 percent |
| Highest Level Found | 136 pCi/L |

FIGURE 2. School Protocol Development Study Screening Results

with elevated radon levels. Figure 2 summarizes these results. This study highlights how prevalent radon is in schools and that it may be especially hazardous to our nation's children.

EPA Administrator, William Reilly, released the results of this study in April, 1989, and recommended that all schools conduct radon measurements. To help school administrators address radon in their buildings, the Agency simultaneously released interim guidance for measuring radon in schools and mitigating buildings with elevated levels.

MAPPING. EPA is cooperating with the United States Geological Survey (USGS) to develop geological maps that indicate areas of potential radon problems in specific regions and throughout the nation. These maps are based on geological data, soil characteristics and indoor air measurements. Additionally, EPA is working with the USGS to conduct land evaluation studies to identify more precisely those geological features which are most useful as indicators of high radon levels.

As a means of determining the radon potential of a specific parcel of land, such as a plot being considered for development, EPA is investigating new methods for measuring soil gases as an indicator of potential radon problems. The radon potential map will be available in the Fall, 1990.

## MITIGATION AND PREVENTION

In the areas of Mitigation and Prevention, EPA has developed cost-effective methods for reducing radon levels in existing structures and for preventing radon entry in new construction. EPA has established programs for transferring these new technologies to the States and the private sector. Post-mitigation monitoring results indicate that the levels in most of the demonstration homes were reduced by more than 90 percent. In many cases, mitigation techniques reduced radon levels by more than 98 percent. EPA is working with the National Association of Homebuilders (NAHB) and nationwide building standard and code organizations to have radon-resistant construction practices incorporated into model building codes. In addition to improving mitigation techniques for residences, EPA is developing techniques that will be useful for remediating radon problems found in schools and other types of structures.

HOUSE EVALUATION PROGRAM. EPA established the House Evaluation Program (HEP) to train State and private sector personnel in radon diagnosis and mitigation and to encourage States to develop their own programs in these areas. As part of this program, State personnel, in cooperation with EPA, evaluate houses with known radon problems and determine the sources of the problem. Course participants then install radon mitigation systems as part of this intensive hands-on training program. Through these activities, States and the private sector receive experience in radon diagnosis and mitigation methods. In addition, EPA uses data from the remediated houses to determine how effectively different mitigation procedures reduce radon levels. EPA is developing a similar program for school radon evaluation.

NEW HOUSE CONSTRUCTION STANDARDS. To reduce the risks from radon exposure, EPA has emphasized preventing radon entry in newly constructed homes. To this end, EPA has developed radon-resistant construction features for use in new homes, has issued interim guidance for new home builders, and has established the New House Construction Evaluation Program to evaluate the effectiveness of these and other features being adopted by homebuilders.

EPA is working closely with the housing industry, particularly the NAHB, to encourage homebuilders to include radon-resistant techniques in new construction activities. EPA is developing a draft standard for new construction that will be published for comments in the Federal Register in Summer, 1990. To ensure a permanent and wide-reaching effect, EPA will work with the national standard and model building code organizations to incorporate radon resistant construction techniques into national building codes.

## CAPABILITY DEVELOPMENT

EPA operates a number of programs to train radon professionals and to evaluate the capabilities of State and private sector companies dealing with radon problems. Radon training courses have taught companies and researchers how to measure radon and mitigate houses. In addition the National Radon Measurement Proficiency Program evaluates the proficiency of radon measurement companies. Last year, EPA developed the Radon Contractor Proficiency Program to evaluate private sector mitigation services, and established three Regional Radon Training Centers to expand the network of training opportunities to State, local, and private communities.

"REDUCING RADON IN STRUCTURES" COURSE. To provide training in assessment and mitigation, EPA, in conjunction with the State of New York, created the diagnosis and mitigation training course entitled "Reducing Radon in Structures." This course has already been offered to over 2,000 Federal and State officials and private contractors across the country and is continually revised and expanded to reflect new technologies.

RADON MEASUREMENT PROFICIENCY PROGRAM. As the number of companies providing measurement services increases, it is important to ensure the reliability and accuracy of the test results. At the request of the States, EPA established the Radon Measurement Proficiency (RMP) Program to assess the capabilities of private radon measurement companies and non-commercial laboratories. Under this program, companies offering measurement services voluntarily demonstrate their measurement capabilities. EPA provides lists of successful participants to the States who distribute them to homeowners upon request. The latest Cumulative Proficiency Report was issued in January, 1990. Last year, EPA also began to blind test measurement companies periodically to make sure they consistently provide accurate results to their customers.

The RMP Program has grown substantially. In 1986, 35 companies were on the list; in 1987, 250 companies were listed, and for the latest test round, the report lists nearly 5,000 companies. Lists of successful participants are currently distributed by all 50 States. Participation in the program remains voluntary; however, several States use the proficiency evaluations as a basis for regulating measurements companies. Other States are considering similar programs.

RADON CONTRACTOR PROFICIENCY PROGRAM. EPA also developed the Radon Contractor Proficiency Program (RCPP), a voluntary program to test the proficiency of mitigation companies that construct and implement measures to reduce radon levels. The first list was published this Spring and includes over 600 individuals who successfully completed the program. EPA has sent States this list to distribute to the public. States also may base mitigator certification programs on the RCPP.

REGIONAL TRAINING CENTERS. To help States and the private sector develop mitigation capabilities, EPA established three Regional Training Centers. These Centers provide a variety of courses on radon health effects, measurement, diagnostics, and mitigation techniques. This Fall, a fourth Training Center will be established in the South.

STATE GRANTS. The Indoor Radon Abatement Act of 1988 established the State Grants Program to help States develop their own radon programs. Federal funds are now being provided for States to conduct surveys, develop public information, develop mitigation programs, purchase radon equipment, and form toll-free hotlines. The legislation authorizes $10 million annually for this program over the next three years, with a declining federal share over that time. This year alone, Congress appropriated $8.7 million for the State Grant Program.

## PUBLIC INFORMATION

Although public concern about the health risks associated with indoor radon has increased over the last few years, relatively few homeowners—about 3 percent—have acted to identify or correct radon problems. As a result, EPA views the development and dissemination of public information as a high priority. EPA is committed to providing the States and the public with current information regarding the risks associated with radon exposure and available methods for assessing and reducing those exposures.

In October, 1989, EPA and the Advertising Council released the first wave of a nationwide Radon Advertising Campaign. The goal of the campaign is to motivate people to test their homes for radon, and to fix homes where elevated levels are detected. The campaign includes TV, radio, newspaper, magazine, outdoor and transit advertising which directs the public to call 1-800-SOS-RADON for information on how to test homes for radon. Direct marketing brochures developed by EPA, the Ad council, and by radon measurement companies with radon test results, are sent in response to hotline calls. Since October, the hotline has received over 50,000 calls, and a second wave of advertising is planned for Fall, 1990.

EPA is using numerous other approaches to disseminate information to the public. Action guidelines for measurement and mitigation have been developed to help citizens determine if there is a problem in their homes, and to present possible courses of action available. EPA has developed, published, and distributed nearly two dozen public information documents, covering a wide range of radon related problems, to educate the public about radon and the health risks associated with exposure. These publications range from concise pamphlets on the basics of radon, to detailed, technical guidance on how to construct or modify buildings with radon problems. These publications include:

- "A Citizen's Guide to Radon," a pamphlet which provides information on radon, such as what radon is, the adverse health effects associated with exposure, and how to reduce the risk of these effects;
- "Radon Reduction Methods: A Homeowner's Guide," which discusses mitigation methods successfully tested by EPA and their costs, installation, limitations, and reduction efficiences;
- "Radon Measurements in Schools, an Interim Report," which suggests various ways to test school buildings, interpret measurements, and select appropriate mitigation techniques; and
- *Radon Reduction Techniques for Detached Houses*, which offers detailed, technical guidance to builders and mitigators about EPA's research on reducing radon levels.

In addition to providing these publications, EPA works with the American Medical Association (AMA) and the American Lung Association (ALA) to educate health care professionals on the risks associated with indoor radon. AMA is conducting a series of workshops for health care professionals and has developed and distributed a brochure on indoor radon. EPA has also awarded grants to 22 ALA chapters in 16 states to promote radon awareness. EPA has conducted several risk communication studies to examine how effectively the Agency communicates with the public and to find alternative ways to present radon risk information and encourage radon testing and mitigation.

## INDOOR RADON ABATEMENT ACT

In October 1988, Congress enacted the Indoor Radon Abatement Act to enhance EPA's existing radon programs and add important new responsibilities. The Act does the following:

- sets a new national goal of reducing indoor radon levels to those found outdoors;
- requires EPA to revise its "Citizen's Guide";
- authorizes EPA to develop model construction codes;
- requires EPA to provide technical assistance to the public;
- establishes a State Grants Program;
- requires EPA to establish three Regional Training Centers;
- authorizes EPA to study radon in schools more extensively; and
- authorizes a Federal Building Study.

The first of these provisions in the Indoor Radon Abatement Act sets a long-term goal that "air within buildings be as free of radon as the ambient air outside." This provision is intended to encourage the development of technology that will be able to reduce radon levels significantly below EPA's current 4 pCi/l guideline.

In its revised "Citizen's Guide," EPA will develop a series of action levels and include information on the risks to both children and smokers. In addition, the new "Citizen's Guide" will discuss measurement options and mitigation techniques.

To develop model construction standards and techniques, EPA is working with several building code and other building organizations. This standard will take into account geographic differences in construction type, geology, weather, and other variables affecting new buildings. The Agency's goal is to offer builders effective radon-resistant construction standards and encourage its adoption at the State and local levels.

EPA's technical assistance efforts currently include both the Radon Measurement and Contractor Proficiency Programs. In addition, the Agency is developing a clearinghouse and database to make information about radon readily available to the public. The State Grants program, a principal component of this technical assistance, is helping States develop their own radon programs, conduct surveys, develop public information, and train mitigators. The Regional

Training Centers, like the State Grants Program, now provide another source of radon information and training at the regional and State levels.

To expand its work on radon in schools, EPA will develop a national list of high risk areas and will survey such areas to determine the radon problem. This effort will also include developing diagnostic and mitigation methods for school buildings. Finally, the Federal Building Study now underway will assess the radon problem in government buildings. EPA is providing technical guidance and information so that Federal agencies can conduct their own surveys of their buildings.

## SUMMARY

EPA has made much progress since the first elevated radon levels were found in Pennsylvania. The Agency has a strong understanding of the national magnitude of the radon problem in homes. Surveys currently underway will further refine estimates of the distribution of radon levels in residences.

The Radon Action Program has developed cost-effective technologies for measuring and fixing radon in homes and has successfully fostered the growth of national industries for measuring radon and mitigating buildings with elevated radon levels. To educate the public, the Agency has published a wide range of radon public information materials, many of which are held as models for other programs in EPA. Influential national organizations and health authorities have joined with EPA to educate people about radon, which has become a household word in many parts of the country.

However, much work remains to be done. The public is not taking action; less than three percent of the nation's homes have been tested. Recent studies show that even those who do test and find elevated radon levels frequently do not mitigate. A further challenge for EPA is to broaden its Radon Action Program to include more complete studies of radon in schools and the workplace.

The Indoor Radon Abatement Act creates new opportunities for State program development and recognizes that collaboration between Federal and State governments is the way to deal with radon most effectively. In addition, universities, research organizations, and the private sector can offer valuable assistance in implementing this new legislation. Only through comprehensive cooperative efforts can we reduce the public health risks of radon.

*Environmental Radon: Occurrence, Control and Health Hazards.* Edited by S. K. Majumdar, R. F. Schmalz and E. Willard Miller © 1990, The Pennsylvania Academy of Science.

# *Chapter Twenty-Six*

# REGULATION OF THE RADON MEASUREMENT AND CONTROL INDUSTRY

**THOMAS M. GERUSKY**
Director
Bureau of Radiation Protection
Pennsylvania Department of Environmental Resources
Harrisburg, PA 17120

## INTRODUCTION

The rapid growth of the radon measurement and control industry following the discovery of extremely high levels of radon in indoor environments in the Reading Prong area of Pennsylvania in December, 1984 was a cause for concern at the federal, state and local government levels. The need to trust the accuracy of measurements led to the establishment of the Radon Measurement Proficiency program of the Environmental Protection Agency. Concern among the states that both testing and mitigation firms were providing service to the public with little or no controls led to the establishment of certification programs at the state level. Federal legislation was passed in 1988 to establish a proficiency program for mitigators. In 1988, with the call for radon testing in all occupiable structures, the need for control of the industry became a top priority at the federal and state government level.

## PRE-1894 ERA

Historically, radon measurement and control in the United States was related almost directly to the uranium mining industry. That industry sprung up during and after World War II and it was realized fairly early that uranium miners were subject to increase lung cancer as a result of exposure to radon daughter products in the confined underground atmosphere of the mine. Standards for radon daughter exposures were established by federal and state governments with the requirement for monitoring and record-keeping. The United States Atomic Energy Commission and later the Department of Energy established laboratory and research capability in radon and radon daughter monitoring.

Indoor radon was found to be a problem in the 1960's in areas where uranium mill tailing were used as fill and in construction materials. In Grand Junction, Colorado, numerous residences were found to have high indoor radon levels. The Federal Government supported the State of Colorado in monitoring and remediating these residences. One engineering firm in Grand Junction became involved in the radon reduction program and, in effect, became the national expert in such techniques.

In the 1970's and early 1980's, the State of Florida became interested in indoor radon because of homes built on reclaimed phosphate mill tailing lands. Phosphate rocks contain higher than normal levels of uranium, thereby providing a source for radon contamination. Indoor radon levels were monitored by the State of Florida and action was taken to reduce levels below a guideline recommended by EPA.

In New York and other states, radiation control programs began monitoring indoor radon levels to determine public exposure.

In the late seventies in Pennsylvania, a nuclear utility, Pennsylvania Power and Light, conducted radon surveys in homes of their employees to determine if radon could be used as an indicator of a home's air-tightness. The utility reported to the Pennsylvania Bureau of Radiation Protection that, while no correlation could be determined between radon and air exchange rates, it nonetheless found a considerable number of homes in Central and Eastern Pennsylvania with elevated radon levels. Pennsylvania attempted to assist these home owners to determine the extent of their problem and suggested ways to mitigate the excessive levels. Requests were also made for inclusion of radon testing in the state budget. USEPA was also contacted for assistance. EPA also attempted to increase their budget to begin a radon evaluation program, but at both the state and federal level, this request fell upon deaf ears. Pennsylvania's priorities for radiation program activities changed dramatically with the Three Mile Island 2 reactor accident in 1979.

In Canada and in Western Europe, indoor radon surveys were being performed and estimates being made of the extent of the population exposure to radon progeny. International and national radiation protection bodies were evaluating the data and recommending actions and guidelines for exposure and remedial measures.

In the U.S., the Department of Energy expanded its activities and began a national indoor radon study to determine the extent of the national radon problem. The Uranium Mill Tailings Recovery Act authorized DOE to carry out mitigation activities at numerous mill tailings sites throughout the country, based mainly on environmental radon levels.

## DECEMBER 1984 AND EXPANSION OF RADON ACTIVITIES

The discovery of the Watras home in eastern Pennsylvania in December 1984 (details presented in another chapter in this book) became the catalyst for intense radon activities throughout the United States. Immediate requests for assistance from the Environmental Protection Agency met with a considerable response. EPA, which had been funding a research and development project into mitigation methods, transferred a major portion of that program into the Reading Prong area of Pennsylvania. EPA provided Pennsylvania with radon survey equipment, with staff assistance, and with consultation and with guidance through the CDC on establishing guidelines for remedial actions.

As Pennsylvania made more and more indoor radon surveys in the geologic region known as the Reading Prong, it became apparent that a considerable number of homes had levels well above the CDC indoor radon guideline level of 4 picoCuries per liter and that a much greater effort into monitoring and mitigation of excessive levels would be required. Studies in areas outside the Reading Prong and in states other than Pennsylvania also revealed widespread radon contamination. Studies in central Pennsylvania, northern New Jersey, Maryland, Ohio, New York, Colorado, and almost every other area where surveys had been conducted began to reveal that the problem was indeed national in scope and not limited to specific geologic areas. It also became apparent that the problem was too large to be handled by government alone and that the private sector including the individual homeowner had to become involved.

## THE NEED FOR PRIVATE SECTOR INVOLVEMENT

In 1984 and early 1985, when Pennsylvania looked for assistance in handling the emerging radon problem in the Reading Prong, they found that there was only one private sector firm that had the capability to provide passive radon detector capability in the quantity needed. Costs to the state were approximately $20 per detector, and there was a significant time delay between testing the home and receiving the survey results. For short term testing, active survey equipment and technical personnel had to be used to carry out these surveys. In spite of these obstacles over 25,000 homes in the Reading Prong were monitored at no cost to the homeowner.

Standard measurement systems were also lacking. EPA provided draft protocols for carrying out radon screening surveys and have since provided protocols for long term screening and for use of a variety of detectors.

Pennsylvania also found that only one United States firm had the capabilities and experience to mitigate radon problems found in private residences, public buildings, and larger commercial buildings. Some firms which had been involved in energy surveys and conservation also expressed interest in attempting to resolve the mitigation problem. Additional involvement of the construction and building industry was required.

Free training courses were set up by EPA, Pennsylvania, New York, and New Jersey to train their own and private consulting personnel in measurement and mitigation. New small companies were emerging with the capabilities needed.

Pennsylvania, under a special act and million dollar appropriation in 1986, carried out a radon gas remedial action research and demonstration project to evaluate and field test various active and passive methods of remedial actions to reduce unsafe levels of naturally occurring radon gas in residential structures, to train new firms in mitigation technique, to provide standards for craftsmanship, to prepare a report on radon mitigation methods and effectiveness, to prepare a homeowners guide for radon contractor selection, and to mitigate radon problems. One hundred and five homes were diagnosed and remediated. The *Final Report of the Pennsylvania Radon Research and Demonstration Project*[1] and the *Homeowners Guidance for Radon Contractor Selection*[2] are available from the Pennsylvania Bureau of Radiation Protection.

## THE QUEST FOR QUALITY IN THE PUBLIC AND PRIVATE SECTORS

As these new companies came forth, especially in the area of radon screening, and as more researchers and state and federal agencies became involved in radon testing, the need arose to assure that measurements taken were consistent and accurate. The Department of Energy, and in particular the Health and Safety Laboratory in New York, was active in operating and maintaining of a radon chamber at its lab and provided an interlaboratory quality assurance program.

## THE EPA RADON MEASUREMENT PROFICIENCY PROGRAM

USEPA Office of Radiation Programs, with the assistance of the DOE Health and Safety Lab, acted in 1985 to establish a Radon Measurement Proficiency Program at their Montgomery, Alabama environmental radiation laboratory. This program is a voluntary one for both public and private sector agencies. To successfully participate in the RMP, a primary company, that is a company which provides laboratory analysis for their measurement program, must meet the following requirements:

1) follow EPA measurement protocols in the EPA documents "Indoor Radon and Radon Decay Product Measurement Protocols" and "Interim Protocols for Screening and Follow-up Radon and Radon Decay Product Measurements,"

2) demonstrate management tracking capability, i.e., the ability to get correct measurement results to the proper homeowner, and

3) demonstrate the ability to measure radon and radon decay products within the established program measurement criteria in an announced or blind test.

A secondary company, one which uses the primary company as their radon measurement laboratory, must meet the requirements in 1 and 2.

Companies must participate in every test to maintain their proficiency ratings. Each primary company enrolls its measurement method for evaluation in the performance test round. EPA conducts either blind tests or instructs a primary company to submit a specific number and type of detectors for exposure to know levels of radon and/or radon decay products. After the detectors are exposed in the EPA radon chambers, EPA returns them to the companies without revealing the radon gas or radon decay product levels. The companies analyze the detectors and report their measurements to EPA. EPA compares the companies' measurements to the know level of exposure. If the results are within twenty-five percent of the known values, the company passes the proficiency test. Companies failing to meet the program's measurement criteria, in either a blind or scheduled test, must wait until the next performance test round before being tested again. No company is allowed to retest.

## THE FEDERAL INDOOR RADON ABATEMENT ACT OF 1988

In late 1988, President Reagan signed the first Federal law directing the Environmental Protection Agency to carry out a national radon program. One major provision of that law is the requirement that EPA establish a voluntary proficiency program "for rating the effectiveness of radon measurement devices and methods, the effectiveness of radon mitigation devices and methods, and the effectiveness of private firms and individuals offering radon-related architecture, design, engineering, measurement, and mitigation services. The proficiency program under this subparagraph shall be in operation within one year after the date of the enactment of this section."

As a result, EPA wll establish a training and proficiency testing program for mitigation firms and individuals by late 1989.

## THE NEED FOR REGULATORY INVOLVEMENT IN PRIVATE SECTOR TESTING AND MITIGATION

With increased publicity concerning radon and with government agencies unable to carry out massive radon screening programs, private firms emerged to provide this service. Quickly, state and federal agencies were contacted by the public for lists of qualified firms. Rumors began to circulate concerning public rip-offs by unscrupulous individuals offering to make radon surveys with empty mayonnaise jars or geiger counters. Specialized equipment must be used to determine if elevated radon levels are to be found. A geiger counter will not detect elevated levels unless the concentrations are severe.

The real estate industry has become the dominant driving force behind radon testing in areas where elevated radon levels have been found. Real estate brokers and agents must divulge, under law and court opinion, any information which may make the sale property unsuitable for occupancy. Real estate, relocation firms, banks and other lending institutions are requiring that radon tests be offered, if not required, prior to any sale.

## STATE CERTIFICATION LAWS AND REGULATIONS

New Jersey, in August 1986, adopted the first law requiring that all testing and mitigation firms be certified by the New Jersey Department of Environmental Protection. New Jersey regulations have not yet been finalized but a voluntary program based upon New Jersey guidelines has been established and New Jersey publishes a list of firms which meet these requirements.

Pennsylvania, in July 1987, enacted a radon certification act based upon the New Jersey legislation. However, the Pennsylvania law required that the Department of Environmental Resources institute a mandatory certification program upon submission of proposed regulations and based upon those regulations. The regulations were proposed and the interim certification program began in May 1988.

New York in 1987 proposed regulations under their general public health authority to certify radon testing individuals and firms. Those regulations have not yet been enacted.

Florida in 1988 passed legislation to certify testing and mitigation firms and individuals. The program became effective in January 1989.

Other states including Iowa and Nebraska have recently announced that certification programs have begun or will begin shortly.

## THE RADON INDUSTRY

The 1989 *Radon Industry Directory*[3] lists over 1300 radon testing firms and organizations and over 400 radon mitigation firms and organizations in the United States. As of July 1, 1989, Pennsylvania alone has certified 269 individuals and 213 firms for testing and 113 individuals and 94 firms for mitigation. In addition, 47 laboratories (firms which provide radon testing devices to other certified testing or mitigation firms) have been certified.

As of January 1, 1989, Florida had certified 60 radon measurement specialists, 141 measurement technicians, 19 mitigation specialists, and 6 mitigation technicians.

## REQUIREMENTS FOR CERTIFICATION

Radon testing and mitigation is not a high-tech business. Education, training, and experience requirements for individuals wishing to be certified have undergone extensive review and debate. College educational requirements are proposed for some classes of certified individuals. However, experience in the subject area and specific training in radon measurement or mitigation is usually required. Training courses given or sponsored by the certification agency are usually required. For testing certification, participation in the EPA Radon Measurement Proficiency Program is required. As an alternative, state sponsored or approved proficiency programs are allowed.

Experience is a problem in an area which is relatively new. Most states authorize at least interim certification based upon minimal experience.

Individual testing is a standard requirement, but requirements vary from state to state.

## THE PENNSYLVANIA RADON CERTIFICATION PROGRAM

The Pennsylvania Radon Certification Act mandates certification of persons who test for and mitigate radon certification. It also requires the Department of Environmental Resources to promulgate regulations which, at a minimum:

1) Establishes a fee schedule which places the costs of the program on the industry,

2) Mandates truthful advertising, and

3) Requires reporting of radon testing and mitigation data by the certified industry to the Department.

The regulations divide testing and mitigation into four categories, tailored to meet the needs of the industry as it exists in Pennsylvania. These are radon testing individuals, radon testing firms, radon mitigation individuals and radon mitigation firms. In addition, radon laboratories and manufacturers of radon testing and mitigation equipment who sell directly to the public must be certified.

Exempted from the requirements for certification are:

1) Persons who perform radon testing or mitigation on their own buildings,

2) Persons who install preventative equipment in new building construction,

3) Persons whose radon-related work is part of their duties as employees of the Department and EPA,

4) Persons conducting scientific research if those persons inform the owner or occupant of the building that they are not certified, that the work is experimental and the results of the work are uncertain,

5) Manufacturers of radon equipment who only deal with persons certified,

6) Construction contractors who build mitigation systems that use plans designed by one certified to mitigate, and if they work under that certified person.

*Requirements for Radon Testing*

Individuals must:

a) Have taken a Department approved course on radon testing and passed a Department approved exam,

b) Had 1 year of professional experience in performing radon measurements,

c) Received a bachelor's degree in the physical or biological sciences or engineering or a related field approved by the Department, or the educational or professional work experience equivalent to a degree, as approved by the Department.

Firms must:

a) employ at least one certified individual,

b) have a quality assurance program,

c) have a continuing education program,

d) have a health and safety program, and,

e) participate in the EPA radon measurement proficiency program or equivalent.

*Requirements for Radon Mitigation*

Individuals must:

a) have taken a Department approved course on radon mitigation and passed a Department-approved exam,

b) had 1 year professional experience or supervised experience in radon mitigation system installment,

c) received a bachelor's degree in the physical sciences, engineering, architecture or related fields as approved by the Department, or the education or professional work experience equivalent to a degree as determined by the Department,

Firms must:

a) employ at least one certified individual,

b) have a health and safety program,

c) have a continuing education program,

d) participate in the EPA radon mitigation proficiency program or equivalent, when available.

*Requirements for Radon Laboratory*

Individuals must:

a) have taken a Department approved course on radon,

b) had 1 year of professional experience in performing radon laboratory analysis,

c) received a bachelor's degree in the physical sciences or engineering or related fields as approved by the Department, or the education or professional work experience equivalent to a degree, as determined by the Department,

Firms must:

a) employ at least one certified individual,

b) have a quality assurance program,

c) have a continuing education program,

d) be successfully enrolled in the EPA Radon Measurement Proficiency Program

*Requirements for Radon Manufacturer*

Any measurement equipment or device cannot be sold directly to the public unless the equipment has received an acceptable rating under the EPA Radon Measurement Proficiency Program or alternative program approved by the Department and the equipment will continue to perform accurately over its lifetime.

Any mitigation equipment or device cannot be sold directly to the public unless the equipment is reliable, can reduce radon levels or prevent radon entry over the expected lifetime of the equipment and performs in the manner in which it is advertised. This requirement does not apply to suppliers of common building materials, such as plastic pipe, electrical cable, fans, sealants, cement and the like.

The requirements for a bachelor's degree and for passing a Department approved test have been waived pending further review and study during the interim certification stage.

## FUTURE ACTIVITIES AND PROBLEMS

Consistency and reciprocity among the certifying states is the major obstacle facing present and future state programs. Laws and regulations are being passed by state legislatures and agencies based upon their individual determination of the problem and what they believe is the best approach to take in solving that problem. Having to become certified in every state by an individual or firm has not yet caused significant problems because of the small number of states now involved in certification.

The states do recognize this problem and are attempting to resolve it before it gets out of hand, working through the Conference of Radiation Control Program Directors, Inc., an organization of state radiation control agencies. In particular, the Conference is working on a Model Radon Certification Act, which can be used by other states now considering certification.

Individual states are also developing individual certification exams, with a variety in quality. One high-quality national certification exam should be established which could be utilized by all states. Again, the Conference of Radiation Control Program Directors could be the agent for accomplishing this with EPA assistance.

## CONCLUSION

As the problem of indoor radon grows throughout the country and as more high-level homes are located, private sector testing and mitigation firms and individuals will grow to fill the void in those areas. The relatively high quality of the radon testing and mitigation industry must be maintained to prevent unnecessary expenditure of private and public funds and to assure that proper measures are taken to protect the public health from this environmental hazard.

## REFERENCES

1. Pennsylvania Department of Environmental Resources—Final Report of the Pennsylvania Radon Research and Demonstration Project; 1988.
2. Pennsylvania Department of Environmental Resources—A Guide to Radon Reduction Contractor Selection; 1988.
3. The Radon Industry Directory—1989; Radon Press, Inc., Alexandria, Virginia

*Environmental Radon: Occurrence, Control and Health Hazards.* Edited by S. K. Majumdar, R. F. Schmalz and E. Willard Miller © 1990, The Pennsylvania Academy of Science.

# *Chapter Twenty-Seven*

# LEGAL ASPECTS OF RADON GAS IN RESIDENTIAL HOME SALES

**FRANK B. CROSS**

Center for Legal and Regulatory Studies
University of Texas
Austin, TX 78712-7450

## INTRODUCTION

There is a growing public recognition that indoor radon exposure presents the greatest environmental health hazard in the United States today, possibly equalling the threat presented by past occupational exposures to asbestos. As this recognition spreads, litigation cannot be far behind. A number of cases seeking some recovery for high indoor radon levels already have been filed.[1] Law reviews have increasingly begun to discuss the prospects for liability in such radon litigation.[2,3,4] Few cases complaining of indoor radon exposure have yet been decided, however, and so the courts may write upon a relatively clean slate.

This chapter discusses some of the legal principles applicable to the expected burgeoning of lawsuits over indoor radon exposure. It will focus on residential exposure, where the problem appears to be greatest. The first section addresses the standards of liability that any plaintiff must meet in order to succeed in a lawsuit, be it under negligence, strict liability, fraud, or implied warranty. Before recovering for radon exposure, a plaintiff must show some such violation of civil law duties. The second section covers the remedies available to a plaintiff who succeeds in establishing such liability. These remedies include any demonstrable property losses, plus possible recovery for prospective health harms. In the course of these sections, various legal difficulties facing plaintiffs are summarized and the proposed approaches to be used by courts in responding to radon litigation.

## PROVING LIABILITY FOR RADON EXPOSURE

For purposes of analyzing potential liability for radon exposure, it is assumed that the prospective plaintiff is a homeowner who is exposed to relatively high levels of radon in his or her home and who was unaware of these prevailing levels at the time that the home was purchased. Such a plaintiff may discover a number of individuals to sue, but the most common potential defendants will be the construction contractor who built the house, the real estate broker who negotiated the sale of the house, the architect who designed the house, the producers and sellers of any radon-contaminated building materials, and the seller of the house. The most likely causes of action available to the plaintiff will be negligence, strict liability, fraud, and violation of an implied warranty of habitability. These theories will now be discussed, in turn.

*Negligence*

Negligence is a traditional and widely known theory of liability that may be available to radon plaintiffs. The requirements of a negligence action are well-established. A negligence plaintiff must demonstrate:

(1) that the defendant owed a duty of care to the plaintiff not to create an unreasonable risk;
(2) that the defendant breached that duty of care; and
(3) that the defendant's breach was the proximate cause of actual injuries suffered by the plaintiff.[5]

In the radon context, it should be easy to prove that most plausible defendants owed the plaintiff a duty of due care. Certainly, sellers, construction contractors, and real estate brokers owe some duty of care to home buyers. The difficult question will be whether the defendants violated their duty of due care in a particular episode by exposing plaintiffs to an unreasonable risk of harm. The answer to this question is based upon how the theoretical "reasonable person" would act. The classic formulation of how this reasonable person behaves was propounded in 1947 by Judge Learned Hand, who wrote:

> Possibly it serves to bring this notion into relief to state it in algebraic terms: if the probability (of harm) be called P; the injury, L; and the burden (of preventing harm), B; liability depends upon whether B is less than L multiplied by P.[6]

This formula requires a cost-benefit analysis of the costs of the harm and its prevention. In practice, however, no judge or jury conducts a formal cost-benefit analysis, and the test for negligence typically includes ethical, as well as economic, considerations. Courts may also consider common industry practice as a guide to how a reasonable person would act.

To prove such a violation of the duty of due care in the radon context, the plaintiff will have three primary burdens: (1) to show that the defendant knew, or should have known, of the potential radon risk; (2) to show that the risk is large enough to be considered potentially unreasonable; and (3) to show that the risk could have been averted without unreasonable expense. With regard to the first burden, the defendant's "duty to know," the timing of the sale will be critical. The knowledge of radon's hazard is now sufficiently widespread that today's transactions probably should all be subject to such a duty to know, which may encompass a duty to inspect and test new housing.[2] EPA has widely disseminated information on the risks of indoor radon, and some states are now integrating radon protection into their building codes and real estate contracts. The National Association of Home Builders is now actively involved in research and informing its members regarding the indoor radon threat. Some banks require radon tests as a condition on mortgages. For today's housing sales, defendants should bear a duty to know of and correct high indoor radon levels.

For past transactions, the case is less clear. In one of the few reported radon liability decisions, a federal appellate court held that a defendant "did not know or have reason to know of the danger of radon gas in 1969."[7] The date at which defendants presumably should have known of the danger of indoor radon gas contamination might be placed in the middle to late 1970s, though courts have not yet ruled on the issue.[2] During this time period, the dangers of indoor radon became known to EPA, though knowledge of the risk was not broadly distributed.

Assuming that a plaintiff can show that a defendant should have known of the potential radon hazard, the plaintiff will next be required to demonstrate that the radon contamination of his or her home is sufficiently high to warrant concern. This determination may involve an elaborate explication of scientific studies and risk assessment, and the rules of legal procedure may confound some plaintiffs' efforts to demonstrate such a risk. Scientific data must be explained to a lay jury. Courts tend to be uncomfortable about their institutional capability to resolve such scientific issues, however, and they may be expected to defer to expert government agency decisions, when available.[6] Thus, if a plaintiff can show indoor radon levels exceeding those deemed "safe" by the Environmental Protection Agency or other expert organizations, the courts are likely to find the risk sufficient to justify a negligence finding. EPA's current "action level" is set at four picocuries per liter of air, and courts may seize upon this level to determine whether a risk is unreasonable.[2] Congress, however, has urged EPA to adopt a "health goal" of two picocuries per liter, and courts will have to struggle with a number of differing recommendations regarding acceptable radon levels. Proof of unreasonable risk will also be complicated by the shortcomings of many of the radon test methods currently in use.

Most plaintiffs should be able to meet the third burden, to show that the risk was or is preventable. In usual cases, relatively inexpensive remedial measures can significantly reduce indoor radon exposures.[2] Even where corrective measures are costly, a plaintiff may still succeed by showing that the house was defectively designed in construction, that the house was negligently sited in a high radon exposure area, or that defendants negligently failed to warn plaintiff of the risk. These latter claims will often be among a plaintiff's strongest.

It is also noteworthy that a defendant's violation of a law or regulation typically represents *per se*, or automatic, negligence. Thus, if a defendant has violated a public safety rule, the plaintiff's task of demonstrating negligence will be eased considerably. Unfortunately, there are still very few government restrictions on indoor radon, and few plaintiffs will be able to avail themselves of this doctrine.

The final requirement of negligence, that the plaintiff prove that a breach was the proximate cause of actual injuries, will depend on the type of injuries alleged. Consequently, I will leave this issue for the second section of this chapter.

### *Strict Liability*

Strict liability is generally perceived as an advantageous theory for plaintiffs, because it eliminates the need to demonstrate negligent misconduct by any defendant and focuses instead on the prevailing risk. Strict liability generally exists for products sold in the marketplace. The availability of a strict liability cause of action for defects in residential homes is still somewhat uncertain and varies by jurisdiction, but New Jersey and California seem to be leading the way in applying strict liability to builders of residential housing. Strict liability will be particularly promising when a radon problem is caused by contaminated building products. Strict liability will be unavailable for some other categories of potential defendants, such as real estate brokers.

If strict liability applies, a plaintiff must prove that the product (house) contains a "defect." The defect may be a defect in construction (such as inadequate sealage or ventilation), a defect in design of the house, or a failure to warn of the potential hazards. In determining whether any given risk rises to the level of a defect, courts typically use a "risk-utility test," in which the magnitude of the risk is compared to the utility of the allegedly defective design. This balancing is roughly comparable to the unreasonable risk standard of negligence, and high radon homes should generally present risks that outweigh the utility of the defect responsible for elevated radon levels. Builders, at minimum, should have some duty to warn purchasers of the presence of high indoor radon exposures.

In most states, however, strict liability is qualified by a "state of the art defense." That is, if the construction, design or warning conformed to the state of the art at the time of production and sale, the defendant will not be liable even in strict liability. Insofar as most builders across the nation have not recognized the potential radon hazard, this state of the art defense may present a considerable obstacle to a plaintiff's recovery for past transactions. Plaintiffs' prospects will be better for current housing sales.

### *Misrepresentation*

Home purchasers may maintain an action for intentional misrepresentation, or fraud, against builders, sellers and brokers in connection with the purchase of a residence with high radon levels. The fraud claim has some strict requirements but offers greater potential recovery for plaintiffs, in the form of punitive

damages. To prove fraud, a plaintiff must show:
(1) a false representation of a fact by a defendant;
(2) the defendant's knowledge that the statement is false;
(3) an intent to induce the plaintiff's reliance on the false statement;
(4) the plaintiff's justifiable reliance on the false statement; and
(5) damage to the plaintiff as a result of such reliance.[5]

The fraud action thus turns on proof of a knowing falsehood.

In the real estate context, a plaintiff may not need to show an actual false statement but may need only prove a failure to disclose the known high radon level. Traditionally, silence could not be fraud, but courts increasingly are imposing a duty on certain parties, such as sellers or brokers, to disclose material facts. Pennsylvania law clearly recognizes a duty to disclose latent defects in housing sales. If the defendant has actual knowledge of high radon levels and fails to inform the purchaser, a court may find fraud. For example, the Texas Association of Realtors directs its members to warn buyers of the presence of radon in a home, even if the seller resists.

If a defendant has no knowledge of high radon levels, there can be no fraud. Such a defendant may still be liable for negligent or innocent misrepresentation, however. Real estate brokers, for example, have some duty to inspect housing for serious hazards, and this requirement may be extended into a duty to test for radon. Failure to test and inform buyers of high radon levels may be deemed negligent misrepresentation and exposure brokers and others to liability.[8]

### *Implied Warranty of Habitability*

While the above claims represent torts, a buyer may also have a contract case for breach of implied warranty of habitability. Most sales of housing are now presumptively subject to this warranty, which promises that the house is reasonably habitable, regardless of whether the seller makes any specific promises regarding the safety of the structure. Pennsylvania has adopted a particularly liberal approach to the implied warranty theory, holding builders liable even when they lacked actual knowledge of the breach. The threshold requirements for this warranty are that the defendant builder-vendor held itself out as expert in construction, that the purchaser has no such particular expertise of his or her own, and that the purchaser relied on the reputation and skill of the defendant. In most instances, these requirements will be easy for a plaintiff to meet.

The plaintiff must next show that the house, as sold, was not habitable. While only limited precedent exists, it seems likely that a home with particularly high radon exposures should be deemed to be not habitable. It is unreasonable to expect a home's residents to risk an unusually high rate of cancer. Once again, this will require the courts to grapple with the question of what radon level warrants the law's concern, and courts will be likely to turn to and rely upon expert determinations, when available.

If a plaintiff can demonstrate a violation of the implied warranty of habitability, the plaintiff must still overcome several defenses. Some jurisdictions require that a buyer notify the seller of a breach of warranty within a reasonable time, or the buyer will not be allowed to bring a claim for damages. Another problem is the contract disclaimer or exculpatory clause, whereby a buyer may expressly waive certain contractual rights. Many current real estate contracts contain provisions waiving the implied warranty of habitability. A buyer who signs such a contract may lose his or her rights to sue for a breach of the warranty. Pennsylvania courts have been relatively reluctant to enforce these disclaimers, however, especially where the seller had greater knowledge and bargaining power than did the buyer. Some states also require privity for an implied warranty case, which means that a buyer can sue only the seller and not other responsible parties, who were not parties to a contract with the plaintiff. Nevertheless, some regard breach of the implied warranty of hàbitability as a radon plaintiff's most promising cause of action.[9]

## RECOVERING DAMAGES FOR RADON EXPOSURE

A plaintiff who can successfully establish liability is only halfway to his or her desired objective in litigation. Some available remedy is required before a plaintiff can benefit from such a victory. Plaintiffs may seek damages for property value or repair, for health harms suffered, for the risk of future health harms, for other damages allegedly linked to radon exposures, and even for punitive damages. Other plaintiffs may wish simply to void the contract and leave the radon-contaminated house that they have purchased unaware of the risk.

### *Property Damage*

The owner of a home with high radon levels may recover for demonstrable property damage under negligence, misrepresentation and warranty theories. The standard measure for property damage to real estate is the reduction in property value traceable to the defendant's action or inaction. In the case of radon, particularly high levels might significantly reduce the value of a home, and some surveys suggest that the price of a house could decline by 30% or more as a result of high-level radon contamination.[2] When provable, a plaintiff should be able to recover these damages.

An exception to the reduced property value measure of damages exists when repair or correction of the problem would be less costly than granting reduced property value. In many cases, radon concentrations can be reduced dramatically through increased or modified ventilation or sealing of entry points, and these procedures are not terribly costly. A plaintiff homeowner suffering high radon concentrations should generally be able to recover the cost of reducing those concentrations below some acceptable limit. Plaintiffs may also recover other miscellaneous monetary damages. In one Pennsylvania case, a plaintiff has sought damages for his expenditure of time and money in detecting the source of the emissions, the expense of surviving winter temperatures while ventilating the house, and various cosmetic expenses, as well as the costs of reducing radon levels. While property damages will generally be available, the amount to be recovered may be relatively small (often in the hundreds or thousands of dollars), and the recovery may not justify the expenses of a lawsuit. For this reason, plaintiffs are likely to seek additional damages, beyond simple property loss.

### *Health Harms Incurred*

Plaintiffs may also sue to recover damages for any health harms already incurred from high radon concentrations. Few plaintiffs will be able to demonstrate such harms, however. The primary health consequence of radon exposure is lung cancer, which may have a twenty-year latency period. Those currently suffering lung cancer may be unable to recover because the defendants had no basis for knowing of the radon threat at the time of the sale and exposure. Those exposed to radon today must wait decades, for cancer to develop, before they may recover for their health harms.

The latency and other characteristics of lung cancer also produce other serious problems for plaintiffs. Plaintiffs will have difficulty proving that their lung cancer was caused by radon exposure, as opposed to tobacco, asbestos, or any number of other carcinogens. Proving this case may be further complicated by rules of evidence and the limited availability of critical data on exposures. Proof of past radon exposures will generally be unavailable. Defendants may be absent or insolvent at the time of trial. In some states, a plaintiff's claim may be barred by statutes of limitations that require plaintiffs to sue within a few years of the action causing their harm. In short, most plaintiffs will find it very difficult to recover for health harms incurred from indoor radon exposures.[6] Some contend that it is all but impossible for cancer victims to recover for their damages under current tort law.

### *Risk of Future Cancer*

In recent years, an increasing number of individuals are suing prior to incurring cancer, seeking recovery for a risk of future cancer caused by some defendant. The courts have traditionally rejected this sort of claim, demanding actual injury before requiring compensation. Some few courts are beginning to recognize that a risk may actually be an actual injury of sorts. Even these courts have tended to require proof of a high probability of future cancer before permitting recovery. Typically, plaintiffs must show a greater than 50% chance of future injury, and even the high radon level homes will seldom present a 50%

risk of future cancer. Consequently, few radon plaintiffs will be able to recover for pure risk of future cancer, without any present health harms.

Courts are more amenable to plaintiffs who have sought recovery for a future risk in tandem with some actual current injury. This rule is a logical outgrowth of the traditional legal requirement that a plaintiff bring all of his or her claims against a defendant in one single suit, rather than claim-splitting. Thus, if a plaintiff has suffered some current harm, he or she may also recover for risks of future harm. While radon may have few obvious current harms that will enable plaintiffs to bootstrap a future risk case, creative plaintiffs have begun to establish favorable precedents. Scientific techniques have made it increasingly possible to identify subcellular chromosome damage in human cells, of the type that may be caused by a carcinogen. Plaintiffs claim that this chromosomal damage is sufficient present injury to justify recovery for a future risk. In one of the few radon cases with a reported decision, a federal district court accepted this argument and permitted plaintiffs to sue for a future risk based on present chromosomal damage. The court found that "experts are able to conclude with a reasonable degree of medical probability" that chromosome damage had occurred and that this damage could "deprive the plaintiffs of a degree of immunity," which in turn meant that the " 'trigger' of a cancer change has been cocked," thereby authorizing recovery.[10] The plaintiffs ultimately settled this case for some undisclosed amount of money. When future risk is coupled with present injury, courts will also be more willing to permit recovery for a less-than-50% probability of future cancer.

While established chromosomal damage may provide plaintiffs with a foot in the door for future risk recovery, they still have a daunting task before actually recovering. They must put forth sufficient scientific evidence to persuade a lay jury of some significant future risk, based upon some risk assessment. The presentation of this evidence may be restricted by rules of evidence. Moreover, the amount of recoverable damages may not be great when risks of future cancer are relatively low.

Individuals living in a home with high radon levels may obtain further recovery for a reasonable fear of future cancer, often called "cancerphobia." While courts tend to be somewhat reluctant to compensate emotional distress, they have recognized cancerphobia recovery in cases involving asbestos, DES, nuclear wastes and medical malpractice. This fear of future cancer is only cognizable when it is a reasonable fear, however, and a plaintiff must demonstrate some meaningful risk of which the plaintiff is afraid. A risk as low as 5% may well be compensable. Cancerphobia may be unavailable in cases brought under strict liability, however, and the magnitude of a damage award may not be great. It is incumbent upon a plaintiff to demonstrate the severity of the fear, and a jury must place some monetary value on the fear. Moreover, some states require proof of a contemporaneous physical injury before permitting recovery for mental distress, such as cancerphobia.[6]

### *Future Medical Surveillance*

Plaintiffs exposed to a risk of future cancer have begun to seek damages to pay for medical surveillance of their condition, in order to detect the possible development of cancer. The general availability of damages for medical surveillance following actual injury is well established in older precedents, and federal regulations emphasize the value of medical surveillance for those exposed to a significant carcinogenic risk. Given the health value of early cancer detection, medical surveillance damages seem a logical remedy for courts to employ.

Courts remain somewhat divided on the availability of these medical surveillance damages for future cancer risk. A major hazardous waste exposure case has granted the damages, while a major asbestos case has denied them. To obtain medical surveillance damages, a plaintiff surely will be required to show a material future risk of cancer, though risks of much less than 50% should justify surveillance. The extent of availability of such damages is still uncertain.

When medical surveillance damages are available, the amount of recovery is likely to be relatively small, under $25,000. While this may not justify litigation for some plaintiffs, the medical surveillance damages conjoin well with other damage claims, including property damage claims, and they may provide an important additional remedy for radon plaintiffs.

### *Punitive Damages*

Many of the largest awards in toxic tort cases have come in the form of punitive damages. Punitive damages differ fundamentally from the above forms of damages, which are designed to compensate a plaintiff for actual losses suffered. Punitive damages go beyond such compensation and provide plaintiffs with something of a windfall recovery, in order to punish the improper conduct of the defendants.

Punitive damages are presently the source of considerable controversy, as defendants complain that these damages have been excessive and even unconstitutional in magnitude. Juries have assessed punitive damages amounting to tens of millions of dollars, even in cases of little actual harm to plaintiffs. Many believe that punitive damages are used by juries to assist sympathetic, suffering plaintiffs who are suing a wealthy, "deep pocket" defendant. Some state legislatures are passing laws to place restrictions on punitive damages, but the remedy remains broadly available.

The prospect for punitive damages in radon litigation is highly uncertain. Traditionally, punitive damages were available for "malicious" conduct by a defendant. If a radon plaintiff succeeds under a fraud claim, punitive damages might be justified. In addition, punitive damages are now awarded when a defendant is considered "reckless" or "grossly negligent." Some negligence cases may thus warrant punitive damages, which are less likely under strict liability and much less likely under a breach of the implied warranty of habitability. It seems unlikely that radon litigation will often justify punitive damages, at least in the early cases. In some extreme cases, when a defendant knew of the radon risk and covered it up, radon plaintiffs may successfully recover punitive damages.

### *Rescission*

Rescission offers an entirely different type of remedy that may appeal to many radon plaintiffs. Rather than recovering monetary damages, the plaintiffs may simply seek to escape the contract and leave the radon-contaminated home without further liability. Such rescission will often be available for cases grounded in fraud, innocent and negligent misrepresentation, or breach of the implied warranty of habitability. Plaintiffs may obtain rescission under still another legal theory, the doctrine of mutual mistake. When the parties to a contract are mistaken regarding a basic assumption of the contract, either party may rescind the contract upon the discovery of the mistake. Extremely high radon levels may qualify as such a mistaken basic assumption, to justify rescission. In the past, rescission of residential housing sales has been granted for mistakes regarding termite infestation, inadequate water pressure, and an inoperable septic system.[2]

The right to rescind under mutual mistake must be qualified, however. Plaintiffs cannot recover when they have expressly assumed the risk of mistake, such as through a contractual disclaimer. Rescission may also be unavailable when the harmful situation can be easily and inexpensively corrected. Nevertheless, mutual mistake offers real promise to plaintiffs, who need not show a defendant's knowledge of the hazard, a defendant's negligence, or even a defect in the product. Proof of high radon levels alone may suffice to provide rescission of housing sales.

## CONCLUSION

The future success of lawsuits seeking damages for indoor radon exposure remains indeterminate. Some authors believe that plaintiffs are unlikely to prevail in these actions.[4] Other authors believe that plaintiffs often will succeed in recovering damages.[2,3] In any event, it seems clear that some plaintiffs will succeed and that their success will inspire many other individuals to initiate legal actions. Increased federal regulatory attention to radon will also spur litigation.

One can reasonably expect a considerable increase in the number of lawsuits filed for indoor radon exposure. The virtually unparalleled magnitude of the health threat of these exposures all but guarantees a "litigation explosion." Experience with another great public health threat, from occupational exposures to asbestos, is not encouraging. Asbestos litigation produced enormous judgments that bankrupted much of an industry. Plaintiffs, however, recovered only a fraction of these judgments, and many deserving plaintiffs have their recoveries stalled by bankruptcy proceedings. The building industry surely fears that radon litigation may produce similar results in their much larger and economically critical industry.

For a number of reasons, radon litigation is likely to be less disruptive than was asbestos litigation. First, while the overall magnitude of the harm may be comparable, the relative risk to any individual plaintiff will usually be less than in the case of asbestos. Second, knowledge of radon risks has been disseminated more promptly than was the information regarding asbestos risks. While society's response to indoor radon often seems painfully slow, it has been more prompt than in past health hazards, such as asbestos. Defendants are more alert to the financial risks of litigation and are quicker to ameliorate health risks. Huge awards in radon litigation should be uncommon, as most plaintiffs should recover only property losses plus some limited damages for medical surveillance, fear of cancer, or future cancer risk. Should this occur, the tort system will help provide a remedy to a major public health problem, without disastrous economic consequences.

Science and scientists will play a prominent role in indoor radon litigation. The optimistic scenario described above depends in part upon accurate, responsible expert testimony, which neither understates nor exaggerates the true risk from indoor radon. The development of engineering knowledge can play a crucial role in remedying the threat of indoor radon, at reasonable expense to builders and homeowners. All the professions share some responsibility for the effective correction of this serious public health problem.

## REFERENCES

1. Galen, Lawyers Grapple With Radon Issue. *Nat'l L.J.,* July 21, 1986, at 1.
2. Cross, F. and P. Murray. 1988. Liability for Toxic Radon Gas in Residential Home Sales. *North Carolina L. Rev.* 66:687-739.
3. Conaway, S. 1986. Grappling with Silent Invaders of the Home: Legal Remedies for Radon Gas Contamination. *Temple Envt'l Law & Tech. J.* 5:36-48.
4. Shuko, C.M. 1986. Radon Gas: Contractor Liability for an Indoor Health Hazard. *Am. J. Law & Medicine.* 12:241-272.
5. Keeton, W.P., D.B. Dobbs, R.E. Keeton, and D.G. Owen. 1984. *The Law of Torts*, (5th ed.). West Publishing Co., St. Paul, MN. pp. 164-165.
6. Cross, F.B. 1989. *Environmentally-Induced Cancer and the Law.* Quorum Press, Westport, CT.

7. Wayne v. Tennessee Valley Auth. 730 F.2d 392, 396 (5th Cir. 1984).
8. Winicour, S. 1987. Clearing the Air on Radon Testing: The Duty of Real Estate Brokers To Protect Prospective Homeowners. *Fordham Urban L.J.* 15:767-799.
9. Waltz, D.A. 1987. Radon Gas: Ramifications for Real Estate Transactions in Pennsylvania. *Dickinson L. Rev.* 91:1113-1155.
10. Brafford v. Susquehanna Corp. 586 F. Supp. 14, 17-18 (D. Colo. 1984).

*Environmental Radon: Occurrence, Control and Health Hazards.* Edited by S. K. Majumdar, R. F. Schmalz and E. Willard Miller © 1990, The Pennsylvania Academy of Science.

# *Chapter Twenty-Eight*

# POTENTIAL IMPACT OF RADON ON REAL ESTATE MARKETING

**LAURA D. KELLER,* ESQUIRE**

Stites & Harbison
2300 Lexington Financial Center
250 West Main Street
Lexington, KY 40507-1758

## INTRODUCTION

As warnings of radon's widespread occurrence and potential health effects move from scientific journals to the mass media, public concern about the problem has begun to impact the real estate market. Many builder-developers in radon-prone areas of the country are seeking legal advice on ways to avoid liability to home purchasers by means of contract provisions and disclosures. Brokers are looking for similar means of minimizing their liabilities. Home buyers more frequently insist on radon testing during the course of their purchase negotiations and refuse to purchase a home until radon problems are corrected or accounted for in the written transaction.

This chapter will discuss the legal duties of builders, real estate agents and even individual sellers of homes to minimize radon contamination and advise prospective home buyers of the potential for radon, and the various theories of liability under which they may be sued for selling an unhealthy home to an unwitting buyer. The chapter will then discuss how sellers and brokers can attempt to avoid such liability and what protections buyers should demand. Applicable Pennsylvania law is singled out where possible.

This chapter is limited in several ways. First, the topic is a broad one and space is restricted, so the chapter will present only an overview; footnotes reference numerous cases and articles which deal with particular subjects in

*Laura D. Keller is an Associate in the law firm of Stites & Harbison, Lexington, Kentucky. She received her J.D. from the University of Kentucky in 1978. The author appreciates the fine editorial assistance provided by Don Pisacano, Stites & Harbison.

greater detail. Second, although a number of cases are pending, there have been no reported opinions concerning liability for selling a home contaminated by naturally occurring radon.[1] Any discussion of such liability must therefore be based on analogous caselaw and a degree of speculation.

## LIABILITY OF BUILDERS/DEVELOPERS FOR RADON CONTAMINATION

A person who has just poured his or her life savings into a home, only to discover that the unseen menace of radon lurks within, is likely to look for a way out of the transaction or for someone to pay for remediation, personal injury and property damages, or compensation for emotional distress. One of the most likely targets, particularly if the home is a new one, is the builder, who is likely to have deep pockets and who clearly owes certain legal duties to a purchaser. Various theories of recovery against a home builder are available, although they have not yet been tested in the radon context.

### *Strict Liability in Tort*[2]

Strict liability, a legal theory available in products liability actions, is a favored theory of recovery for plaintiffs, because misconduct by the defendant need not be proved and disclaimers and other contract defenses do not apply. Strict liability is the offspring of public policy considerations: legislatures and courts have decided that it is preferable to place the onus of an injury caused by a product on its manufacturer and suppliers, parties in a better position to prevent injury and to spread losses, than upon an injured individual who relied upon their competence. The theory is explained in the Restatement (Second) of Torts, as follows:

> (1) One who sells any product in a defective condition unreasonably dangerous to the user or consumer or to his property is subject to liability for physical harm thereby caused to the ultimate user or consumer, or to his property, if
> (a) the seller is engaged in the business of selling such a product, and
> (b) it is expected to and does reach the user or consumer without substantial change in the condition in which it is sold.
> (2) The rule stated in Subsection (1) applies although
> (a) the seller has exercised all possible care in the preparation and sale of his product, and
> (b) the user or consumer has not . . . entered into any contractual relation with the seller.[3]

For this theory to apply in the radon context, the aggrieved homeowner must establish that a home is a "product." Some courts have accepted the premise. The seminal case for application of products liability doctrine to real estate was *Schipper v. Levitt & Sons, Inc.*[4] involving a builder's failure to install a mixing valve to regulate water temperature, resulting in the scalding of the homeowner's child. The court decided that the policy considerations for strict liability applied to mass-produced homes as well as mass-produced goods and found the builder strictly liable: "The public interest dictates that if such injury does result from the defective construction, its cost should be borne by the responsible developer who created the danger and who is in the better economic position to bear the loss rather than the injured party who justifiably relied on the developer's skill and implied representation."[5] Courts have extended application of the theory since *Schipper.* A California court applied strict liability to the siting of a home, finding a developer strictly liable for damages suffered by an owner as a result of construction on an unstable hillside, although the home itself was not defective.[6] A New Jersey court held that products liability theory applied even where the defendant was not a mass builder but had only developed six homes for sale.[7]

Even if a buyer can establish that a home constitutes a product, he must also prove that the source of the radon contamination is an "unreasonably dangerous defective condition." He might do this by demonstrating that contamination is the result of a construction flaw (e.g., cracks in the foundation, poor ventilation, excessive insulation, contaminated building material) or the result of bad design (e.g., failure to construct subslab ventilation; use of concrete block for basement walls). It is also possible that, even if a court determined that the home's design was not defective, it might find the builder strictly liable for failure to warn the purchaser of the existence and risk of excessive radon if he knew or should have known of its presence.[8]

Although strict liability does not require proof of fault, a fault-related defense to the action does exist in many jurisdictions. If the builder can show that the home conformed to the "state-of-the-art" when it was sold, he may have a defense against a charge of design defect or failure to warn.[9] A builder might thus escape liability by demonstrating that, due to a lack of scientific knowledge at the time the product was made, he had no reason to know of the danger. Radon as a naturally occurring health hazard was largely unknown to the general public until the 1980's. It may therefore be difficult for a homeowner to prevail in a strict liability suit against a builder for pre-1980 construction based on design defect or failure to warn.

The public policy considerations which engendered products liability law apply in the context of radon in the home. The home builder is in a better position to avoid radon contamination by construction techniques than is the homeowner attempting to remedy the problem after the home is built. The builder holds himself out as an expert in home construction and site selection, and the home buyer relies heavily upon his expertise. Radon is not readily detectable by a buyer upon inspection. Public health and safety could best be served by establishing an economic incentive for the builder to prevent excessive indoor radon levels altogether, thereby promoting lower overall exposure to the substance. Finally, home builders are in a better position to spread the cost of preventing entry of radon into the homes, by passing the increased cost to the buyer or purchasing liability insurance.

Products liability will be most viable for a home purchaser if he can prove that a particular construction flaw resulted in high radon levels in his home or if he can demonstrate that the house was constructed recently and the home builder failed to design the home to avoid radon problems and to warn the buyer of its hazards. The plaintiff's chances of recovery will be increased if the builder is a mass producer of homes and radon concentrations are so high that the house is clearly unsuitable for its intended use.

### *Implied Warranty of Habitability*[10]

The doctrine of *caveat emptor* ("buyer beware") has been abandoned in a majority of states with regard to construction of new homes. Most states have now adopted some type of implied warranty of habitability,[11] based upon a number of public policy reasons similar to those for strict liability. A builder-vendor of a new home normally holds a bargaining position superior to that of the buyer. He has superior knowledge about the condition of the home, is better able to prevent hazards and to repair defects, holds himself out as an expert, and in addition is more able to spread the cost of remediation or injury than an individual buyer.[12]

Courts have held that a builder-vendor of a new home impliedly warrants that the dwelling was constructed in a "workmanlike manner and using suitable materials."[13] Despite the disparity of opinions as to the required level of workmanship,[14] one could reasonably argue, based upon EPA publications about health hazards,[15] that excessive radon contamination renders a home uninhabitable and thus constitutes a breach of the implied warranty. A builder might counter that the home itself was structurally sound, made of suitable materials, and built in a workmanlike manner, and that the cause of the problem was not construction-related, but the result of geology. In many states, this argument might well succeed. This sort of argument was rejected by the Pennsylvania Supreme Court, however, in *Elderkin v. Gaster,* in which the court found a builder responsible for a home's unhealthy water supply, a problem due not to faulty construction of the home or well, but to an inadequate water source, a problem over which the builder had no control.[16] Because of the interplay of building construction and natural geological conditions in determining radon levels in a home, it is likely that a homeowner in Pennsylvania could prevail on an implied warranty theory on the ground of inadequate construction or poor site selection, at least with regard to recent construction.

Several hurdles must be overcome by a homeowner asserting an implied warranty of habitability. First, the standard of workmanship ordinarily required is that of ordinary care, or the care a typical builder in the community would exercise.[17] Since radon hazards were largely unknown before 1984, a builder might well not be liable under the warranty theory for pre-1984 failure to use protective building methods and radon-free materials, even in a radon-prone area. Second, some courts require that the plaintiff give notice of any defect to the builder so that he can repair the defect.[18] Third, many states require privity of contract as an element of contract-based liability, although a number of courts have rejected the traditional privity requirement, particularly for latent defects.[19] The trend, although still in the minority, is to allow subsequent purchasers to recover from original builders.[20] A final hurdle is that many courts enforce contract disclaimers as a bar to actions for breach of implied warranty, at least where the disclaimer is specific enough to advise the buyer of exactly what he is waiving.[21] Thus a homebuilder could state in the sales contract that he made no warranty that the home was free of radon, and would likely thereby avoid liability.

As in the case of strict liability in tort, a plaintiff homeowner will have a better chance of success if he can point to a particular construction defect causing the increased levels of radon. Chances of recovery will be greatly improved if the construction occurred recently, that is after radon had become a widely known health hazard, and where the concentrations are so high that there will be no issue as to the habitability of the home.

### *Express Warranty*

Although most home sales agreements do not include an express warranty covering latent defects, an exception may be found in sales of homes financed by the Federal Home Administration or the Veterans Administration, for which express "blanket" warranties are required.[22] Any such express warranty would offer an avenue of relief for a homeowner in the event the warranty was breached.

### *Fraud or Intentional Misrepresentation*

Fraud, by definition, is a false representation of a fact by one with knowledge that the statement is false (or awareness that he lacks sufficient basis for making the statement), intended to induce another party to rely on the information. To recover, a plaintiff must demonstrate that he was damaged as a result of his justifiable reliance on that information.[23] The false representation of fact may be in the form of non-disclosure; that is, silence may be considered fraud. Concealment by a seller of material defects not readily detectable by a buyer who has relied upon the seller is a ground for liability.[24] There is little doubt that a builder knowing of excessive levels of radon in a home has a duty to disclose this information concerning a latent hazard to potential buyers.

Fraud is often difficult to prove because of the intent requirement. A number of courts, however, have relaxed the requirement and recognized a cause of action for negligent misrepresentation.[25] Thus, if a builder makes a statement about a home, he may be liable for it, even if he believes it to be true, as discussed in the next subsection.

### *Negligence*[26]

Another legal theory of liability of a builder-vendor is negligence, a theory grounded on the concept that each person has a duty to protect others from unreasonable risk. In order to succeed in a negligence action, the plaintiff must show that the defendant had a duty to act toward him with a certain standard of care, that the defendant breached that duty, that the plaintiff was somehow injured, and that the defendant's act or omission proximately caused the plaintiff's injury. The defendant is judged by the conduct of a "reasonable person." If the defendant holds himself out as having special knowledge or skill, then he is measured by the skill and learning commonly possessed by members of his profession in good standing.[27] A court will examine the custom or general practice of an industry, but that custom does not conclusively establish what a reasonable person would do in the circumstances.[28]

A builder-vendor has a duty of care in a number of different areas, including site selection, design and construction of the home, and use of safe materials. In the radon context, a plaintiff might argue that the builder had breached his duty by using contaminated products, such as concrete blocks made of radioactive phosphate slag,[29] or that he negligently chose a dangerous home site,[30] left cracks in the foundation or openings around electrical and plumbing fixtures, excessively insulated, failed to ventilate or in some other way failed to use proper precautions to prevent entry of radon into the building. A builder also has an obligation to disclose to a buyer any natural or artificial condition which presents an unreasonable risk of injury, so long as he knew or had reason to know of the condition and the buyer could not readily discover it.[31]

A builder would not be liable for radon contamination unless he knew or should have known, as a member of his profession, that the structure would be dangerous if he did not take precautions or warn the purchaser. A plaintiff would certainly have a reasonable argument, now that the hazards of radon have been widely publicized, that any builder should be aware of the problems associated with radon, particularly builders in radon-prone areas. Although no magical date exists when builders would suddenly be held to have knowledge about radon, it might be difficult for a plaintiff to prevail on a radon negligence case for a home constructed prior to the 1980's.

## LIABILITY OF A SELLER WHO IS NOT A BUILDER

A number of the avenues of relief available to a homeowner against a builder-vendor are probably not viable against an individual homeowner. The policy reasons in favor of recognizing actions for strict liability and implied warranty of habitability against builders do not apply to a single seller of a home. The author is unaware of any court applying those theories of liability to an individual homeowner. A number of lower courts in Pennsylvania have expressly rejected such an extension of the implied warranty of habitability.[32]

The most viable theories of liability against an individual homeowner are negligence and fraud. A seller would clearly be liable for fraud if, knowing of high radon levels in his home, he denied radon contamination. Likewise, he could be held liable for representing that radon levels were low when he had conducted no tests in his home, or for failing to advise a buyer of excessive radon levels about which he knew. Concealment by a seller of material defects, not readily detectable by a buyer who has relied on the seller, is a ground of liability for a noncommercial seller as well as a builder.[33] As a defense, the seller might argue that the buyer also knew of the contamination and therefore did not rely on the seller's representations.

With regard to negligence, it is conceivable that a seller of a home might be sued for negligently having allowed radon entry, for example, by creating cracks in the basement foundation or by overinsulating. It is far more likely, however, that a negligence cause of action against an individual seller would be based on the seller's failure to warn the buyer of the presence of radon gas when the seller should have known of the potential for radon. In other words, a buyer may have an action for negligent misrepresentation even if the seller did not know of radon contamination, but should have known based upon information available to him, that is, if he had constructive knowledge of its presence. An individual would be less likely to be held to this constructive knowledge standard than a builder-vendor or a real estate agent, because he is not a professional, and does not hold himself out as having knowledge superior to that of the buyer. Unlike real estate agents or builders, who are likely members of trade associations which disseminate information to their members, the seller could more persuasively argue that he had no reason to know of the potential hazards of radon, or that the buyer was just as likely as the seller to have constructive knowledge of such risks and should, on theories of contributory negligence or comparative negligence, share in the liability.

Finally, if the seller expressly warranted that radon levels were within an acceptable range, and it subsequently became apparent that they were not, the buyer would have a cause of action against the seller for breach of express warranty. The same principle would apply if the seller had generally warranted that the home was free of latent defects.

## LIABILITY OF REAL ESTATE AGENT[34]

Brokers have traditionally been considered to be agents of sellers, who pay their commissions, rather than of buyers, and to owe their primary loyalty to the seller. Recent case law, however, has held that brokers owe a fiduciary duty to purchasers as well, based largely on buyers' reliance upon their brokers' expertise. In keeping with this principle, courts are following a trend toward greater liability of brokers.

### *Fraud or Intentional Misrepresentation*

Like a seller, if a broker knows of a material fact, that is, one which would affect the purchaser's decision, and he misrepresents the fact or withholds information about it, he can be liable for intentional fraud. In *Roberts v. Estate of Barbagallo,*[35] a Pennsylvania court granted rescission of a purchase contract, based upon the broker's fraudulent concealment of the presence of ureaformaldehyde foam insulation in a home. The broker was aware of the existence of the insulation and that it was a potential health hazard, but did not disclose its presence. This case lends support for a broker's liability for damages due to radon. A broker could try to distinguish the case by arguing that foam insulation is man-made, whereas radon is a naturally occurring substance, but the distinction is a weak one.

### *Unintentional Misrepresentation*

Traditionally real estate agents, like sellers of homes, have enjoyed the protection of the doctrine of caveat emptor. Under this traditional standard, still followed by about half of the states,[36] the broker's duty to the buyer is only to disclose latent defects of which he knows or should know.[37] In the states following this negligence standard, the broker is not considered a guarantor of the principal's assurances.[38] The broker has no duty to verify independently statements made by the seller unless he has some reason to know that the seller's statements are false.[39] Courts following this negligence standard would declare that the broker's only duty is to inquire about and review aspects of the property about which there is a reasonable basis for doubt or concern, to speak truthfully, to obtain necessary information before making any representations, to make no representations in the absence of such information, and to disclose all material defects known to the broker and unobservable to the buyer.[40]

A broker has a number of defenses available to him if sued for negligent misrepresentation. The broker may escape liability if he can show that the purchaser was negligent in relying upon his statements,[41] or that the purchaser already knew of the radon levels or should have known, based upon newspaper articles or common knowledge that radon levels were high in the neighborhood. A broker could attempt to show that representations about radon were not material to the real estate transaction, although it is doubtful this defense would work, since EPA publications warn of radon's severe health hazards and a purchaser might well not have bought the house if he had known of the dangers. Some courts have accepted a broker's defense that his representations were those of the seller, where the broker had stated or implied that he was merely relying on the seller's statements and had no personal knowledge of their veracity.[42] A disclaimer clause to that effect could be included in the purchase agreement.

Courts have noted an "increasing trend of state courts to impose liability on real estate brokers for any kind of misrepresentation,"[43] including innocent misrepresentation. Such jurisdictions recognize a fiduciary duty from the broker to the buyer, even though the broker is legally considered the seller's agent. The trend is essentially grounded in public policy: the broker has the greater expertise and professional knowledge and the broker is in a better position to verify the seller's statements than is the buyer.[44]

Courts have followed this trend to varying degrees. Some would require that the broker perform a diligent inspection and disclose the results, but the broker would not be liable for failure to disclose conditions which were not apparent through such an inspection. For example, in *Tennant v. Lawton,*[45] a broker was found negligent when, based on a seller's statement, she innocently misrepresented to the buyer that the property had passed a percolation test and had an approved septic tank site. (The buyer had specifically conditioned the transaction on such approval). The court held that "the broker is required to employ a reasonable degree of effort and professional expertise to confirm or refute information from the seller which he knows, or should know, is pivotal to the transaction from the buyer's perspective."[46] In *Easton v. Strassburger,*[47] listing agents were found liable for negligent misrepresentation because they never informed the buyer of potential instability problems, although they were aware the property had been filled and that fill areas may be unstable. The court held that the broker had a duty to conduct a reasonably competent inspection of the property in order to discover any defects. If he did discover such defects, he had a further duty to recommend to the buyer that the buyer solicit professional help to investigate the potential defects. In *Gouveia v. Citicorp Person-to-Person Financial Center, Inc.,*[48] a listing broker was held to have a duty to inspect and disclose defects of which the broker knew or should have known through the exercise of reasonable care, even though the broker never had any direct contact with the purchaser, but was only a listing service.[49]

Other courts have carried the broker's duty to the purchaser farther, finding a broker strictly liable for misrepresentations, even if he conducted a diligent inspection and found no latent defects.[50] Based on public policy considerations, such courts would find a real estate agent liable for failure to disclose any material latent defect in the property, regardless of the agent's knowledge of the condition or ability to detect the condition or even if he knew of no facts suggesting that the condition might exist.

In addition to applying common law concerning a broker's fiduciary duties to a buyer, courts may use real estate licensing statutes or consumer protection statutes to set standards for real estate agents.[51] Some courts have used the statutes to impose liability, finding that the statute created a higher duty to which a real estate agent must be held, above that of a lay vendor of property.[52] Others have used the statutes to limit liability, finding that the statute only punished knowing misrepresentation and that innocent misrepresentations should therefore not be actionable.[53]

The above principles of liability apply not only to inaccurate statements, but also to silence, or failure to disclose, assuming that radon contamination would be deemed a material latent defect. Liability for nondisclosure would depend on the standard applied in the jurisdiction as well as the knowledge available to the broker. Under the traditional rule, a broker in an area where radon was not prevalent would probably have no liability for failure to disclose the presence of radon in a particular home. In such a jurisdiction, he would have no duty to inspect unless radon "red flags" existed, and it is likely that a trier of fact would not find that he should have known of the problem. On the other hand, if the home were located in a radon-prone area, the broker could easily be found to have constructive knowledge of high radon levels, accompanied by a duty to disclose this fact to a prospective purchaser. In a jurisdiction requiring a diligent inspection, the question would be whether a reasonably competent inspection should include a radon test. It is likely that a court would require such a test in a high radon area, but not in an area of low radon levels. In a strict liability jurisdiction, on the other hand, the result would be the same in either case: the broker would be strictly liable for failure to disclose a latent defect, even though he did not know and had no reason to know of excessive radon levels.

The difficulties for a broker with regard to radon are many. First of all, a broker is not an environmental specialist, but many buyers will look to the broker for advice and quite possibly lead him to make statements about which he does not have knowledge. Furthermore, because of the possibility of unrepresentative results with short-term radon tests and the risk of losing a deal where a long-term test is required, brokers may hesitate to recommend testing. Furthermore, brokers generally feel a responsibility to the seller, who is paying their commission, rather than to the buyer, and may hesitate to lose a sale through their own disclosures. The cases establishing broker liability are in flux, vary according to the jurisdiction and the circumstances, and do not set out any bright-line standards for a broker to follow. Finally, there are absolutely no reported cases setting forth the liability of a broker with regard to radon.

## REMEDIES AGAINST BUILDERS, BROKERS AND SELLERS

A complete discussion of the possible damages available to a homeowner/plaintiff exceeds the scope of this article.[54] Upon a showing of radon contamination, a homeowner might well obtain from the court an order rescinding his purchase contract and requiring the seller to return the deposit. As compensation for property damages, a homeowner might receive the difference between the fair market value of the property without radon and the fair market value of the home with radon, or in the alternative he might receive the cost of remediation.

Ordinarily, plaintiffs receive greater awards for personal injury than for property damages. In the radon context, however, personal injury will be far more difficult to establish. Because naturally occurring indoor radon is a relatively new "discovery," risk assessments concerning the effects of low levels of radon on humans are inconclusive. Experts disagree about what is a "safe" level or to what extent a certain radon level increases one's chances of cancer. Medical evidence suggests that the latency period for radon-caused cancer is 20 years or more. Meeting the causation burden of demonstrating that a plaintiff's personal injury (cancer) was caused by radon exposure, as opposed to smoking, exposure to a myriad of carcinogenic chemicals present in our lives today, or some other cause, might be impossible. Furthermore, after 20 years, the defendant may be insolvent, evidence may be lacking, and statutes of limitation may bar recovery. Courts often award damages for the increased risk of cancer, but as a prerequisite the plaintiff must ordinarily demonstrate that he already has a physical injury. Some courts have accepted, as proof of current physical injury, evidence from medical experts that the plaintiff has suffered a cellular or subcellular change which would increase his vulnerability to cancer. In addition, compensation may be available for emotional distress or anxiety over the prospect of contracting cancer as a result of exposure (cancerphobia).[55]

## PROTECTING THE PARTIES

As set forth above, brokers, sellers and builders are potentially liable under a number of theories, although none of those theories has been fully tested with respect to the radon situation. Based on experience with other kinds of latent defects, however, attorneys can make recommendations to protect potentially responsible sellers or agents, or on the other hand, advise buyers of measures to avoid unknowing purchase of a home containing high levels of radon. The following sections discuss some of these recommended measures. The recommendations for contractual language and other protections contained in this outline are made in the absence of particular facts and are intentionally general in order to be helpful to a nationwide audience. They should not be relied upon for legal protection.

## *Protections for the Seller or Builder*

Because a seller must disclose to a potential buyer the existence of a latent defect of which he knows and which the buyer cannot readily discover, it may be advisable for a seller not to test for radon, in order not to have such knowledge. If the seller is a builder or developer in the business of selling homes, however, it is not unlikely that he would be held to have constructive knowledge of radon levels, particularly in an area where radon is commonly found in homes. In such a case, even though he did not know of excessive levels, he might be liable for negligent misrepresentation because he should have known. Therefore, in high radon areas, it is probably advisable for a builder-vendor to test for radon and to disclose the results of the test and the potential health hazards, or to recommend that the purchaser conduct his own test. Of course, one further way for a builder-developer to avoid liability would be to employ radon-reducing construction techniques at the outset, especially where the radon levels are high.

Various contractual provisions may provide some protection for a seller. A clause providing that the house is sold "as is" might shield a seller from warranty liability, but courts disfavor disclaimers. In Pennsylvania, mere recitation on an "as is" clause in the deed, stating that the buyer has inspected the property and agrees to purchase it in its current condition, does not constitute a defense under the implied warranty of habitability applied to a builder-vendor. In *Tyus v. Resta,*[56] the superior court decided that the implied warranty had been adopted by the court for public policy reasons, and public policy demanded that it not be waivable by broad "as is" terms in the deed. Furthermore, the latent defects at which the implied warranty is aimed would not be discoverable by an ordinary purchaser's inspection, so it would be inequitable to allow an inspection clause to extinguish the warranty.[57]

A more effective protection than a broad "as is" clause would be a specific disclaimer concerning radon levels in the home, so long as the disclaimer can be honestly made. The *Tyus* opinion affirmatively stated that a clear and unambiguous disclaimer clause is enforceable in a contract for sale of a new home. In order to constitute a defense to the implied warranty, such a disclaimer must refer to specifically designated potential latent defects.[58] A specific disclaimer about radon levels would likely meet this test.

In the event that a buyer desires to perform radon tests, one way to protect the seller from inaccurate testing would be to add a provision giving the seller a specified time after he receives notice of the buyer's results to retest in order to demonstrate that the buyer's test was inaccurate or not representative.

## *Protecting the Broker*

Like sellers, a broker must decide at the outset whether to test or remain ignorant. Since many jurisdictions lack specific guidelines on the subject in the form of court opinions, there is no clear answer. In an area high in radon, especially in a jurisdiction holding that a broker has a fiduciary duty to a prospective purchaser, the realtor would be well advised to test and disclose the results or to specifically advise the purchaser that high levels of radon have been found in the area and testing is recommended. The New Jersey Association of Realtors wants to add a clause to its standard real estate sales contract noting that high levels of radon have been found in the area and that, if the buyer bears the expense of the radon tests, the seller will be obliged to provide remediation;[59] or to add a disclaimer of knowledge as to radon's presence in the house and a recommendation that the purchaser perform his own test.[60] The Pennsylvania Board of Realtors has developed a radon disclaimer form to be attached to all contracts stating whether tests were performed and what the results were.[61] Obviously a potential problem with testing and disclosing the test results is that the results may not be representative of year-round radon levels. A broker might wish to include in its statement of the results a description of the conditions under which the test was made, along with a statement that test results may vary depending upon a number of circumstances, and suggesting that the buyer consult with a professional for advice.

For a broker in a low-risk area, there is always the question as to whether it would be reasonable to test for radon, and whether a broker would be negligent if he did not test. Even in such an area, the broker may wish to add a disclaimer in the purchase agreement stating that he has no knowledge of radon levels.

A broker should document and limit his representations. If the broker intends to rely on testing performed by the seller, the broker should clearly advise the purchaser that his statements are made in reliance upon the seller's representations, and that he has no personal knowledge of their accuracy. A further protection would be for the broker to require the seller to sign a statement listing all the representations that will be made, certifying that they are true, and agreeing to indemnify the broker for any damages if they turn out not to be true, even if liability is based on the broker's negligence.[62] A broker representing a buyer as opposed to a seller might include in a brokerage agreement provisions eliminating any obligation on his part to investigate the property for radon independently and acknowledging that the buyer will not rely on the broker to perform such an investigation. A broker should, when initially examining any property, be alert to any defects in the property which he should disclose to a potential buyer. Finally, he should review his errors and omissions insurance policy to determine whether its coverage would extend to liability associated with radon.

*Protections for the Purchaser*

A buyer may ask for warranties and representations by the seller that a home does not have excessive radon levels, accompanied by an indemnification clause whereby the seller agrees to pay for any costs of remediation. Because of the difficulty of enforcing such provisions, especially if the seller moves out of state, relying solely on such clauses is not recommended.

Although a seller may have reasons not to test for radon, a buyer's interest is best served by requiring testing. Various approaches are available. The purchase contract could provide that the purchaser will conduct a radon test before the closing. The contract should specify a certain number of days in which the buyer will perform the test, obtain the results, and notify the seller and broker.[63] The provision should guarantee the buyer or his consultants access to the residence and should provide that the seller agrees not to tamper with the test procedures and agrees to keep all windows and doors closed except for regular ingress and egress. The contract might specify a particular level of radon (usually 4 picocuries per liter of air) and provide that if that level is exceeded, the buyer may void the contract or opt for remediation.[64] Remediation cost allocation is open to negotiation. The contract might provide that the seller would correct the hazardous condition before the closing and that the effectiveness of the work would be verified by another test after remediation. In the alternative, the contract might provide for certain financial or other concessions by the seller to enable the purchaser to select a consultant to do the remedial work after he takes possession. Variations of these provisions might be that an upper limit of remediation cost be established, and if estimates for remediation exceeded that level, either of the parties could back out.[65] On the other hand, the buyer could use positive test results to persuade the seller to lower his price and accept the home in its current condition, "as is."

The above recommendations provide that, if testing indicates excessive levels of radon, the parties may then decide whether to void the contract or mitigate the hazardous condition according to a certain agreement. In the alternative, the parties could, before testing, agree to a specific concentration level and tie themselves in to one option or the other. That is, if the level exceeded 8 picocuries per liter, for example, the purchaser could back out, but if the level were lower, the purchaser would be tied in and the seller would undertake corrective action.

All of the above discussion pertains to pre-closing testing, that is, testing done between the signing of the purchase contract and the closing. Obviously a disadvantage for a buyer in this situation is that he has no control over the test; the seller with occupancy of the home could tamper with the radon test. In addition, the length of time required for a more accurate representative sample could delay the closing beyond the time frame desired by the parties. Therefore, in some cases, post-closing testing may be preferred. Since EPA suggests that remediation of most homes can ordinarily be effected for less than $5,000, the parties might provide in their purchase contract that the seller would place into escrow $5,000 with which the buyer could conduct remediation. The contract would then provide for a specified time after closing during which the purchaser would test, and if radon levels exceeded a certain amount, the escrowed money could be used for remediation. As one option, the contract might provide that any money not spent for remediation would be split between the seller and the buyer. This option would encourage the buyer to limit his costs. Another option would be for all of the money to be returned to the seller.[66] Obviously there are disadvantages in leaving the matter to post-closing testing, as it complicates the agreement and delays final completion of the transaction.

## CONCLUSION

Although the author is unaware of any cases against builders, sellers, or brokers for exposure to naturally-occurring radon having been taken to trial, the number of lawsuits in this area is bound to proliferate. Although radon is not a man-made hazard, its concentrations are closely tied to man-made conditions and can be avoided by careful building techniques and later corrective action. Although proving that cancer was caused by radon will be difficult, plaintiffs will undoubtedly seek to combine personal injury actions with property damage actions, because such actions, when successful, yield higher judgments. Property damage lawsuits concerning high concentrations of radon in newly purchased homes will be easier for a plaintiff to win, and will therefore be more commonplace. The upshot of the inevitable litigation over radon in residential homes is that brokers' agreements and purchase agreements will likely be modified to provide for automatic testing in some parts of the country and for boilerplate disclaimers and disclosures. Already a number of state realtor associations have proposed such language. Hopefully, the discovery of the effects of radon in homes will not lead solely to more legal escape clauses in the various documents associated with the transaction, but will force all parties to the transaction to investigate radon levels in homes and to allocate remediation costs in a fair and effective manner. The end result will hopefully be more than increased employment of attorneys; it should cause all parties to become more informed and force the elimination of this particular hazard in the home.

## FOOTNOTES

1. Radon related cases have been reported. *See, e.g., Wayne v. TVA,* 730 F.2d 392 (5th Cir. 1984), *cert. denied,* 469 U.S. 1159 (1985) (product liability and negligence action by homeowner against producer of phosphate slag and manufacturer and seller of concrete blocks incorporating slag, used to construct plaintiff's home; suit was barred by statute of limitations); *Brafford v. Susquehanna,* 586 F. Supp. 14 (D. Colo. 1984) (suit for forcible eviction and personal injury, increased risk of cancer and mental anguish for exposure to radon from radioactive mill tailings placed around foundation

of home by defendant mill; dispute was subsequently settled); *Nobel v. Marvin E. Kanze, Inc.,* Civ. No. 83-05253 (Montgomery Co. Court of Common Pleas, Pennsylvania, filed April 13, 1983) (suit against air conditioning contractor by homeowner for radon contamination entering cracked ventilation system under an implied warranty of contract theory; not reported).

2. For a thorough discussion of strict liability as applied to builders, see Frank Cross and Paula Murray. 1988. Liability for Toxic Radon Gas in Residential Home Sales. *N.C.L. Rev.* 66:687, 703. Readers are cautioned to examine the applicable strict liability law in their own states, as such laws differ from state to state.
3. Restatement (Second) of Torts, § 402A (1965).
4. 207 A.2d 314 (N.J. 1965).
5. *Id.* at 326. *See generally* Annot., *Recovery Under Strict Liability in Tort for Injury or Damages Caused by Defects in Building or Land,* 25 A.L.R. 4th 351 (1983). Cases applying strict liability to real estate defects include *Bastian v. Wausau Homes,* 620 F. Supp. 947 (N.D. Ill. 1985); *Blagg v. Fred Hunt Co.,* 272 Ark. 185, 612 S.W.2d 321 (1981); *Kriegler v. Eichler Homes, Inc.,* 269 Cal. App. 2d 224, 74 Cal. Rptr. 749 (1969); *State Stove Mfg. Co. v. Hodges,* 189 So.2d 113 (Miss. 1966), *cert. denied,* 386 U.S. 912 (1967). Pennsylvania appears to be on the verge of viewing a newly constructed home as a "product." Donald Waltz. 1987. Radon Gas: Ramifications for Real Estate Transactions in Pennsylvania. *Dick. L. Rev.* 91:1113, 1132. The Waltz article is also recommended for a more thorough discussion of other theories of liability under Pennsylvania law.
6. *Avner v. Longridge Estates,* 272 Cal. App. 2d 607, 77 Cal. Rptr. 633 (1969).
7. *Patitucci v. Drelich,* 153 N.J. Super. 177, 379 A.2d 297 (Law Div. 1977).
8. *See generally* Cross and Murray, *supra* note 2, at 708-711.
9. *See* W. Prosser and W. Keeton. 1984. *The Law of Torts.* § 99 at 697-98. West Publishing Co., St. Paul, MN. (5th Ed.) (hereinafter Prosser and Keeton on Torts).
10. A more complete treatment of implied warranty of habitability may be found in Peter J. Shedd. 1980. The Implied Warranty of Habitability: New Implications, New Applications. *Real Est. L.J.* 8:291; Paul G. Haskell, 1965. The Case for an Implied Warranty of Quality in Sales of Real Property. *Geo. L.J.* 53:633; Warren H. McNamara, 1972. The Implied Warranty in New-House Construction. *Real Est. L.J.* 1:43. For a discussion of Pennsylvania law, *see* Note. 1973. Elderkin v. Gaster—The Pennsylvania Experience with Implied Warranties in Sales of New Homes. *Temp. L.Q.* 47:172.
11. Shedd, *supra* note 9, Table 1 at 303-06.
12. *See* Annot., *Liability of Builder-Vendor or Other Vendor of New Dwelling for Loss, Injury, or Damage Occasioned by Defective Condition Thereof,* 25 A.L.R. 3d 383, 392 (1969).

13. *Crawley v. Terhune,* Ky., 437 S.W.2d 743, 745 (1969); *accord, Hartley v. Ballou,* 286 N.C. 51, 209 S.E.2d 776 (1974).
14. *Wagner Construction Co. v. Noonan,* 403 N.E.2d 1144, 1148 (Ind. Ct. App. 1980) (workmanship sufficient so as not to substantially impair the enjoyment of the residence); *Petersen v. Hubschman Construction Company,* 76 Ill.2d 31, 42, 389 N.E.2d 1154, 1159 (1979) (workmanship sufficient to ensure the house is reasonably suited for its intended use); *Klos v.Gockel,* 87 Wash.2d 567, 554 P.2d 1349, 1351 (1976) (workmanship sufficient to ensure the house is safe for the buyer's intended purpose of living in it).
15. U.S. Environmental Protection Agency. 1986. *A Citizen's Guide to Radon: What It Is and What to Do About It.*
16. *Elderkin v. Gaster,* 447 Pa. 118, 288 A.2d 771 (1972). The Pennsylvania Supreme Court found an implied warranty not only that the home was built in a workmanlike manner, but also that it is reasonably suited for human habitation.
17. Note. 1982. Extension of Implied Warranties to Subsequent Purchases of Real Property; Insurance Company of North America v. Bonnie Built Homes. *Ohio St. L.J.* 43:951, 958. In Pennsylvania, however, the Supreme Court in *Elderkin* essentially applied a strict liability (liability without fault) standard, based upon policy reasons. The opinion has been criticized for making the builder an insurer of both the home and the land on which it sits. Note. 1972. Vendor and Purchaser—Abrogation of Caveat Emptor in New Home Sales by Builder-Vendors. *U. Rich. L. Rev.* 7: 399, 403-04. It is quite possible that the Pennsylvania Supreme Court, faced with the prospect of finding builder-vendors liable for excessive levels of radon in thousands of homes built in conformity with community standards when radon was not recognized as a residential hazard, would carve out an exception in the *Elderkin* holding.
18. *Wagner Construction Co. v. Noonan,* 403 N.E.2d 1144 (Ind. Ct. App. 1980).
19. *Redarowicz v. Ohlendorf,* 92 Ill. 2d 171, 441 N.E.2d 324 (1982); *Barnes v. Mac Brown & Co.,* 264 Ind. 227, 342 N.E.2d 619 (1976); *Gupta v. Ritter Homes, Inc.* 646 S.W.2d 168 (Tex. 1983); *Moxley v. Laramie Builders, Inc.,* 600 P.2d 733 (Wyo. 1979).
20. But *see Coburn v. Lenox Homes, Inc.* 173 Conn. 567, 572-73, 378 A.2d 599, 601 (1977); *Oliver v. City Builders, Inc.,* 303 So.2d 466, 468 (Miss. 1974) ("It would be strange indeed, if, when the original purchaser conveyed the property to another, that his vendee could resort to the builder for damages for deficiencies in workmanship or materials which the original purchaser...accepted.")
21. *See* Waltz, *supra* note 5, at 1146-49, for a discussion of a disclaimer's effect on an implied warranty.
22. P. Goldstein, Real Estate Transactions 176 (2d ed. 1985).
23. Prosser and Keeton on Torts § 105, at 728.

24. *Id.* § 106, at 738; *see, e.g., Lingsch v. Savage,* 213 Cal. App. 2d 729, 735-36, 29 Cal. Rptr. 201, 204-05 (1963); *Hall v. Carter,* 324 S.W.2d 410 (Ky. 1959); *Shane v. Hoffman,* 227 Pa. Super. 176, 324 A.2d 532 (1974); *Quashnock v. Frost,* 299 Pa. Super. 9, 445 A.2d 121 (1982).
25. Prosser and Keeton on Torts § 107, at 740-49. A transaction tainted by intentional fraud is voidable by the buyer even if the misrepresentation was not material. Any innocent misrepresentation, however, must be material, i.e., a significant factor in the transaction, in order to warrant avoidance of the transaction. *See Delahanty v. First Pa. Bank,* 318 Pa. Super. 90, 464 A.2d 1243 (1983).
26. For a more thorough discussion of negligence liability of builders with regard to radon, *see* Steven F. Conaway, 1986. Grappling with Silent Invaders of the Home: Legal Remedies for Radon Gas Contamination. *Temple Envt'l. Law and Tech. Journal* 5:36, 38-43 (1986).
27. Prosser and Keeton on Torts § 32, at 185.
28. *The T.J. Hooper,* 60 F.2d 737 (2d Cir. 1932), *cert. denied,* 287 U.S. 662, 53 S.Ct. 220, 77 L. Ed. 571 (1932).
29. *Wayne v. Tennessee Valley Authority,* 730 F.2d 392 (5th Cir. 1984), *cert. denied,* 469 U.S. 1159 (1985).
30. *ABC Builders, Inc. v. Phillips,* 632 P.2d 925 (Wyo. 1981).
31. Restatement (Second) of Torts, Section 353 (1965) states: § 353. Undisclosed Dangerous Conditions Known to Vendor.

    (1) A vendor of land who conceals or fails to disclose to his vendee any condition, whether natural or artificial, which involves unreasonable risk to persons on the land, is subject to liability to the vendee and others upon the land with the consent of the vendee or his subvendee for physical harm caused by the condition after the vendee has taken possession if

    (a) the vendee does not know or have reason to know of the condition or the risk involved, and

    (b) the vendor knows or has reason to know of the condition, and realizes or should realize the risk involved, and has reason to believe that the vendee will not discover the condition or realize the risk.

    (2) If the vendor actively conceals the condition, the liability stated in Subsection (1) continues until the vendee discovers it and has reasonable opportunity to take effective precautions against it. Otherwise the liability continues only until the vendee has had reasonable opportunity to discover the condition and to take such precautions.
32. *See, e.g., Boozell v. Bollinger,* 30 Pa. D.E.C.3d 247 (1983); *Estill v. Vandernat,* 90 York 133 (1976).
33. *See* cases cited *supra* note 24.
34. For other discussions of real estate agent liability in the radon context, see Ralph Holmen. 1988. Radon—Legal Issues for the Real Estate Agent. *Probate & Property.* 2:51; Note. 1987. Clearing the Air on Radon Testing: The

Duty of Real Estate Brokers to Protect Prospective Homebuyers. *Fordham Urban L.J.* XV: 767.

35. 366 Pa. Super. 551, 531 A.2d 1125 (1987).
36. Mary Rose Kornreich. 1987. Dealing with the Invisible Trap of Radon Liability. *The Practical Real Estate Lawyer.* 3:17, 29.
37. Restatement (Second) of Torts § 353. For a discussion of broker liability, *see also* Holmen, *supra* note 34. For a Pennsylvania case, see *Smith v. Renant*, 564 A.2d 188 (Pa. Super. 1989) "There can be no liability upon the seller or his agent. . .for failure to disclose unknown condition or one whose risk is not realized." *Id.* at 193).
38. *Speigner v. Howard,* 502 So.2d 367, 374 (Ala. 1987).
39. *Provost v. Miller,* 144 Vt. 67, 473 A.2d 1162, 1164 (1984).
40. Kornreich, *supra* note 36, at 28.
41. Restatement (Second) of Torts, § 552A at p. 140; *Aztlan Lodge No. 1 v Ruffner,* 745 P.2d 611 (Ariz. App. 1987); *Evans v. Teakettle Realty,* 736 P.2d 472 (Mont. 1987).
42. *See* Robert S. Treece and Carolyn B. Clawson. 1988. The Real Estate Broker's Duty to Investigate. *For the Defense.* 30:11
43. *Hoffman v. Connall,* 108 Wash. 2d 69, 736 P.2d 242, 247 (1987).
44. *Id.* at 248. ("Between innocent purchasers, who may justifiably rely on the broker's knowledge and expertise, and a broker, who is in a far better position to check the accuracy of any purported boundary lines, I believe the broker should be liable for any misrepresentation as to the boundary's location. Equity demands such a result.") *See also* Restatement (Second) of Torts § 551(1).
45. 26 Wash. App. 701, 615 P.2d 1305 (Wash. 1980).
46. *Id.* at 1310.
47. 152 Cal. App. 3d 90, 199 Cal. Rptr. 383 (1984).
48. 686 P.2d 262 (N.M. App. Div. 1984).
49. *See also Lynn v. Taylor,* 642 P.2d 131 (Kan. App. 1982); *Reed v. King,* 193 Cal. Rptr. 130 (1983).
50. *Bevins v. Ballard,* 655 P.2d 757 (Alaska 1982).
51. Kornreich, *supra* note 36, at 29.
52. *Dugan v. Jones,* 615 P.2d 1239 (Utah 1980).
53. *Hoffman v. Connall,* 108 Wash.2d 69, 736 P.2d 242, 245 (1987).
54. For a more complete review of remedies for residential radon contamination, *see* Cross and Murray, *supra* note 2, at 724-38.
55. *Id.*
56. 328 Pa. Super. 11, 476 A.2d 427 (1984).
57. *Id.* at 434.
58. *Id.* at 432.
59. Quinn, Andy. 1988. Radon: An Environmental Problem that is Too Close to Home. *J. of Contemp. Health Law and Pol.* 4:415, 426.

60. Stephen L. Kass and Michael B. Gerrard. 1987. Radon and Real Estate Transactions: The Risks and the Legal Implications. *Toxics L. Rep.* 2:289, 292.
61. *Id.*
62. Kornreich, *supra* note 36, at 30.
63. *Id.* at 23.
64. *Id.*
65. *Id.*
66. *Id.*

*Environmental Radon: Occurrence, Control and Health Hazards.* Edited by S. K. Majumdar, R. F. Schmalz and E. Willard Miller © 1990, The Pennsylvania Academy of Science.

# *Chapter Twenty-Nine*
# LEGAL IMPLICATIONS OF RADON GAS

**BENJAMIN N. HENSZEY**

Professor of Business Law
The Pennsylvania State University
University Park, PA 16802

## INTRODUCTION

Although artificially created radon gas has been the subject of governmental regulation and litigation for at least two decades, it took an incident in Boyertown, Pennsylvania to bring naturally created radon gas to national attention. In December, 1984, an employee at the Limerick Nuclear Power Plant began setting off radiation detectors at the plant. Since the employee was not in close proximity to radioactive material at the time the detectors sounded an alarm, his house was tested and was found to have levels of cancer-causing radon nearly a thousand times higher than the four picocuries a liter the EPA uses as a threshold for concern.[1] Predictably, shortly after the Limerick incident, reports were issued by a variety of governmental agencies speculating about the gravity of the naturally occurring radon problem. Testimony was taken that identified radon as posing a greater health hazard than asbestos or cigarettes. It was estimated that as many as 20,000 lung cancer deaths may be caused each year by naturally occurring radon gas and that as many as ten percent of the eighty-five million houses in the nation may have hazardous radon levels.[2]

Practitioners and scholars in the legal community quickly recognized the numerous potential radon-related liability problems. As a result, legal issues related to naturally occurring radon gas have been the subject of a growing body of recent educational and scholarly programs and literature.[3] What has emerged is a rather consistent pattern of thought relating to potential defendants and legal theories of liability. The various legal theories that relate to naturally occurring radon gas have not been tested in the courts to date.[4] But undoubtedly, when an appropriate opportunity arises, they will be vigorously pursued in the future. It remains to be seen with what success and in what number.

This chapter will define artificially and naturally created radon gas. It will identify potential defendants in radon gas related cases and theories of liability on which they might be sued. Because the potential for liability is greatest in naturally occurring radon gas cases (hence the surge in publicity and national interest) the majority of the chapter will be devoted to it. In conclusion, an assessment will be made on the number of radon related cases that will be processed through the legal system and their potential for success based upon the strengths and weaknesses of the legal theories as discussed.

## DEFINITION

Radon has been generally defined as a colorless, odorless gas produced naturally in the ground as a radioactive byproduct of decaying uranium. It is found in small amounts in all soil and in larger amounts in granite, shale, phosphate and pitchblende. Radon breaks down into radioactive particles called "radon daughters" which, when inhaled, may cause lung cancer. One of the highest concentrations of radon gas is found in a geologic formation called the "Reading Prong" which runs through parts of Pennsylvania, New Jersey and New York. The unit of measurement is picocuries per liter (pCi/l) where, in accordance with Environmental Protection Agency (EPA) standards, a reading in excess of 4 pCi/l creates a potential health hazard.[5]

Nontechnical as this definition may be, with respect to some legal theories, it goes to the very heart of establishing or defending against legal liability resulting from radon gas intrusions. For example, does a builder or an owner of a house have a duty to disclose to a buyer the presence of an odorless, invisible gas whether or not he is aware of its presence? Because of the vast amount of publicity given to radon gas since the Limerick incident, is one presumed to have knowledge of radon gas despite its inability to be detected by the senses? If disclosure is required, should a house owner be held to the same disclosure standard as a builder? Once discovered, under what circumstances is disclosure required? These are among the many questions that directly tie the physical characteristics of radon gas to a legal theory.

A distinction needs to be made between artificially created and naturally created radon gas because certain legal theories will be more applicable to one than the other. For purposes of this chapter, radon gas which is highly concentrated in a specific area by an act of man will be referred to as artificially created radon gas. A common example is mine tailings or solid waste used to fill vacant land or make products such as block used in the construction of houses. Naturally occurring radon gas comes from the ground itself without additional material being added or from sources of water or natural gas. The general definition is the same, but the source is different. It is naturally occurring radon gas that presents the greatest possible legal problem since, by definition, it has potential legal implications for each and every owner of land in states having radon gas in high concentrations.

## LEGAL THEORIES

Landowners injured by the effects of radon gas have at their disposal several legal theories. The most commonly recognized are strict liability, breach of implied warranty of habitability, negligence and misrepresentation. Case law indicates that one or more of these theories apply to artificial radon cases, but to date it is not clear that any of them apply to naturally occurring radon cases. Each theory will be examined in detail as it applies to both artificial and natural radon gas. However, because the law has not been established in the naturally occurring radon cases and because it is the focal point of public interest and concern as indicated above, most of the analysis and speculation that follows will deal with it.

## STRICT LIABILITY

### *General application*

Although strict liability is often referred to as liability without fault, perhaps more accurately it means "liability that is imposed on an actor apart from either (1) an intent to interfere with a legally protected interest without a legal justification for doing so, or (2) a breach of a duty to exercise reasonable care, i.e. actionable negligence."[6] The attractiveness of the theory lies in the fact that a plaintiff does have to prove that the defendant intended to cause harm or in the alternative that the defendant was negligent. In addition, because strict liability is a tort theory, it does not give rise to a variety of contract defenses including disclaimers and failure to give notice in the event of a contract breach.

### *Applied to artificially created radon gas—ultrahazardous and/or abnormally dangerous conditions*

As originally conceived, strict liability applies to ultrahazardous and abnormally dangerous things and activities. Artificially created radon gas clearly falls into that category. In a recent New Jersey case, the appellate court held that defendant successors to a radium processor which disposed of a waste product on property were absolutely liable to plaintiff, a subsequent purchaser of the property, for damages caused by gamma radiation and radon gas from radium tailings. The court reached this conclusion even though a building constructed by an intervening party contributed to the dangerous concentration of radon gas. The court noted that as a matter of law, the New Jersey Supreme Court has concluded that the handling of toxic waste is an abnormally dangerous activity, hence bringing it automatically within the confines of strict liability. Specifically, it is abnormally dangerous because: it creates a high degree of risk to the person or property of another; there is a likelihood that the harm that results from it will be great; the risk cannot be eliminated by the exercise of reasonable care; the activity is not a matter of common usage; the activity is not appropriate to the place where it is carried on; and the extent of its value to the community is outweighed by its dangerous attributes.[8] "The issue of toxic waste is a sensitive issue of paramount concern in today's society. Yet, many of today's problems due to toxic waste are a result of yesterday's mistakes. These are mistakes for which yesterday's perpetrators must be held responsible."[9]

On the Federal level, the legal concept of strict liability is also applicable to artificially created radon cases. The Comprehensive Environmental Response, Compensation and Liability Act (CERCLA)[10] subjects the "owner" or "operator" of a facility (real property) strictly liable for removal and other costs involved in remedying sites where hazardous substances were disposed. The liability extends to current and certain past owners of real property on which the hazardous substance was deposited.

### *Applied to naturally created radon gas—Restatement of Torts Section 402A*

Radon gas that comes naturally out of the ground is not abnormally dangerous and therefore traditional strict liability principles applicable to ultrahazardous or abnormally dangerous conditions will not apply. In short, there is no intervening act of man that concentrates or in any other way interferes with naturally occurring radon gas. Nevertheless, naturally occurring radon gas tends to seep into houses and other buildings, remain there and concentrate at such levels as to present a health hazard. Either the current owner or a subsequent owner cannot recover from the seller, contractor or anyone else in the chain based upon the argument that radon gas under the circumstances is abnormally dangerous. But an owner may be able to argue that the "product" purchased (the house) was defective and consequently strict liability principles relating to the sale of a product applies. Section 402A of the Second Restatement of Torts which has been widely adopted by the courts states as follows as it applies to the sale of products:

(1) One who sells any product in a defective condition unreasonably dangerous to the user or consumer or to his property is subject to liability for physical harm thereby caused to the ultimate user or consumer, or to his property, if
   (a) the seller is engaged in the business of selling such a product, and
   (b) it is expected to and does reach the user or consumer without substantial change in the condition in which it is sold.
(2) The rule stated in Subsection (1) applies although
   (a) the seller has exercised all possible care in the preparation and sale of his product, and
   (b) the user or consumer has not bought the product from or entered into any contractual relation with the seller.[11]

Similar to strict liability as it applies to ultrahazardous or abnormally dangerous activities, Section 402A eliminates the need to prove the elements of negligence and does not require a contractual relationship between a plaintiff (past or present possessor or occupant of a residential or other building) and defendant (builder/vendor). But it applies only to sellers "engaged in the business of selling such a product;" therefore, potential defendants are builders and generally not owners of residential or other buildings in a chain of title.

Under the language of Section 402A, defendant-builders will argue that Section 402A does not apply to them because: (a) a residential or other building is not a "product;" (b) radon gas in a residential or other building does not create a "defective condition;" and (c) radon gas is not "unreasonably dangerous." Given the tremendous public interest in and concern about radon gas, it is reasonable to assume that the courts will be under some pressure to find that a building is a product, that radon gas in a building can be a defect and that it is unreasonably dangerous. Certainly good arguments can be made for reaching that conclusion.

a. Houses or other buildings as a "product"

In response to the first argument, from inception, courts have tended to be expansive in their application of Section 402A. Defendants are not exclusively "sellers" and "products" are not exclusively manufactured goods. Courts have construed the seller-product requirement to apply to leases and bailments of chattels, use of chattels by business invitees and licensees, services and real estate sales.[12]

The case that first applied strict tort liability to a builder-vendor was *Schipper v. Levitt & Sons*[13] The defendant, a well-known mass developer and builder of residential houses, sold a house to the Kreitzer family in 1958. At the time of the sale, the defendant warned the Kreitzers that the hot water in the house would be hotter than in the ordinary house. The Kreitzers subsequently leased the house to plaintiff-Schipper in 1960. In the same year, plaintiff's sixteen month old son was severely scalded by hot water drawn from a bathroom faucet in a hot water system that was defectively designed by plaintiff. In dispensing with defendant's arguments that plaintiff had to prove negligence and that strict tort liability relating to the sale of products did not apply to the sale of houses, the court stated: "We consider that there is no meaningful distinctions between Levitt's mass production and sale of automobiles and that the pertinent overriding policy considerations are the same. That being so, ...strict liability principles (applicable to the sale of products) ...should be carried over into the realty field, at least as it is dealt with here."[14] To hold otherwise, the court argued, would be to give credence to ancient distinctions that make no sense in today's society.

It has been argued that the *Schipper* conclusion should not apply to custom houses because custom houses, relative to mass produced houses, are unique, owner-supervised and generally have greater longevity than products produced by the manufacturing process.[15] However, the better reasoning, as supported by case law,[16] makes no distinction between mass and custom builders. As one commentator argues, "the plaintiff's injuries are life-threatening, and public policy will not be well served by categorically exempting from liability builders who are not mass-producers."[17]

b. Radon gas in a building as a "defective condition"

Comments g. and h. to Restatement of Torts Section 402A purport to define the meaning of "defective condition." Although vague, emphasis is placed upon a product being in a "safe condition" at the time it leaves the seller's hand. A safe condition includes proper precautions taken to create the safe condition and, where there is reason to believe that the product may be dangerous, providing adequate warning of the danger. A builder's responsibility to transfer a building in a safe condition fits nicely into this definition, particularly as it applies to radon gas.

Radon gas enters residential houses and other buildings by three primary routes. First, through the soil gaining entry into the structure through cracks in a concrete slab, cracks between a poured concrete slab and blocks, pores and cracks in concrete blocks, slab-footing joints, exposed soil, weeping tile, mortar joints and loose fitting pipes. Second, through building material, primarily granite. Third, through water. The EPA has suggested a variety of ways to take simple yet effective action to reduce concentration levels. For example, the sealing and closing of all pores, voids, open joints, and exposed earth, and the effective use of a variety of ventilation systems.[18] The presumption is that if a builder has not effectively sealed and ventilated a newly constructed house, especially when the presence of radon gas was known or should have been known, he has transferred a defective product to a buyer. At the very least, where the builder knew or should have known of the presence of radon gas, he has a duty to warn a potential buyer of the danger presented. Furthermore, because of the scope of public awareness, it is becoming increasingly difficult for a builder to escape liability based upon lack of knowledge. In a sense, he is presumed to know of the problem.

c. Radon gas as an "unreasonably dangerous" condition

Comment i. to Restatement of Torts Section 402A treats as unreasonably dangerous anything purchased by a consumer which exceeds common knowledge of its characteristics. Tobacco, for example, is not unreasonably dangerous because smoking is harmful, but tobacco that contains marijuana may be unreasonably dangerous. By analogy, a house by itself or a house that contains a low concentration of radon gas is not unreasonably dangerous, but a house that contains high concentrations of radon, if the EPA Radon Reports are to be believed, is unreasonably dangerous. A builder might argue that because of public exposure, most buyers have knowledge of the dangerous characteristics of radon gas and therefore it cannot be treated as an unreasonably dangerous condition. The practical answer is that actual knowledge does not exist because radon gas is colorless and odorless, and if a buyer is knowledgeable about the potential problem, the buyer will invariably inquire. If mislead or lied to, a buyer can always bring an action for fraud or misrepresentation.

In summary, strict liability of former owners for artificially created radon is clear because it creates an abnormally dangerous condition. In the case of naturally occurring radon gas, it remains to be seen whether strict liability, under Restatement of Torts Section 402A, is a viable legal theory that can be invoked by a buyer against a builder. At least one commentator argues that this theory "offers the most promise for homeowners exposed to excessive indoor radon levels."[19]

## WARRANTIES

### *General Application*

In the context of a commercial sale, a warranty is a statement that relates to the quality of property being sold. It can be expressly stated or it can be presumed or implied.[20] In either case, contrary to strict liability, it arises out of contract, hence it is subject to all contract requirements and defenses. There must be privity of contract, that is, both parties have some relationship to the same contract, and contract disclaimers are applicable, that is, a buyer may agree not to hold a seller responsible for a defect if one arises.

Historically, the common law concept of *caveat emptor* (let the buyer beware) was universally applied to the sale of housing. Thus, only with an express contract provision warranting against defects was a buyer protected. Express warranties continue to be applicable. Many have taken the form of a Home Owners Warranty (HOW) which was created in 1974 by the National Association of Home Builders. It is a form of private insurance whereby claims sent by injured buyers to the builder may be indemnified by HOW.[21] Contractually, any seller (builder or owner) can expressly warrant a house to be free of defects including radon gas and such a warranty will be enforced. However, since sellers are not motivated to give express warranties, the more difficult question arises as to whether or not implied warranties are applicable to the sale of houses generally, and to radon gas specifically.

### *Implied warranty of habitability*

As the law has evolved, *caveat emptor* no longer protects a builder from latent or undisclosed defects. Now, an implication exists, commonly referred to as an implied warranty of habitability, that a house will be free of unreasonable defects and suitable for habitation. Therefore, even though not specifically stated in a contract, the buyer of a defective house or other building may have a cause of action based upon breach of implied warranty of habitability.

A majority of states have accepted some form of the implied warranty of habitability, most by judicial decision.[22] In Pennsylvania, for example, the implied warranty of habitability theory was first applied to a latent defect in the 1972 case of *Elderkin v. Gaster.* [23] Elderkins purchased a lot and house from Gaster. Shortly after moving into the house, Elderkins discovered that the well water he used in his house tested for nitrates four times greater than the standard promulgated by the United States Department of Health, Education and Welfare which made the water unfit for human consumption. In response to application of the implied warranty of habitability theory, Gaster argued that it extends only to completing a functional well and does not extend to the purity of the water produced by the well. On appeal, the appellate court discounted this argument by stating: "The developer holds himself out, not only as a construction expert, but as one qualified to know what sorts of lots are suitable for the types of homes to be constructed. Of the two parties to the transaction, the builder-vendor is manifestly in a better position than the normal vendee to guard against defects in the home site and if necessary to protect himself against potential but unknown defects in the projected home site."[24]

Although it is apparent that the implied warranty of habitability theory applies to the sale of residential houses and other buildings, there are a number of legal obstacles that could easily prevent its application.

First is contract disclaimer. Since the relationship between a builder-vendor and buyer arises out of contract, the seller may insert a disclaimer clause. Generally, contract disclaimers are not favorably viewed by the courts. Therefore, not only will they be construed strictly against the builder-vendor, but they must meet certain requirements. For example, relative to a warranty of habitability, a Pennsylvania court recently stated that disclaimer language must be clear, unambiguous, understandable and sufficiently particular to give the new home purchaser adequate notice of the implied warranty that is being waived. It must "refer to its effect on specifically designated, potential latent defects."[25] To be legally effective, it appears therefore that a disclaimer should specifically mention radon gas as a possible latent defect.

Second is lack of privity. If an owner is injured who did not purchase directly from a builder-vendor, the builder-vendor will argue that there is no contractual relationship between himself and the injured owner and therefore an action based on contract (implied warranty) cannot exist. Many courts have done away with the privity requirement. In a case of first impression involving a basement that leaked and walls that were cracked, an Indiana court concluded that "Our technology is increasingly complex. The traditional requirement of privity between a builder-vendor and a purchaser is an outmoded one."[26] The court also pointed out that the rule applied to latent defects and the burden of proof is upon the claimant to show that the defect was caused by the original builder-vendor. Various commentators have also noted with approval the application of implied warranty of habitability to the sale of used housing.[27]

Third is the reasonableness test. The builder-vendor would argue that the presence of radon gas is a reasonably expected condition and not a defective condition. The *Schipper* case teaches that the test of defectiveness is "reasonableness rather than perfection." Under the holding of *Elderkin,* the admission of nitrates in well water is factually identical to the admission of radon gas from well water. It seems logical to conclude therefore that the presence of radon gas in a house or other building makes it unreasonably defective.[28]

In summary, express warranty relating to the existence of radon gas is a clear, uncontestable theory on which an injured owner could reply if privity of contract exists. Although not as certain, it appears that implied warranty of habitability is also a viable legal theory for an injured owner to use against a vendor-builder unless it has been effectively disclaimed or, in some jurisdictions, privity does not exist.

## NEGLIGENCE

### *General Application*

Strict liability depends upon the nature of the activity. It does not depend upon the breach of a duty to exercise reasonable care. Warranty, whether express or implied, generally depends upon the existence of a contractual relationship. Negligence, on the other hand, does not relate to the nature of the activity or a contractual relationship. Rather, it relates to any "conduct which falls below the standard established by law for the protection of others against unreasonable risk of harm."[29]

Negligence requires four basic elements: (1) duty imposed on a defendant to conform to a certain standard of conduct established for the benefit of the plaintiff; (2) failure to conform to a standard of conduct resulting in breach of the duty; (3) harm to plaintiff proximately caused by defendant's failure to conform to an established standard of conduct; and (4) failure to conform to the standard of conduct resulted in injury or harm. From an injured plaintiff's point of view, it is more expedient to base a cause of action on strict liability or warranty because they eliminate the need to prove each of the four elements required for negligence. Nevertheless, negligence is still a viable theory on which to proceed. In particular, negligence is a legal theory that would probably be used in conjunction with strict liability and warranty by an owner of real property against a builder-vendor. It is also possible for an owner of property to use it in a suit against another owner (not the builder-vendor) or his agent in a situation where, for example, a real estate broker forgot to have a radon test made in a house he is selling and high concentration of it is discovered after the sale is completed.

*Applied to naturally created radon gas*

Based upon the four basic elements, the first element will invariable create the greatest obstacle to recovery where radon gas is discovered in a residential house or other building. One question is whether a seller (whether a builder-vendor or anyone else in the chain) has a duty to test for and disclose the presence of radon gas. As time passes and the presence of radon gas and its health implications becomes more a matter of public knowledge, the answer may move in the direction of being positive. At this point, however, there appears to be no reason, short of statutory mandate, for imposing such a duty. A different question arises with respect to a builder-vendor and the sighting and construction process. If the presence of radon gas is known, then the builder-vendor arguably has a duty to conform to a standard of care in sighting and constructing the house or other building. The EPA, for example, has provided certain techniques for reducing entrance of radon gas into a building. If the builder-vendor has not conformed to the established standard of care and it was the proximate cause of the harm, recovery may be possible.

In their national publications, the building industry has recognized the problem, one writer speculating that: "a home built before radon became known as an indoor air pollutant should not provide a basis to sue a builder under a negligence theory. However, now that radon is known to cause problems, there could be a litigation explosion engulfing homes built now and in the future."[30]

In summary, it may be difficult for an injured plaintiff to meet the burden of proof required to establish a certain standard of care when none exists now and to prove that the standard was not met. Furthermore, it may be difficult to prove that there was a connection between the harm or potential harm and something which occurred naturally in the ground. It appears therefore that the strict liability and warranty theories will prove to be more plaintiff friendly.

## MISREPRESENTATION

*General Application*

During the negotiation of a contract, the parties often make statements of fact concerning the subject matter of the contract. If a statement is influential in bringing about an agreement, it is called a representation. If the statement is false, it is called a misrepresentation. Misrepresentation in turn can be either fraudulent or negligent. A fraudulent misrepresentation, which can also be a tort, crime or both, occurs when the representer does not honestly believe that the fact represented was true.[31] Factually, a fraudulent misrepresentation can arise in a variety of situations such as express lies, partial information, nondisclosure when there is a duty to disclose and from acts or conduct. A negligent misrepresentation occurs when the representer honestly believes that the fact represented was true, but in reality was not. Generally, the injured party can recover for both types of misrepresentation.

*Applied to naturally created radon gas*

Potential defendants in fraudulent and negligent misrepresentation radon gas cases include vendor-builders, owner-sellers, real estate brokers and professionals or other persons who certify radon gas levels in residential houses or other buildings. The following examples deal with potential misrepresentation situations.

*a. Fraudulent misrepresentation*

A builder-vendor who has tested for radon gas in a housing development and found high levels to exist at all his building sites, may have made a fraudulent misrepresentation to a buyer if: he tells the buyer that radon gas is not present in the area; he tells the buyer that radon gas is present but in insignificant amounts; he says nothing to the buyer about the known presence of radon gas; or if he interferes with a device that records radon levels in such a way as to favorably distort the reading. The same basic factual pattern can be applied to a real estate broker or any seller who does or does not do the same thing. Real estate brokers in particular need to be extremely careful about their representations in selling a house. They are professionals whose standard of conduct relative to the disclosure of radon gas may rise higher than the seller for whom he is acting as an agent.

*b. Negligent misrepresentation*

A builder-vendor who has not tested for radon and honestly believes that the house he offers for sale is free of radon, may be guilty of negligent misrepresentation if high levels of radon gas are subsequently discovered. A cause of action based upon negligent misrepresentation is more likely to be successful against a builder-vendor than an owner because the former is in the business of selling the product and thus would be held to a higher disclosure standard. If an owner unintentionally interferes with a radon test causing an unrealistically low reading, and a buyer relies on that reading, the owner may be guilty of negligent misrepresentation.

A number of states have recognized the importance of accurate radon testing procedures. Pennsylvania, for example, recently enacted the Radon Certification Act[32] which provides for certification of persons who perform radon testing and radon remediation. Should a certified person not conform to the standard created thereunder, the person would be negligent per se.

In summary, to date it is not entirely clear what the extent of the vendor-builder's or seller's duty is to disclose known or unknown information about radon gas. On one end of the spectrum is a builder-vendor who has constructed a house and knows that high levels of radon gas exist in the house. A strong argument can be made that a builder-vendor should accurately disclose available information relating to radon gas concentrations. On the other end of the spectrum is the seller of a used house who has little or no knowledge of the potential harm caused by radon gas. If the seller is not aware of the presence of radon gas, it would appear that he is not negligent in failing to disclose its existence. However, it would be tantamount to fraud if the seller actively concealed the presence of high levels of radon gas. The real estate industry is already aware of the misrepresentation problem in the sense that it is now standard practice in many known radon gas areas to either insert a disclaimer clause in a contract of sale and/or require a radon test before the real estate contract becomes effective.

## CONCLUSION

At this time, the legal implications of naturally occurring radon gas are uncertain. Recovery for harm caused by radon gas based upon any one of the suggested theories (strict liability, warranty, negligence and misrepresentation) is clearly possible. Nevertheless, the predicted flood of law suits, particularly related to residential houses, has not occurred and may never occur to the extent predicted. One explanation is the extensive coverage and exposure the issue has received in the national press and the academic, professional and trade literature. For example, builder-vendors undoubtedly are aware of notification requirements and construction standards to eliminate the problem. As a result, builders will take appropriate steps to protect themselves through notification, disclaimers or hopefully, acceptable construction practices. Real estate brokers in known radon areas commonly advise buyers to include a radon certification clause in a contract of sale. Although there is no evidence to confirm the fact, it is probably safe to say that in most cases the problem is resolved short of litigation by price adjustments and corrective measures. Certainly, as this chapter has shown, injured plaintiffs are not without potential legal remedies which will be tested in the courts. But perhaps the better question is whether a great number of law suits will materialize in the first place.

## REFERENCES AND FOOTNOTES

1. *National Geographic* is the most recent national publication to report the Limerick incident. See C.E. Cobb, Jr. 1989. "Living With Radiation," *National Geographic,* Vol. 175, No. 4, p. 425.
2. Since the Limerick incident, Congress has held a number of hearings dealing with the radon gas issue. See for example *Hearings on Radon and Indoor Air Pollution Before the Subcommittee on Natural Resources, Agriculture Research and Environment of the House Committee on Science and Technology,* 99th Cong., 1st Sess. 1 (1985) and *Joint Hearing on Radon Gas Issues Before the Subcommittees on Environmental Protection and Committee on Environmental and Public Works,* 100th Cong., 1st Sess. 1 (1987) [hereinafter referred to as Radon Reports].
3. See for example Pennsylvania Bar Institute, *What Real Estate Lawyers Should Know About Environmental Laws.* 1988. Pub. No. 1988-424, p. 120; M.R. Kornreich. 1987. Dealing with the Invisible Trap of Radon Liability, *The Practical Real Estate Lawyer,* Vol. 3 No. 5, pp. 17-23; S.L. Kass and M.B. Gerrard. 1987. Radon and Real Estate Transactions: The Risks and the Legal Implications. *Toxics Law Reporter,* Vol. 2, No. 10, pp. 289-293; Comment 1987. Radon Gas: Ramifications for Real Estate Transactions in Pennsylvania. *Dick L. Rev.,* Vol. 91:1113-1155; F.B. Cross and P.C. Murray. 1988. Liability for Toxic Radon Gas in Residential Home Sales, *N.C.L. Rev.,* Vol. 66:687-739.
4. Although the author could find no reported cases dealing with naturally occurring radon gas, it is his understanding that a number of cases have been filed in the lower courts.
5. The previously referenced literature and Radon Reports define radon gas in a consistent manner. See for example Kans and Gerrard, Radon and Real Estate Transactions: The Risks and the Legal Implications, p. 289 (1987). In addition, see M. Galen. 1986. Lawyers Grapple with Radon Issue, *The National Law Journal, Reduction Methods: A Homeowner's Guide.* August 1986, Report No. OPA-86-005 and U.S. Environmental Protection Agency, *A Citizen's Guide to Radon: What it is and What to do About it,* August 1986, Report No. OPA-86-004.
6. Prosser, W. and W.P. Keeton. 5th ed. 1984. *Prosser and Keeton on The Law of Torts,* Sec. 75, p. 534.
7. T & E Ind. v. Safety Light Corp. 546 A.2d 570 (1988).
8. Restatement (Second) of Torts Sec. 520, p. 36 (1976).
9. T & E Ind., 546 A.2nd at 575.
10. 42 U.S.C. Sec. 9601.
11. Restatement (Second) of Torts Sec. 402A (1965).
12. Noel, D.W. and J.J. Phillips. 1974. Products Liability. West Publishing Co., St. Paul, Minn. pp. 85-90.

13. 207 A.2d 314 (1965).
14. *Id.* at 325. The same rationale was applied in a California case which overturned prior California law that applied strict liability only to manufacturers, retailers and suppliers of personal property and not real estate. Kriegler v. Eichler Homes, Inc., 74 Cal. Rptr. 749, 752 (1969).
15. See Hiner 1985. Strict Liability and the Building Industry, *J. Prod. Liab.,* Vol. 8:381-389.
16. See Papitucci v. Drelich, 379 A.2nd 297 (1977).
17. Cross and Murray, *Liability for Toxic Radon Gas in Residential Home Sales,* p. 707.
18. U.S. Environmental Protection Agency, *Radon Reduction Techniques for Detached Houses,* June 1986, EPA/625/5-86/019.
19. Cross and Murray, *Liability for Toxic Radon Gas in Residential Home Sales,* p. 703.
20. Black's Law Dictionary (abridged 5th ed. 1973) p. 818.
21. For a detailed explanation of HOW, see R.T. Strickland. 1984. Implied Warranties in New Home Sales—Is the Seller Defenseless?, *S.C.L. Rev.* Vol. 35:491-492.
22. See R.T. Strickland. 1984. Implied Warranties in New Home Sales—Is the Seller Defenseless? *S.C.L. Rev.,* Vol. 35: p. 469 and P. Shedd. 1980. The Implied Warranty of Habitability: New Implications, New Applications, *Real Est. L.J.* Vol. 8: pp. 303-306.
23. 288 A.2d 771 (1972).
24. *Id.* at 777.
25. Tyus v. Resta, 476 A.2nd 427, 432 (1984).
26. Barnes v. Mac Brown & Co., 323 N.E. 2nd 619, 620 (1976).
27. See for example B.N. Henszey. (1978). Caveat Emptor in Full Retreat—Application of Implied Warranty to the Sale of Used Housing, *Am. Bus. L.J.* Vol. 15: p. 351-363.
28. In agreement is Cross and Murray, Liability for Toxic Radon Gas in Residential Home Sales, pp. 718-719.
29. Restatement (Second) Sec. 282 (1976).
30. Ethier, W. 1986. Some Legal Issues Touching on Radon Gas Problems. *Nation's Building News.* Vol. II.
31. Prosser & Keeton, Sec. 135 p. 728.
32. 63 P.S. Sec. 2001 - 2014.

*Environmental Radon: Occurrence, Control and Health Hazards.* Edited by S. K. Majumdar, R. F. Schmalz and E. Willard Miller © 1990, The Pennsylvania Academy of Science.

# *Chapter Thirty*

# RADON RISKS: PEOPLE'S PERCEPTIONS AND REACTIONS*

**ANN FISHER[1] and LENNART SJÖBERG[2]**

[1]Manager
Risk Communication Program
Office of Policy Planning and Evaluation, PM-221
U.S. Environmental Protection Agency
Washington, D.C. 20460
and
[2]Director
Center for Risk Research
Stockholm School of Economics
Box 6501
113 83 Stockholm, Sweden

## INTRODUCTION

Radon is a colorless, odorless, tasteless gas that is radioactive. It escapes from the ground, and can accumulate in buildings. Radon became a media issue in early 1985 after Stanley Watras set off radiation detectors on his way into work at the Philadelphia Electric Company's Limmerick generating station. It turned out that the radiation came from very high radon levels in his Boyertown, Pennsylvania home, not from the nuclear power plant. Since then, Pennsylvania, other states, and the U.S. Environmental Protection Agency (EPA) have discovered that many homes and schools have elevated radon.[1] EPA estimates that about 20,000 lung cancer deaths per year are attributable to radon.[2] There have been a variety of efforts to make people aware of the potential threat from radon, but less than 5 percent of homes have been tested, and only a small share of those with elevated levels have been mitigated. This paper examines how people think about radon, and why testing and mitigation are not more widespread.

*A preliminary version of this paper was presented to the Risk Group at the University of Wisconsin, April 21, 1989. The views are of the authors' and should not be attributed to their employers.

Recent research is summarized on how to communicate effectively about the risks from radon. The concluding section provides some recommendations for a more effective program to reduce the risks from radon.

## REGULATION VERSUS INDIVIDUAL CHOICE

The risks from many environmental contaminants are controlled through regulation. But regulation of radon may not be a viable solution to this risk problem. It also may not be the most efficient solution.

Regulation may not be viable because most of the exposure to radon occurs in people's homes. One potential form of regulation could be to require testing of every home and mitigation whenever the tests exceed a designated level. However, the federal government has been reluctant to tell people what they must do inside their homes. EPA also lacks clear regulatory authority with respect to radon.

The health risk estimates indicate that radon is a larger health threat than any others being faced by EPA.[3] There is more confidence in the health risk estimates for radon than for most other environmental pollutants. This is because the risk estimates are based on human data rather than on extrapolations from animal data. EPA's risk estimation techniques are conservative, in the sense that they are not likely to underestimate the risk even though they could be substantial overestimates of the risk. Even if the EPA estimates are too high by an order of magnitude, radon accounts for more risk than any other environmental pollutant in the United States. Setting a standard below which exposures would be safe is not feasible on the basis of present scientific understanding. A non-threshold dose-response relationship is presumed for cancer-causing substances such as radon. This means that there is some increase in cancer risk even at very low exposure levels.

Even if regulation were viable, it might not be the most efficient way to reduce the risks from radon. Early data from the portion of Pennsylvania near the Watras home indicated that radon levels vary widely even in the same neighborhood.[4] Thus a requirement to mitigate every home (or to use radon-proof construction techniques for all new buildings) would use society's scarce resources for mitigation that sometimes would not reduce risk because many of the houses had very low risks to start with. Data from state surveys also show that some large areas apparently do not have the potential for elevated radon levels. Thus a requirement that all homes be tested similarly could be argued to use resources unnecessarily.

EPA's present radon policy is clearly a compromise. It has set *action guidelines* rather than a standard. These action guidelines recommend that people mitigate within a few years if their readings exceed 4 pCi/l, within a few months if they are over 20 pCi/l, and within a few weeks if they are over 200 pCi/l. Lifetime exposure to 4 pCi/l is estimated to have an excess lung cancer risk of

1-3/100, which is about the same as that associated with smoking 10 cigarettes per day. This risk level is much higher than the 1/10,000 to 1/1,000,000 risk levels often chosen for EPA's safety standards. But the EPA action guidelines are not health-based standards; they account for the potential radon reductions judged to be feasible at reasonable cost. That judgment was made in 1986, when there was not much information available about mitigation methods, their costs, or their effectiveness.

EPA has supplemented the information program for the public with technical assistance to the states. One type of assistance has been materials and courses so that states can train contractors in effective mitigation techniques. EPA also has provided the resources for state surveys to determine whether individual states have radon "hot spots."

In September 1988, EPA and the Public Health Service issued a joint public health advisory. The advisory recommends that all homes be tested, unless they are above the second floor of a building.[5] In April 1989, EPA issued guidance for testing in schools. (See reference in endnote 1.) In 1988, Congress directed EPA to provide more information about the health risks from radon exposures below 4 pCi/l, and set a national indoor radon goal of ambient outdoor levels, which are estimated to be 0.2 to 0.7 pCi/l.[6] The changes will be incorporated in a revision of the *Citizen's Guide,* which is expected to be available in 1990.

Thus EPA's program has relied on individual action rather than regulation. Individual choice is consistent with democratic principles, and has the potential for maximizing the efficient use of society's resources. However, this efficient use of resources depends on the realization of informed consumer decisions about whether to test for radon and whether to mitigate if test results show unacceptably high radon levels. The crucial question, then, is whether a risk communication program will result in informed choices on the part of the public.

## CHALLENGES OF COMMUNICATING ABOUT RADON

Increasing evidence makes it clear that the public's perceptions of the seriousness of a particular risk typically are at odds with the scientific estimates of the magnitude of the risk. Some researchers stress the difficulty with senders, channels and receivers in explaining why people do not understand small probability risks.[7] Others argue that these shortcomings in explaining and understanding small probabilities are only part of the problem. Figure 1 modifies Peter Sandman's explanation of this phenomenon.[8] It emphasizes that risk has many characteristics, and that people care about characteristics in addition to the magnitude of the risk.

There are several characteristics of radon that make it unlikely to arouse concern. Because it is a colorless, odorless, tasteless gas, there are no sensory cues to remind people of this potential hazard. People often perceive a benign experience with radon because they have lived in their homes for some time, (e.g., I don't have any ill effects and I have lived here for 10 years). Of course, this is related to the fact that effects do not become apparent until after a 20-40 year latency period. This latency also makes it easy to delay action. Home is viewed as a haven, and often as an extension of one's self.[9] It is hard to view one's self or one's haven as being threatening. People recognize that radon is a natural contaminant, so there is no villain to blame. Its effects occur singly and without drama; no one can say they knew someone who died from radon exposure.

Even when there is awareness, some characteristics make it less likely that people will act. Research at Carnegie-Mellon University has shown that people have misperceptions about radon. Because most gases are lighter than air, some people think radon concentrations will be greatest in the upper floors of a

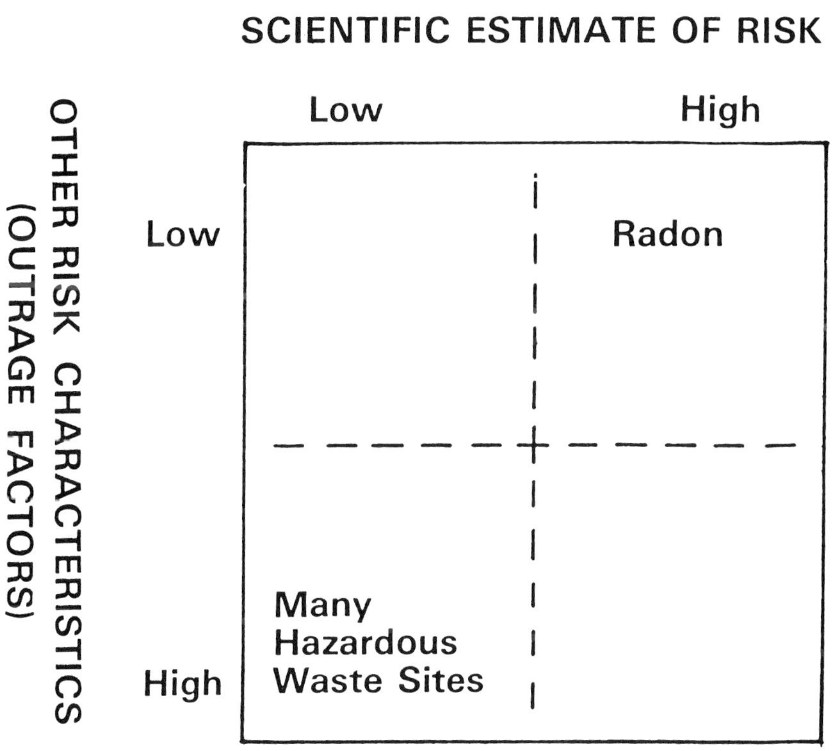


FIGURE 1. Perceived Seriousness Depends on Both Size of Risk and Other Characteristics of the Risk.

building. In fact, concentrations tend to be greatest on the lowest floor because of dilution in the upper floors. Some people think that if a building has high radon levels, then the only way to get rid of it is to remove all of the "contaminated" furnishings. In fact, radon already in the house dissipates quickly by normal ventilation, so preventing additional radon from coming in quickly lowers exposures. Many people also seriously overestimate the cost of mitigation.

On the other hand, some characteristics of radon make it more likely to arouse concern. It is radioactive, and people tend to fear anything radioactive. It causes lung cancer, and cancer is a dreaded disease. Unlike health threats such as smoking, radon confers no benefits, so that it is not psychologically difficult to give up. Testing is easy and inexpensive ($10-$50), and will relieve anxiety for many people. Mitigation typically is no more expensive than ordinary home repairs, and is essentially a one-shot effort. (However, it is important to retest immediately after mitigation and periodically to make sure these measures remain effective.[10])

## WHAT DO WE KNOW ABOUT RADON RISK PERCEPTIONS AND ACTIONS?

Several studies have examined what people know about radon, how concerned they are about it, and what they have done to lower their concern. The most revealing of these studies are discussed in this section.

*Maine:* Early observations of reality and research were discouraging. Over half of a sample in Maine reported some form of mitigation, but much less than half of them had radon levels above the 4 pCi/l action guideline.[11] This suggests too many resources were being devoted to mitigation. Further analysis showed that there was no relation between mitigation and their radon levels. Thus, resources were being used to reduce some very low risks while others were tolerating very high risks. These results did not disappear when corrected for family size and other variables that might affect people's feelings about taking risks.

*New Jersey:* Officials in New Jersey became concerned when high radon levels were discovered in Pennsylvania's Reading Prong, because that geological formation extends through a large part of New Jersey. They commissioned research on how to handle the public panic they anticipated in response to their announcements about the potential radon threats in New Jersey. The research showed that although most people in the New Jersey portion of the Reading Prong were aware of radon, apathy was a far more typical reaction than panic.[12]

*Pennsylvania:* Pennsylvania reacted to the public health threat by offering free radon test kits to all homeowners in its portion of the Reading Prong. Less than 40 percent took advantage of the free offer.[13]

*Maryland:* The marketing literature contributed to the design of an experiment in Maryland to evaluate a community-based "multiple hit, multiple channel" effort to increase radon awareness and testing.[14] A "Test Now and Be Sure" theme was combined with an emphasis on the state's toll-free radon hotline number. Posters, radio public service announcements, and inserts in utility bills were used in two communities. One community's mayor and city council sponsored a radon awareness week, and health officials presented slide shows about radon. A third community received no special attention, and was examined for comparison purposes. Telephone interviews collected baseline data before these outreach activities. Follow-up data were collected three months after the interventions had started.

Three months may not be long enough to measure the full impact of an information program, and success means different things to different evaluators. Four success measures were chosen for the Maryland study: awareness, attitudes, knowledge, and testing. Testing is the most important measure from a public health perspective. However, from the informed choice perspective, it is important to know whether people consciously decided that they did not need to test (compared with never having thought about testing).

The two test communities showed improvements in awareness, attitudes, and knowledge. People still perceived that it would be very costly to fix a radon problem, and they still were weak on such technical issues as what factors can affect a radon test reading. In the community with the full range of activities (Frederick), the share of people testing tripled from 5 percent to 15 percent (with essentially no change in the other two communities). This is a tremendous success from a marketing perspective, but disappointing from a public health perspective.

The Maryland data suggested an increase in the share of people who had made informed decisions about testing. The share of uninformed nontesters decreased, and especially in Frederick there was a big increase in the number of people who thought it was important to test but just had not gotten around to it. These procrastinators may have tested after the data were collected, but factors mentioned in the "challenges" section may have made it easier for them to delay action indefinitely.

The Maryland study was complicated by the sort of external event that can occur in any social science field experiment. Shortly after the joint EPA/Maryland interventions began in the two test communities, television channel WJLA worked with a local supermarket chain to initiate a major campaign for encouraging radon testing. All three study communities are in the WJLA viewing area, but the people in Frederick are more likely to watch this channel. Special analysis accounted for the impact of the WJLA campaign, and it did seem to affect awareness, attitudes, and knowledge in Frederick. However, the analysis did not detect any effect on testing. At least over a longer time period, some synergism can be expected between the EPA/Maryland activities and the

WJLA campaign.

*Washington, D.C. area:* EPA decided to evaluate the impact of the WJLA campaign on testing and mitigation. The WJLA activities combined public service announcements, nightly news coverage for a week, advertising of their special coverage and the "Radon Awareness Watch" on WJLA, and a follow-up about a month later. Their TV coverage repeatedly mentioned the availability of test kits through the supermarket chain.

The research team analyzed the steps leading from deciding to test to deciding to mitigate, as shown in the middle column of Figure 2.[15] Approximately 100,000 test kits were sold as a result of the WJLA activities, at a price about half the usual price for the particular test kit used. The demand exceeded expectations, so more would have been sold if the supermarket chain had been able to keep them in stock. Only about 55 percent of the test kits were returned for analysis, which is substantially lower than the 90-95 percent usually returned. The 55 percent of kits returned amounts to only 3.6 percent of the single-family homes in the Washington area, as shown in the right-hand column of Figure 2. Following the analysis of steps necessary to mitigate, credible mitigation occurred in only about 0.1 percent of homes in the Washington area. This can be compared with 41 percent of the homes in their weighted sample having readings above the 4 pCi/l action guideline. A potential shortcoming of this study is that EPA recommends against taking mitigating action on the basis of a single short-term test of the type used in the WJLA campaign. However, the data showed that few people had conducted confirmatory tests, regardless of whether they mitigated. The data were collected about 9 months after the WJLA campaign, so there should have been enough time to retest and take appropriate action.

*Colorado:* The same research team conducted a preliminary pilot study of testing and mitigation associated with housing transactions. They interviewed 100 recent home buyers in Boulder, Colorado, where there has been no equivalent to the WJLA campaign. Forty-three percent of these buyers had their homes tested as part of the purchase agreement. (See the left-hand column of Figure 2). IBM recently transferred a large number of employees to its Boulder facility. Its employees can participate in the IBM housing "buyback" program (which helps when they are transferred away from Boulder) only if their homes are below the 4 pCi/l action guideline. Some of the real estate agents in the area also are knowledgeable about radon. The pilot study results suggest mitigation of about 15 percent of the homes in the Boulder area, compared with .1 percent in the Washington area.

The sample size is small in the Boulder study, and not representative. Even so, it suggests the need for further research to explore the effectiveness of at least two time-of-sale options. One would be state or local requirements for radon testing in areas known to have "hot spots." This would be similar to the requirements for termite inspections in susceptible areas. Another option would be information programs targeted to realtors, who can suggest a testing clause as part of the purchase contract. EPA currently is working with the National Association of Realtors to develop materials for them to distribute to potential buyers.

*Rutgers Study:* In a separate study, the research team at Rutgers compared alternative ways of expressing the risk of radon and the risk of asbestos.[16] Several of their experimental formats used a risk "ladder." The action guideline for radon was placed in the center of its ladder. Similarly, the action guideline for asbestos was placed in the center of its ladder. Participants received only one information treatment, and they assigned the same degree of seriousness to the asbestos guideline as to the radon guideline. However, scientific estimates of the risk associated with the radon guideline are about 25 times larger than corresponding estimates of the risk associated with the asbestos guideline. More research is needed, but the implication seems to be that people would show more concern about radon if both guidelines appeared on the same ladder.

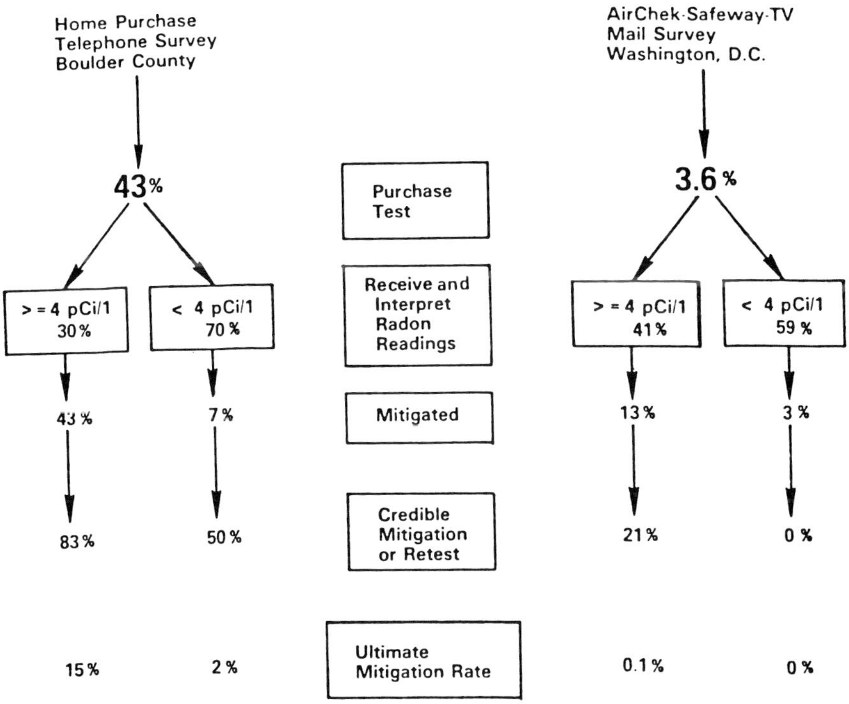


Source: Adapted from Doyle, et al. "An Evaluation of Strategies for Solving the Radon Problem", Figure 10.1.

FIGURE 2. Steps in Testing and Mitigation.

*New York:* A long-term study has been underway to examine how people form and modify their perceptions of radon risk and what actions they take in response to those risk perceptions.[17] This study involves about 2000 homeowners who participated in a study to assess the extent of elevated radon levels in New York State (NYS). People were asked to participate in the testing, so no information is available about motivations for testing, only for risk perceptions and mitigation. EPA worked with NYS to develop and test alternative ways of expressing information about the risk from radon. A group of experts in marketing, journalism, radon, cancer, psychology, communication, and economics suggested a long list of factors that might influence the effectiveness of alternative information treatments. Their opinions were most divided on the effects of a) quantitative versus qualitative information and b) whether people were told what to do rather than given information and asked to decide what they want to do. Four experimental booklets were developed, varying along these two dimensions. The EPA *Citizens' Guide* and a one-page fact sheet also were tested. Baseline data were collected in the summer of 1986, with follow-up data collected in December 1986-January 1987, October-December 1987, and March 1989.

The first follow-up data were collected after people received a 2-3 months reading for their basement, along with one of six sets of information about radon risk. When asked whether they liked the materials they received, and whether the materials were easy to understand and answered their questions, the ratings were very similar across all six information treatments. However, as in the Maryland study, multiple measures of effectiveness were designed into the experiment: learning about radon, ability to form a perception of risk consistent with the objective radon reading, and whether people felt they had enough information to decide about additional action. The results for these measures showed marked differences. The one-page fact sheet clearly was inferior. It had been sent only to a subset of the sample with readings lower than 1 pCi/l, one-fourth of the EPA action guideline. Yet these people were the most concerned about their radon, even though they should have been reassured. (They were sent one of the other information treatments in later phases of the study.) There was no single winner among the five booklets. Which one worked best depended on which effectiveness measure was used. The qualitative version that showed how to adjust the risk ladder for personal exposure differences was best for learning about radon. The two quantitative versions were best for helping people form realistic risk perceptions. The versions that told people what to do and the *Citizens' Guide* were best at helping people feel they had enough information to make a decision about further action.

The rest of the data were collected after people received their readings for year-long monitoring. The monitoring data yielded good news with respect to the state's original question: Only 16 percent of the homes had basement readings above the 4 pCi/l action guideline, and only 5.3 percent had living area readings above that guideline. The first question of interest is whether the right people are concerned and the rest are reassured. Only preliminary data are available at this time (May 1989), but they are suggestive.[18] Figure 3 shows that a substantial share have shifted into the categories indicating that they do not perceive their risk from radon to be serious. This is consistent with a large share being below the action guideline. It also is comforting that the share who cannot judge the seriousness of their risk has fallen from nearly one-fourth to 3 percent. There still is a small percentage viewing their risk as serious or very serious. This also is consistent with the shares above the action guideline, and the fact that participants probably included some of the factors in the "challenges" section in their own definitions of seriousness (rather than considering only the magnitude of the risk estimates corresponding to their readings).

The other question this study hopes to answer is whether the right people acted effectively. That is, accounting for socioeconomic characteristics and preferences about risk-taking, were those at higher radon levels more likely to mitigate and reduce their exposure? (Recall that this was not the case in the Maine study.) Figure 4 shows the responses to the question: "What have you done since receiving your annual readings?" Only 13 percent have mitigated, which is somewhat lower than the proportion above the action guideline. However, another 12 percent still might mitigate after getting more information—if they resist those characteristics of radon risk that tend to encourage

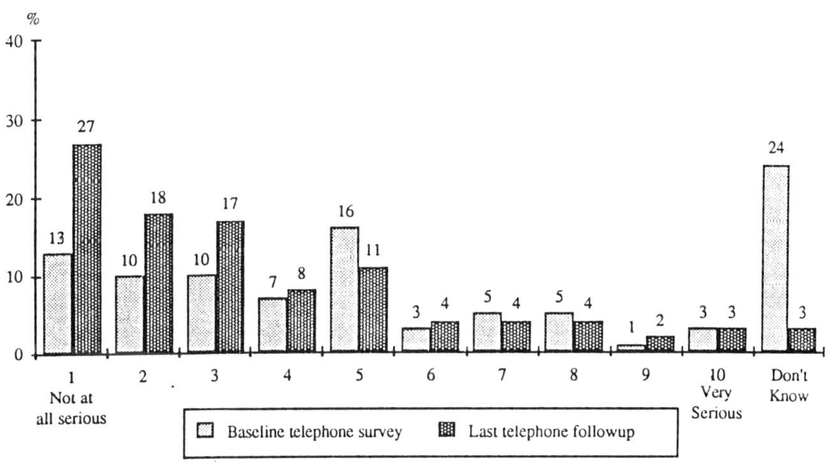


SOURCE: Desvousges, William H., V. Kerry Smith, and Hillery H. Rink, III. 1989. Risk Communication Effectiveness.

FIGURE 3. Perceived Risk from Radon.

procrastination. Figure 5 shows that those who mitigated did have higher radon readings than those who did not. (These differences are statistically significant except for the four percent of the sample who said they wanted to think more before deciding whether to act.) Detailed analysis of their motivations is not yet available, but 52 percent of the mitigators said that health risk was the most important factor. Only 5 percent said that property values or the cost and effectiveness of mitigation were the most important factors.

## CONCLUSIONS

The common result of all this research is that it may be unreasonable to expect a risk communication program alone to reduce radon risk. This is partly

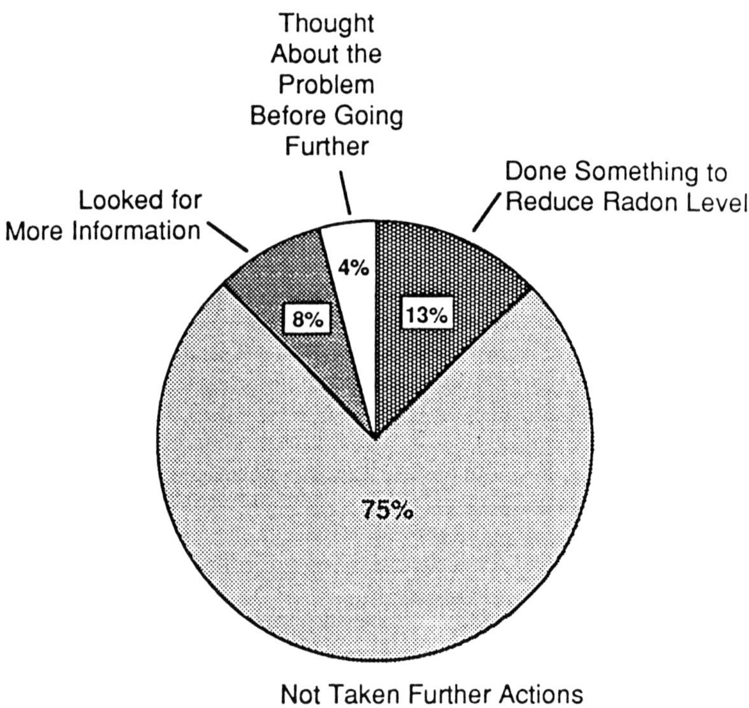


SOURCE: Desvousges, William H., V. Kerry Smith, and Hillery H. Rink, III. 1989 Risk Communication Effectiveness.

FIGURE 4. Responses to Question: "What Have You Done Since Receiving Your Annual Readings?"

because of the tension between the policy maker's desire to protect public health on the basis of what the experts "know is best," versus the choices made by informed homeowners. Citizens who truly understand radon risk may decide that it is less important than other things that they worry about and spend their resources to reduce these other worries. Unfortunately, those who take no action also include homeowners who do not understand the risk, so it may be difficult to distinguish the ill-informed from the well-informed on the basis of behavior.

Another difficulty in determining the effectiveness of a radon risk communication program is that a result judged to be a big success from the marketing perspective typically appears inadequate using public health standards. Any use of an information program to promote public health goals inevitably encounters ethical issues associated with moving along the spectrum from informing to persuading to manipulating.

There has been more research on radon risk communication than for any other environmental health risk. (And possibly more research than for any other health risk with the exception of smoking and the cancer prevention program.) Even so, improved radon policy decisions need additional interdisciplinary research. Dunwoody and Neuwirth observe that researchers tend not to use the communication literature to "inform the relationship between information and risk judgment."[19] The cost-effectiveness of present radon risk communication efforts has not been estimated. Potential improvements in effectiveness might

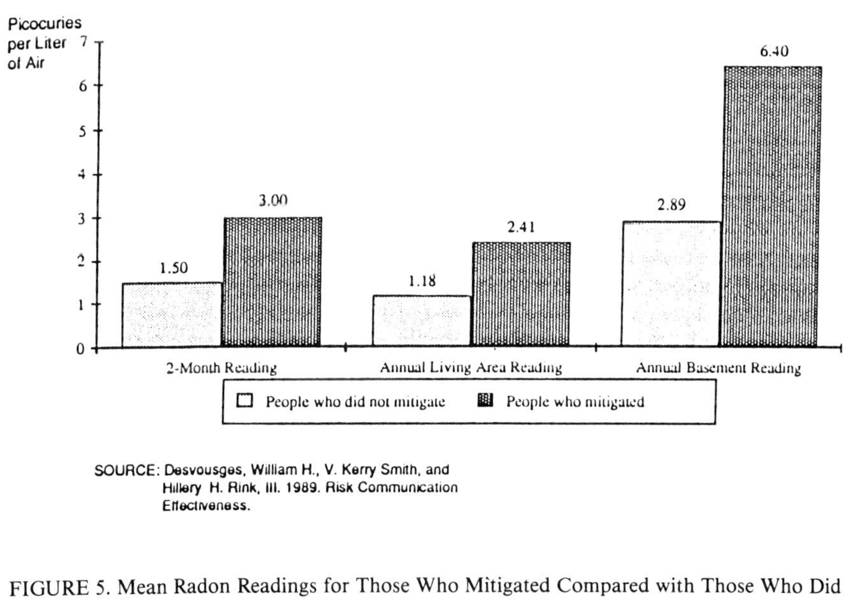


FIGURE 5. Mean Radon Readings for Those Who Mitigated Compared with Those Who Did Not (or have not yet).

be enhanced by combining a multi-faceted risk communication program with point-of-sale requirements for homes and other buildings. These requirements could be imposed by real estate agents and state or local ordinances as suggested by the Colorado study, or by bankers in transactions for mortgages or refinancing.

On balance, the radon risk communication research is consistent with a more general need for convergence between the scientists' and the public's perceptions of the seriousness of many hazards. The public needs a better appreciation of the magnitude of the risk as estimated by the scientists. At the same time, the scientists and policy makers need to recognize that judged seriousness of the risk depends on multiple characteristics of the risk, not just its size.

## REFERENCES

1. U.S. Environmental Protection Agency. September 1988. *Radon Action Report,* EPA 520/1-88-022, Office of Radiation Programs, Washington, D.C. U.S. Environmental Protection Agency. March 1989. *Radon Measurements in Schools: An Interim Report.* EPA 520/1-89-010, Office of Radiation Programs, Washington, D.C.
2. U.S. Environmental Protection Agency and U.S. Department of Health and Human Services. August 1986. *A Citizens' Guide to Radon.* OPA 86-004, Office of Air and Radiation and Centers for Disease Control.
3. U.S. Environmental Protection Agency. February 1987. *Unfinished Business: A Comparative Assessment of Environmental Problems.* Office of Policy Analysis, Washington, D.C.
4. Gerusky, T.M. 1987. Protecting the Home Front. *Environment.* 29: March, 12 ff.
5. U.S. Environmental Protection Agency. 1988. EPA and Assistant Surgeon General Call for Radon Home Testing. September 12 News Release. Office of Public Affairs.
6. Amendments to the Toxic Substances Control Act, Title III. 1988. Section 301. Also see references in endnote 1.
7. Covello, V.T., Detlof von Winterfeldt and Paul Slovic. October 1986. Risk Communication: A Review of the Literature. *Risk Abstracts.* 3:4, 171-182.
8. Sandman, Peter. 1986. Explaining Environmental Risk. U.S. Environmental Protection Agency, TSCA Assistance Office, Washington, D.C.
9. Sjöberg, Lennart. 1989. Radon Risks, Attitudes, Perceptions and Actions, prepared for U.S. Environmental Protection Agency, Office of Policy Analysis, Washington, D.C. March draft.
10. U.S. Environmental Protection Agency. September 1987. *Radon Reduction Methods.* OPA-87-010. Office of Research and Development, Washington, D.C.

11. Johnson, F. R. and R.A. Luken. 1987. Radon Risk Information and Voluntary Protection. *Risk Analysis.* 7:1, 97-107.
12. Weinstein, N.D., P.M. Sandman and M.L. Klotz. 1987. *Public Response to the Risk from Radon, 1986.* January 1987 Final Report to the Division of Environmental Quality, New Jersey Department of Environmental Protection.
13. Personal communication with Tim Hartman, Bureau of Radiation Protection, Pennsylvania Department of Environmental Resources.
14. Desvousges, W.H., V.K. Smith and H.H. Rink, III. 1988. *Communicating Radon Risk Effectively: Radon Testing in Maryland.* Final Report. EPA-230-03-89-048. U.S. Environmental Protection Agency, Office of Policy Analysis, Washington, D.C. March 1989.
U.S. Environmental Protection Agency. 1988. *Region 3/OPPE/State of Maryland Radon Risk Communication Project.* Office of Policy, Planning, and Evaluation, Washington, D.C., November.
15. Doyle, J.K., S.R. Elliott, P.A. Locke, G.H. McClelland, G.W. Russell and W.D. Schulze. 1989. *An Evaluation of Strategies for Solving the Radon Problem,* draft report to Office of Policy, Planning and Evaluation, U.S. Environmental Protection Agency, Washington, D.C.
16. Weinstein, N.D., P.M. Sandman and N.E. Roberts. 1988. *Communicating Effectively about Changes in Risk.* November progress report to U.S. Environmental Protection Agency, Office of Policy Analysis, Washington, D.C.
17. Smith, V.K., W.H. Desvousges, A. Fisher and F.R. Johnson. 1987. *Communicating Radon Risk Effectively: A Mid-Course Evaluation.* EPA-230-07-87-029. U.S. Environmental Protection Agency, Office of Policy Analysis, Washington, D.C.
18. Desvousges, W.H., V.K. Smith and H.H. Rink, III. 1989. Risk Communication Effectiveness: Radon Mitigation Decisions. Working paper at Research Triangle Institute, Research Triangle Park, NC.
19. Dunwoody, S. and K. Neuwirth. 1988. Coming to Terms with the Impact of Communication on Scientific and Technological Risk Judgments. December 1988 Symposium on Science Communication: Environmental and Health Research, Annenberg School of Communications, University of Southern California, Los Angeles, CA.

*Environmental Radon: Occurrence, Control and Health Hazards.* Edited by S. K. Majumdar, R. F. Schmalz and E. Willard Miller © 1990, The Pennsylvania Academy of Science.

# *Appendix-One*

# TITLE IV—RADON GAS AND INDOOR AIR QUALITY RESEARCH

## SEC. 401. SHORT TITLE.

This title may be cited as the "Radon Gas and Indoor Air Quality Research Act of 1986".

## SEC. 402. FINDINGS.

(1) High levels of radon gas pose a serious health threat in structures in certain areas of the country.

(2) Various scientific studies have suggested that exposure to radon, including exposure to naturally occurring radon and indoor air pollutants, poses a public health risk.

(3) Existing Federal radon and indoor air pollutant research programs are fragmented and underfunded.

(4) An adequate information base concerning exposure to radon and indoor air pollutants should be developed by the appropriate Federal agencies.

## SEC. 403. RADON GAS AND INDOOR AIR QUALITY RESEARCH PROGRAM.

(a) DESIGN OF PROGRAM.—The Administrator of the Enrivonmental Protection Agency shall establish a research program with respect to radon gas and indoor air quality. Such program shall be designed to—

(1) gather data and information on all aspects of indoor air quality in order to contribute to the understanding of health problems associated with the existence of air pollutants in the indoor environment;

(2) coordinate Federal, State, local, and private research and development efforts relating to the improvement of indoor air quality; and

(3) assess appropriate Federal Government actions to mitigate the environmental and health risks associated with indoor air quality problems.

(b) PROGRAM REQUIREMENTS.—The research program required under this section shall include—

(1) research and development concerning the identification, characterization, and monitoring of the sources and levels of indoor air pollution, including radon, which includes research and development relating to—

(A) the measurement of various pollutant concentrations and their strengths and sources,

(B) high-risk building types, and

(C) instruments for indoor air quality data collection;

(2) research relating to the effects of indoor air pollution and radon on human health;

(3) research and development relating to control technologies or other mitigation measures to prevent or abate indoor air pollution (including the development, evaluation, and testing of individual and generic control devices and systems);

(4) demonstration of methods for reducing or eliminating indoor air pollution and radon, including sealing, venting, and other methods that the Administrator determines may be effective

(5) research, to be carried out in conjunction with the Secretary of Housing and Urban Development, for the purpose of developing—

(A)methods for assessing the potential for radon contamination of new construction, including (but not limited to) consideration of the moisture content of soil, porosity of soil, and radon content of soil; and

(B) design measures to avoid indoor air pollution; and

(6) the dissemination of information to assure the public availability of the findings of the activities under this section.

(c) ADVISORY COMMITTEES.—The Administrator shall establish a committee comprised of individuals representing Federal agencies concerned with various aspects of indoor air quality and an advisory group comprised of individuals representing the States, the scientific community, industry, and public interest organizations to assist him in carrying out the research program for radon gas and indoor air quality.

(d) IMPLEMENTATION PLAN.—Not later than 90 days after the enactment of this Act, the Administrator shall submit to the Congress a plan for implementation of the research program under this section. Such plan shall also be submitted to the EPA Science Advisory Board, which shall, within a reasonable period of time, submit its comments on such plan to Congress.

(e) REPORT.—Not later than 2 years after the enactment of this Act, the Administrator shall submit to Congress a report respecting his activities under this section and making such recommendations as appropriate.

## SEC. 404. CONSTRUCTION OF TITLE.

Nothing in this title shall be construed to authorize the Administrator to carry out any regulatory program or any activity other than research, development, and related reporting, information dissemination, and coordination activities specified in this title. Nothing in this title shall be construed to limit the authority of the Administrator or of any other agency or instrumentality of the United States under any other authority of law.

## SEC. 405. AUTHORIZATIONS.

There are authorized to be appropriated to carry out the activities under this title and under section 118(k) of the Superfund Amendments and Reauthorization Act of 1986 (relating to radon gas assessment and demonstration program) not to exceed $5,000,000 for each of the fiscal years 1987, 1988, and 1989. Of such sums appropriated in fiscal years 1987 and 1988, two-fifths shall be reserved for the implementation of section 118(k)(2).

# Appendix-Two

# TITLE III—INDOOR RADON ABATEMENT

PUBLIC LAW 100-551—OCT. 28, 1988

Public Law 100-551
100th Congress

## An Act

To amend the Toxic Substances Control Act to assist States in responding to the threat to human health posed by exposure to radon.

Oct. 28, 1988
[S. 744]

*Be it enacted by the Senate and House of Representatives of the United States of America in Congress assembled,*

**SECTION 1. INDOOR RADON ABATEMENT.**

(a) AMENDMENT OF TOXIC SUBSTANCES CONTROL ACT.—The Toxic Substances Control Act (15 U.S.C. 2601 and following) is amended by adding after title II the following new title:

### "TITLE III—INDOOR RADON ABATEMENT

**"SEC. 301. NATIONAL GOAL.** 15 USC 2661.

"The national long-term goal of the United States with respect to radon levels in buildings is that the air within buildings in the United States should be as free of radon as the ambient air outside of buildings.

**"SEC. 302. DEFINITIONS.** 15 USC 2662.

"For purposes of this title:

"(1) The term 'local educational agency' means—

"(A) any local educational agency as defined in section 198 of the Elementary and Secondary Education Act of 1965 (20 U.S.C. 3381);

"(B) the owner of any nonprofit elementary or secondary school building; and

"(C) the governing authority of any school operated pursuant to section 6 of the Act of September 30, 1950 (64 Stat. 1107), relating to impact aid for children who reside on Federal property.

"(2) The term 'nonprofit elementary or secondary school' has the meaning given such term by section 202(8).

"(3) The term 'radon' means the radioactive gaseous element and its short-lived decay products produced by the disintegration of the element radium occurring in air, water, soil, or other media.

"(4) The term 'school building' has the meaning given such term by section 202(13).

15 USC 2663. **"SEC. 303. EPA CITIZEN'S GUIDE.**

"(a) PUBLICATION.—In order to make continuous progress toward the long-term goal established in section 301 of this title, the Administrator of the Environmental Protection Agency shall, not later than June 1, 1989, publish and make available to the public an updated version of its document titled 'A Citizen's Guide to Radon'. The Administrator shall revise and republish the guide as necessary thereafter.

"(b) INFORMATION INCLUDED.—

"(1) ACTION LEVELS.—The updated citizen's guide published as provided in subsection (a) shall include a description of a series of action levels indicating the health risk associated with different levels of radon exposure.

"(2) OTHER INFORMATION.—The updated citizen's guide shall also include information with respect to each of the following:

"(A) The increased health risk associated with the exposure of potentially sensitive populations to different levels of radon.

"(B) The increased health risk associated with the exposure to radon of persons engaged in potentially risk-increasing behavior.

"(C) The cost and technological feasibility of reducing radon concentrations within existing and new buildings.

"(D) The relationship between short-term and long-term testing techniques and the relationship between (i) measurements based on both such techniques, and (ii) the actions levels set forth as provided in paragraph (1).

"(E) Outdoor radon levels around the country.

15 USC 2664. **"SEC. 304. MODEL CONSTRUCTION STANDARDS AND TECHNIQUES.**

"The Administrator of the Environmental Protection Agency shall develop model construction standards and techniques for controlling radon levels within new buildings. To the maximum extent possible, these standards and techniques should be developed with the assistance of organizations involved in establishing national building construction standards and techniques. The Administrator shall make a draft of the document containing the model standards and techniques available for public review and comment. The model standards and techniques shall provide for geographic differences in construction types and materials, geology, weather, and other variables that may affect radon levels in new buildings. The Administrator shall make final model standards and techniques available to the public by June 1, 1990. The Administrator shall work to ensure that organizations responsible for developing national model building codes, and authorities which regulate building construction within States or political subdivisions within States, adopt the Agency's model standards and techniques.

15 USC 2665. **"SEC. 305. TECHNICAL ASSISTANCE TO STATES FOR RADON PROGRAMS.**

"(a) REQUIRED ACTIVITIES.—The Administrator (or another Federal department or agency designated by the Administrator) shall develop and implement activities designed to assist State radon programs. These activities may include, but are not limited to, the following:

"(1) Establishment of a clearinghouse of radon related information, including mitigation studies, public information materials, surveys of radon levels, and other relevant information. Public information.

"(2) Operation of a voluntary proficiency program for rating the effectiveness of radon measurement devices and methods, the effectiveness of radon mitigation devices and methods, and the effectiveness of private firms and individuals offering radon-related architecture, design, engineering, measurement, and mitigation services. The proficiency program under this subparagraph shall be in operation within one year after the date of the enactment of this section.

"(3) Design and implementation of training seminars for State and local officials and private and professional firms dealing with radon and addressing topics such as monitoring, analysis, mitigation, health effects, public information, and program design.

"(4) Publication of public information materials concerning radon health risks and methods of radon mitigation.

"(5) Operation of cooperative projects between the Environmental Protection Agency's Radon Action Program and the State's radon program. Such projects shall include the Home Evaluation Program, in which the Environmental Protection Agency evaluates homes and States demonstrate mitigation methods in these homes. To the maximum extent practicable, consistent with the objectives of the evaluation and demonstration, homes of low-income persons should be selected for evaluation and demonstration.

"(6) Demonstration of radon mitigation methods in various types of structures and in various geographic settings and publication of findings. In the case of demonstration of such methods in homes, the Administrator should select homes of low-income persons, to the maximum extent practicable and consistent with the objectives of the demonstration.

"(7) Establishment of a national data base with data organized by State concerning the location and amounts of radon.

"(8) Development and demonstration of methods of radon measurement and mitigation that take into account unique characteristics, if any, of nonresidential buildings housing child care facilities.

"(b) DISCRETIONARY ASSISTANCE.—Upon request of a State, the Administrator (or another Federal department or agency designated by the Administrator) may provide technical assistance to such State in development or implementation of programs addressing radon. Such assistance may include, but is not limited to, the following:

"(1) Design and implementation of surveys of the location and occurrence of radon within a State.

"(2) Design and implementation of public information and education programs.

"(3) Design and implementation of State programs to control radon in existing or new structures.

"(4) Assessment of mitigation alternatives in unusual or unconventional structures.

"(5) Design and implementation of methods for radon measurement and mitigation for nonresidential buildings housing child care facilities.

"(c) INFORMATION PROVIDED TO PROFESSIONAL ORGANIZATIONS.—The Administrator, or another Federal department or agency designated by the Administrator, shall provide appropriate information concerning technology and methods of radon assessment and mitigation to professional organizations representing private firms involved in building design, engineering, and construction.

"(d) PLAN.—Within 9 months after the date of the enactment of this section and annually thereafter, the Administrator shall submit to Congress a plan identifying assistance to be provided under this section and outlining personnel and financial resources necessary to implement this section. Prior to submission to Congress, this plan shall be reviewed by the advisory groups provided for in section 403(c) of the Superfund Amendments and Reauthorization Act of 1986 (42 U.S.C. 7401 note).

"(e) PROFICIENCY RATING PROGRAM AND TRAINING SEMINAR.—

"(1) AUTHORIZATION.—There is authorized to be appropriated not more than $1,500,000 for the purposes of initially establishing the proficiency rating program under subsection (a)(2) and the training seminars under subsection (a)(3).

"(2) CHARGE IMPOSED.—To cover the operating costs of such proficiency rating program and training seminars, the Administrator shall impose on persons applying for a proficiency rating and on private and professional firms participating in training seminars such charges as may be necessary to defray the costs of the program or seminars. No such charge may be imposed on any State or local government.

"(3) SPECIAL ACCOUNT.—Funds derived from the charges imposed under paragraph (2) shall be deposited in a special account in the Treasury. Amounts in the special account are authorized to be appropriated only for purposes of administering such proficiency rating program or training seminars or for reimbursement of funds appropriated to the Administrator to initially establish such program or seminars.

"(4) REIMBURSEMENT OF GENERAL FUND.—During the first three years of the program and seminars, the Administrator shall make every effort, consistent with the goals and successful operation of the program and seminars, to set charges imposed under paragraph (2) so that an amount in excess of operation costs is collected. Such excess amount shall be used to reimburse the General Fund of the Treasury for the full amount appropriated to initially establish the program and seminars.

"(f) AUTHORIZATION.—(1) There is authorized to be appropriated for the purposes of carrying out sections 303, 304, and this section an amount not to exceed $3,000,000 for each of fiscal years 1989, 1990, and 1991.

"(2) No amount appropriated under this subsection may be used by the Environmental Protection Agency to administer the grant program under section 306.

"(3) No amount appropriated under this subsection may be used to cover the costs of the proficiency rating program under subsection (a)(2).

15 USC 2666. "SEC. 306. GRANT ASSISTANCE TO STATES FOR RADON PROGRAMS.

"(a) IN GENERAL.—For each fiscal year, upon application of the Governor of a State, the Administrator may make a grant, subject to such terms and conditions as the Administrator considers appropriate, under this section to the State for the purpose of assisting the State in the development and implementation of programs for the assessment and mitigation of radon.

"(b) APPLICATION.—An application for a grant under this section in any fiscal year shall contain such information as the Administrator shall require, including each of the following:

"(1) A description of the seriousness and extent of radon exposure in the State.

"(2) An identification of the State agency which has the primary responsibility for radon programs and which will receive the grant, a description of the roles and responsibilities of the lead State agency and any other State agencies involved in radon programs, and description of the roles and responsibilities of any municipal, district, or areawide organization involved in radon programs.

"(3) A description of the activities and programs related to radon which the State proposes in such year.

"(4) A budget specifying Federal and State funding of each element of activity of the grant application.

"(5) A 3-year plan which outlines long range program goals and objectives, tasks necessary to achieve them, and resource requirements for the entire 3-year period, including anticipated State funding levels and desired Federal funding levels. This clause shall apply only for the initial year in which a grant application is made.

"(c) ELIGIBLE ACTIVITIES.—Activities eligible for grant assistance under this section are the following:

"(1) Survey of radon levels, including special surveys of geographic areas or classes of buildings (such as, among others, public buildings, school buildings, high-risk residential construction types).

"(2) Development of public information and educational materials concerning radon assessment, mitigation, and control programs.

"(3) Implementation of programs to control radon in existing and new structures.

"(4) Purchase by the State of radon measurement equipment or devices.

"(5) Purchase and maintenance of analytical equipment connected to radon measurement and analysis, including costs of calibration of such equipment.

"(6) Payment of costs of Environmental Protection Agency-approved training programs related to radon for permanent State or local employees.

"(7) Payment of general overhead and program administration costs.

"(8) Development of a data storage and management system for information concerning radon occurrence, levels, and programs.

"(9) Payment of costs of demonstration of radon mitigation methods and technologies as approved by the Administrator, including State participation in the Environmental Protection Agency Home Evaluation Program.

"(10) A toll-free radon hotline to provide information and technical assistance.

"(d) PREFERENCE TO CERTAIN STATES.—Beginning in fiscal year 1991, the Administrator shall give a preference for grant assistance under this section to States that have made reasonable efforts to ensure the adoption, by the authorities which regulate building construction within that State or political subdivisions within States, of the model construction standards and techniques for new buildings developed under section 304.

"(e) PRIORITY ACTIVITIES AND PROJECTS.—The Administrator shall support eligible activities contained in State applications with the full amount of available funds. In the event that State applications for funds exceed the total funds available in a fiscal year, the Administrator shall give priority to activities or projects proposed by States based on each of the following criteria:

"(1) The seriousness and extent of the radon contamination problem to be addressed.

"(2) The potential for the activity or project to bring about reduction in radon levels.

"(3) The potential for development of innovative radon assessment techniques, mitigation measures as approved by the Administrator, or program management approaches which may be of use to other States.

"(4) Any other uniform criteria that the Administrator deems necessary to promote the goals of the grant program and that the Administrator provides to States before the application process.

"(f) FEDERAL SHARE.—The Federal share of the cost of radon program activities implemented with Federal assistance under this section in any fiscal year shall not exceed 75 percent of the costs incurred by the State in implementing such program in the first year of a grant to such State, 60 percent in the second year, and 50 percent in the third year. Federal assistance shall be made on the condition that the non-Federal share is provided from non-Federal funds.

"(g) ASSISTANCE TO LOCAL GOVERNMENTS.—States may, at the Governor's discretion, use funds from grants under this section to assist local governments in implementation of activities eligible for assistance under paragraphs (2), (3), and (6) of subsection (c).

"(h) INFORMATION.—(1) The Administrator may request such information, data, and reports developed by the State as he considers necessary to make the determination of continuing eligibility under this section.

"(2) Any State receiving funds under this section shall provide to the Administrator all radon-related information generated in its activities, including the results of radon surveys, mitigation demonstration projects, and risk communication studies.

"(3) Any State receiving funds under this section shall maintain, and make available to the public, a list of firms and individuals within the State that have received a passing rating under the Environmental Protection Agency proficiency rating program referred to in section 305(a)(2). The list shall also include the address and phone number of such firms and individuals, together with the proficiency rating received by each. The Administrator shall make such list available to the public at appropriate locations in each State which does not receive funds under this section unless the State assumes such responsibility.

"(i) LIMITATIONS.—(1) No grant may be made under this section in any fiscal year to a State which in the preceding fiscal year received a grant under this section unless the Administrator determines that such State satisfactorily implemented the activities funded by the grant in such preceding fiscal year.

"(2) The costs of implementing paragraphs (4) and (9) of subsection (c) shall not in the aggregate exceed 50 percent of the amount of any grant awarded under this section to a State in a fiscal year. In implementing such paragraphs, a State should make every effort, consistent with the goals and successful operation of the State radon program, to give a preference to low-income persons.

"(3) The costs of general overhead and program administration under subsection (c)(7) shall not exceed 25 percent of the amount of any grant awarded under this section to a State in a fiscal year.

"(4) A State may use funds received under this section for financial assistance to persons only to the extent such assistance is related to demonstration projects or the purchase and analysis of radon measurement devices.

"(j) AUTHORIZATION.—(1) There is authorized to be appropriated for grant assistance under this section an amount not to exceed $10,000,000 for each of fiscal years 1989, 1990, and 1991.

"(2) There is authorized to be appropriated for the purpose of administering the grant program under this section such sums as may be necessary for each of such fiscal years.

"(3) Notwithstanding any other provision of this section, not more than 10 percent of the amount appropriated to carry out this section may be used to make grants to any one State.

"(4) Funds not obligated to States in the fiscal year for which funds are appropriated under this section shall remain available for obligation during the next fiscal year.

"(5) No amount appropriated under this subsection may be used to cover the costs of the proficiency rating program under section 305(a)(2).

"SEC. 307. RADON IN SCHOOLS. 15 USC 2667.

"(a) STUDY OF RADON IN SCHOOLS.—

"(1) AUTHORITY.—The Administrator shall conduct a study for the purpose of determining the extent of radon contamination in the Nation's school buildings.

"(2) LIST OF HIGH PROBABILITY AREAS.—In carrying out such study, the Administrator shall identify and compile a list of areas within the United States which the Administrator determines have a high probability of including schools which have elevated levels of radon.

"(3) BASIS OF LIST.—In compiling such list, the Administrator shall make such determinations on the basis of, among other things, each of the following:

"(A) Geological data.

"(B) Data on high radon levels in homes and other structures nearby any such school.

"(C) Physical characteristics of the school buildings.

"(4) SURVEY.—In conducting such study the Administrator shall design a survey which when completed allows Congress to characterize the extent of radon contamination in schools in each State. The survey shall include testing from a representative sample of schools in each high-risk area identified in paragraph (1) and shall include additional testing, to the extent resources are available for such testing. The survey also shall include any reliable testing data supplied by States, schools, or other parties.

"(5) ASSISTANCE.—(A) The Administrator shall make available to the appropriate agency of each State, as designated by the Governor of such State, a list of high risk areas within each State, including a delineation of such areas and any other data available to the Administrator for schools in that State. To assist such agencies, the Administrator also shall provide guidance and data detailing the risks associated with high radon levels, technical guidance and related information concerning testing for radon within schools, and methods of reducing radon levels.

"(B) In addition to the assistance authorized by subparagraph (A), the Administrator is authorized to make available to the appropriate agency of each State, as designated by the Governor of such State, devices suitable for use by such agencies in conducting tests for radon within the schools under the jurisdiction of any such State agency. The Administrator is authorized to make available to such agencies the use of laboratories of the Environmental Protection Agency, or to recommend laboratories, to evaluate any such devices for the presence of radon levels.

"(6) DIAGNOSTIC AND REMEDIAL EFFORTS.—The Administrator is authorized to select, from high-risk areas identified in paragraph (2), school buildings for purposes of enabling the Administrator to undertake diagnostic and remedial efforts to reduce the levels of radon in such school buildings. Such diagnostic and remedial efforts shall be carried out with a view to developing technology and expertise for the purpose of making such technology and expertise available to any local educational agency and the several States.

"(7) STATUS REPORT.—On or before October 1, 1989, the Administrator shall submit to the Congress a status report with respect to action taken by the Administrator in conducting the study required by this section, including the results of the Administrator's diagnostic and remedial work. On or before October 1, 1989, the Administrator shall submit a final report setting forth the results of the study conducted pursuant to this section, including the results of the Administrator's diagnostic and remedial work, and the recommendations of the Administrator.

"(b) AUTHORIZATION.—For the purpose of carrying out the provisions of paragraph (6) of subsection (a), there are authorized to be appropriated such sums, not to exceed $500,000, as may be necessary. For the purpose of carrying out the provisions of this section other than such paragraph (6), there are authorized to be appropriated such sums, not to exceed $1,000,000, as may be necessary.

15 USC 2668.

**"SEC. 308. REGIONAL RADON TRAINING CENTERS.**

Schools and colleges. Grants.

"(a) FUNDING PROGRAM.—Upon application of colleges, universities, institutions of higher learning, or consortia of such institutions, the Administrator may make a grant or cooperative agreement, subject to such terms and conditions as the Administrator considers appropriate, under this section to the applicant for the purpose of establishing and operating a regional radon training center.

"(b) PURPOSE OF THE CENTERS.—The purpose of a regional radon training center is to develop information and provide training to Federal and State officials, professional and private firms, and the public regarding the health risks posed by radon and demonstrated methods of radon measurement and mitigation.

Schools and colleges.

"(c) APPLICATIONS.—Any colleges, universities, institutions of higher learning or consortia of such institutions may submit an application for funding under this section. Such applications shall be submitted to the Administrator in such form and containing such information as the Administrator may require.

"(d) SELECTION CRITERIA.—The Administrator shall support at least 3 eligible applications with the full amount of available funds. The Administrator shall select recipients of funding under this section to ensure that funds are equitably allocated among regions of the United States, and on the basis of each of the following criteria:

"(1) The extent to which the applicant's program will promote the purpose described in subsection (b).

"(2) The demonstrated expertise of the applicant regarding radon measurement and mitigation methods and other radon-related issues.

"(3) The demonstrated expertise of the applicant in radon training and in activities relating to information development and dissemination.

"(4) The seriousness of the radon problem in the region.

"(5) The geographical coverage of the proposed center.

"(6) Any other uniform criteria that the Administrator deems necessary to promote the purpose described in subsection (b) and that the Administrator provides to potential applicants prior to the application process.

"(e) TERMINATION OF FUNDING.—No funding may be given under this section in any fiscal year to an applicant which in the preceding fiscal year received funding under this section unless the Administrator determines that the recipient satisfactorily implemented the activities that were funded in the preceding year.

"(f) AUTHORIZATION.—There is authorized to be appropriated to carry out the program under this section not to exceed $1,000,000 for each of fiscal years 1989, 1990, and 1991.

"SEC. 309. STUDY OF RADON IN FEDERAL BUILDINGS. 15 USC 2669.

"(a) STUDY REQUIREMENT.—The head of each Federal department or agency that owns a Federal building shall conduct a study for the purpose of determining the extent of radon contamination in such buildings. Such study shall include, in the case of a Federal building using a nonpublic water source (such as a well or other groundwater), radon contamination of the water. Water.

"(b) HIGH-RISK FEDERAL BUILDINGS.—(1) The Administrator shall identify and compile a list of areas within the United States which the Administrator, in consultation with Federal departments and agencies, determines have a high probability of including Federal buildings which have elevated levels of radon.

"(2) In compiling such list, the Administrator shall make such determinations on the basis of, among other things, the following:

"(A) Geological data.

"(B) Data on high radon levels in homes and other structures near any such Federal building.

"(C) Physical characteristics of the Federal buildings.

"(c) STUDY DESIGNS.—Studies required under subsection (a) shall be based on design criteria specified by the Administrator. The head of each Federal department or agency conducting such a study shall submit, not later than July 1, 1989, a study design to the Administrator for approval. The study design shall follow the most recent Environmental Protection Agency guidance documents, including 'A Citizen's Guide to Radon'; the 'Interim Protocol for Screening and Follow Up: Radon and Radon Decay Products Measurements'; the 'Interim Indoor Radon & Radon Decay Product Measurement Protocol'; and any other recent guidance documents. The study design shall include testing data from a representative sample of Federal buildings in each high-risk area identified in subsection (b). The study design also shall include additional testing data to the extent resources are available, including any reliable data supplied by Federal agencies, States, or other parties.

"(d) INFORMATION ON RISKS AND TESTING.—(1) The Administrator shall provide to the departments or agencies conducting studies under subsection (a) the following:

"(A) Guidance and data detailing the risks associated with high radon levels.

"(B) Technical guidance and related information concerning testing for radon within Federal buildings and water supplies.

"(C) Technical guidance and related information concerning methods for reducing radon levels.

"(2) In addition to the assistance required by paragraph (1), the Administrator is authorized to make available, on a cost reimbursable basis, to the departments or agencies conducting studies under subsection (a) devices suitable for use by such departments or agencies in conducting tests for radon within Federal buildings. For the purpose of assisting such departments or agencies in evaluating any such devices for the presence of radon levels, the Administrator is authorized to recommend laboratories or to make available to such departments or agencies, on a cost reimbursable basis, the use of laboratories of the Environmental Protection Agency.

"(e) STUDY DEADLINE.—Not later than June 1, 1990, the head of each Federal department or agency conducting a study under subsection (a) shall complete the study and provide the study to the Administrator.

"(f) REPORT TO CONGRESS.—Not later than October 1, 1990, the Administrator shall submit a report to the Congress describing the results of the studies conducted pursuant to subsection (a).

15 USC 2670. **"SEC. 310. REGULATIONS.**

"The Administrator is authorized to issue such regulations as may be necessary to carry out the provisions of this title.

15 USC 2671. **"SEC. 311. ADDITIONAL AUTHORIZATIONS.**

"Amounts authorized to be appropriated in this title for purposes of carrying out the provisions of this title are in addition to amounts authorized to be appropriated under other provisions of law for radon-related activities.".

(b) TECHNICAL AMENDMENTS.—The Toxic Substances Control Act (15 U.S.C. 2601 and following) is amended by adding at the end of the table of contents in section 1 the following:

"TITLE III—INDOOR RADON ABATEMENT

"Sec. 301. National goal.
"Sec. 302. Definitions.
"Sec. 303. EPA Citizen's guide.
"Sec. 304. Model construction standards and techniques.
"Sec. 305. Technical assistance to States for radon programs.
"Sec. 306. Grant assistance to States for radon programs.
"Sec. 307. Radon in schools.
"Sec. 308. Regional radon training centers.
"Sec. 309. Study of radon in Federal buildings.
"Sec. 310. Regulations.
"Sec. 311. Additional authorizations.".

Approved October 28, 1988.

---

LEGISLATIVE HISTORY—S. 744 (H.R. 2837):

HOUSE REPORTS: No. 100-1047 accompanying H.R. 2837 (Comm. on Energy and Commerce).
SENATE REPORTS: No. 100-91 (Comm. on Environment and Public Works).
CONGRESSIONAL RECORD:
Vol. 133 (1987): July 8, considered and passed Senate.
Vol. 134 (1988): Oct. 5, H.R. 2837 considered and passed House; proceedings vacated and S. 744, amended, passed in lieu.
Oct. 7, Senate concurred in House amendments.

# Subject Index

# ACADEMY OFFICERS 1990-1992

SHERMAN S. HENDRIX
*President*
Department of Biology
Gettysburg College
Gettysburg, Pennsylvania 17325

LEONARD M. ROSENFELD
*President-Elect &*
*Recording Secretary*
Department of Physiology
Jefferson Medical College of
Thomas Jefferson University
Philadelphia, Pennsylvania 19107

KURT C. SCHREIBER
*Immediate Past-President*
Professor Emeritus
Duquesne University
1812 Wightman Street
Pittsburgh, Pennsylvania 15217

GEORGE C. SHOFFSTALL
*Past-President & Executive Secretary*
502 Misty Drive, Suite 1
Lancaster, Pennsylvania 17603

SHYAMAL K. MAJUMDAR
*Past-President*
*Editor of the Journal*
Professor of Biology
Lafayette College
Easton, Pennsylvania 18042

JAN MIKESELL
*Treasurer*
Department of Biology
Gettysburg College
Gettysburg, Pennsylvania 17325

LARRY R. SHERMAN
*Assistant Treasurer*
Department of Chemistry
University of Scranton
Scranton, PA 18510

HOWARD S. PITKOW
*Corresponding Secretary*
Department of Physiology
Pennsylvania College of Podiatric Medicine
Philadelphia, Pennsylvania 19107

FRED J. BRENNER
*Editor of the Newsletter*
Department of Biology
Grove City College
Grove City, Pennsylvania 16127

J. ROBERT HALMA
*Historian*
Department of Biology
Cedar Crest College
Allentown, Pennsylvania 18104

JOSEPH DURKIN
*Director*
*PA Junior Academy of Science*
Valley View High School
Archbald, Pennsylvania 18403

HEINZ G. PFEIFFER
*Chairman, Advisory Council*
Allentown, Pennsylvania 18103